STERILE PRODUCT FACILITY DESIGN AND PROJECT MANAGEMENT

Second Edition

STERILE PRODUCT FACILITY DESIGN AND PROJECT MANAGEMENT

Second Edition

Jeffery N. Odum

CRC Press is an imprint of the
Taylor & Francis Group, an **informa** business

CRC Press
Taylor & Francis Group
6000 Broken Sound Parkway NW, Suite 300
Boca Raton, FL 33487-2742

First issued in paperback 2019

ISBN-13: 978-0-8493-1874-0 (hbk)
ISBN-13: 978-0-367-39440-0 (pbk)

Library of Congress Cataloging-in-Publication Data

Odum, Jeffery N.
Sterile product facility design and project management / Jeffery N. Odum -- 2nd ed.
p. cm.
Includes bibliographical references and index.
ISBN 0-8493-1874-2 (alk.paper)
1. Biotechnology laboratories--Design and construction. 2. Pharmaceutical industry--Management. 3. Clean rooms--Design and construction. 4. Project management. 1. Title.

TP248.24.O38 2004
660.6′078—dc22 2003064023

Library of Congress Card Number 2003064023

Visit the Informa Web site at
www.informa.com

and the Informa Healthcare Web site at
www.informahealthcare.com

Preface

In 2003, as in 1997 when this work was first published, the delivery of facility projects remains a very complex series of integrated activities. In today's marketplace, there remains a strong emphasis on managing risk: risk associated with advancing technologies, regulatory change, increased financial pressures, and tighter time lines.

The pharmaceutical and biotechnology industries have remained a constant in tough economic times. The drug discovery process has made tremendous advances, allowing many new companies to enter the marketplace with new drugs and delivery methods for those drug products. The drug pipeline remains strong, maybe stronger than it has ever been. The promise of biotechnology has become reality, as new biotech drugs begin to enter the market and become the new "blockbuster" drugs.

At the same time, however, there is recognition that the industry must not only look ahead, but also ensure that current facilities meet the needs and expectations of the industry and the regulators that oversee it. The challenges that are being met in design and construction of sterile product facilities continue to focus on compliance, flexibility, and manufacturing optimization.

This edition of *Sterile Product Facility Design and Project Management* takes a new look at some of the changes that have occurred over the past six years. Not surprisingly, there also remain many constants. I do not feel this will change, because many fundamental principles of successful project management are not impacted by technology, economic, or social change. They are simply good common sense.

Every project team that must deliver a compliant facility on time, and within budget, still must address the strategic nature of project execution. This fundamental principle has not changed. It is my hope that the updated information in this book will again be a reference source to each person in need of the current thinking within the industry.

While our world is changing, it is nice to know that some constants remain.

Acknowledgments

Revising a previous work is not easy. It takes a focused effort to update and refine material that has stood the test of time. I want to thank Dena for helping to create the new material. I could not have done this without her. And I want to thank Bo for keeping me focused on the task.

And again, thanks to all of my friends and peers in the industry. Your dedication and commitment is a continued breath of fresh air.

Acknowledgments

Contents

chapter 1

Introduction to facility project management

What is a project? A simple question with a not-so-simple answer. The response that you receive will depend on the person being asked and the industry with which he or she is familiar. A project can be defined as a collection of both human and financial resources that are focused on the achievement of a specific set of goals and objectives. Others may see a project as the opportunity to produce a new product or provide a new service that must meet predetermined specifications and standards. This effort must be completed within an established set of parameters that include a fixed time, a defined budget, available assets, and limited human resources.

The term project is used loosely by most individuals. Cleaning out the garage is referred to as a project. Painting a deck is seen as a project. Hanging new wallpaper in the bathroom is a project. But to the biotech/pharmaceutical company that is embarking on a capital spending program to provide new or additional facilities, a project has many more complex characteristics; many individual tasks must be defined, understood, and managed in order for the project to be seen as a success upon completion.

For our discussion, a project is defined as the performance of specific work activities, having a defined starting point and end point, resulting in meeting the technical and financial objectives of the client organization. A project has a defined life cycle (Figure 1.1) that results in a discrete deliverable or end product; in this case, a new facility, a renovated space, or an upgraded process to produce biological or pharmaceutical products.

Project management

Project management is the practiced discipline that integrates the process of producing the end product with the processes of planning, managing, and controlling the tasks that lead to its completion. It is both a science and an "art." It is a science because of the technical tools that are used to document

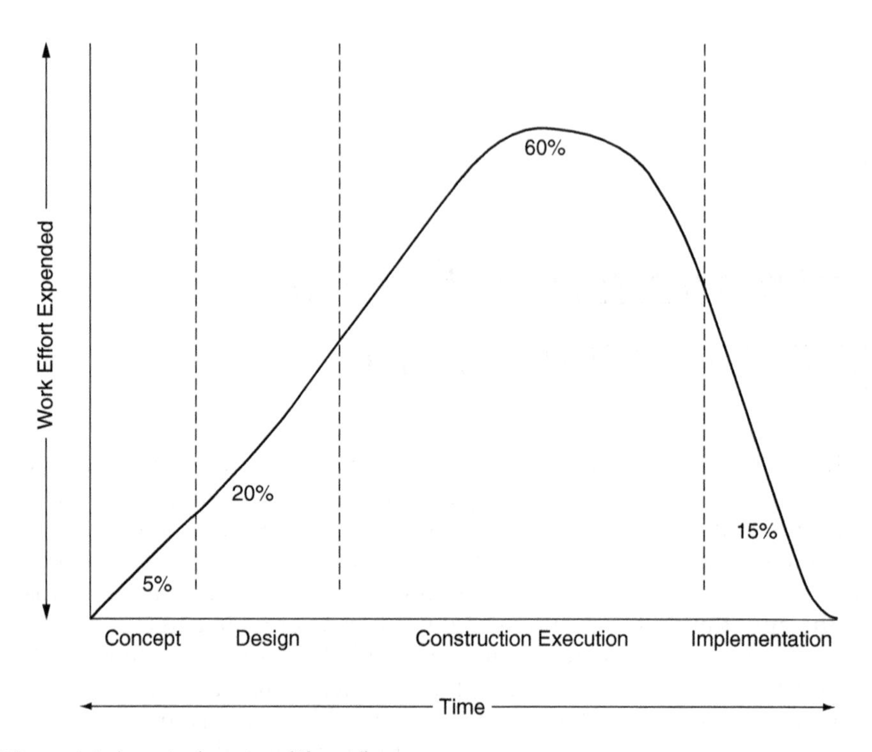

Figure 1.1 A typical project life cycle.

information. It is an "art" because it is driven by communication, negotiation, and conflict resolution.

Project management begins when the decision is made to devote human and financial resources to an effort—the execution of a project. It does not end until the desired goal has been achieved—successful validation of the facility. The skills with which the project staff uses the methods, tools, and techniques of effective management between these two points determines the success of the project.

How does all this fit in with the execution of a project to design and build a sterile manufacturing facility for drug production? Like all projects, parenteral facility projects have the following basic task requirements:

- Assemble a project team.
- Define the project goals and objectives.
- Plan the project.
- Manage changes to the project scope.
- Control the project so that it is completed on time and within budget.

In addition, there are other issues that the project manager and his or her staff must address while executing the above mentioned tasks.

- Technology issues
- Regulatory compliance
- Risk management
- Quality management
- Safety

Taking all of these factors and putting them into a flowchart depicting the project life cycle (Figure 1.2) provides a good "snapshot" of what project management is all about.

Project philosophy

Every project, either by design or by default, has a controlling philosophy that directly influences every decision made during the life of the project. Overall corporate philosophy related to project execution, the experiences and methods of the sponsor or project manager, or the backgrounds of the principle project participants may become the primary driving force for project management and execution.

In discussing philosophy, the following key issues will become important to each and every project, regardless of the size, cost, complexity, or schedule for completion:

- Planning issues
- People issues
- Control issues
- Contracting issues

Planning

No activity will have a greater impact on the final outcome of a sterile product manufacturing project in terms of cost, schedule, quality, and operability than the planning phase of the project. In managing a project, it is essential to have the proper information available to make informed decisions regarding the definition of the project and the direction it will take. With this information, owners and clients can address risk and commit resources that will maximize the chances for a successful project. Doing this in an effective manner requires diligence in the planning process.

I once heard a frustrated client representative say, "We never have time to do it right, but we always seem to find time to do it over." This single statement says volumes about the value of planning. Spending the time up front can reap tremendous benefits in the long run. It is an acknowledged fact that the best "bang for the buck" in developing a project comes during the planning stage (Figure 1.3). The Construction Industry Institute (CII) has documented studies that show an effective planning effort can reduce project design and construction costs by as much as 20 percent on average versus authorization estimates, and can reduce the project design and construction

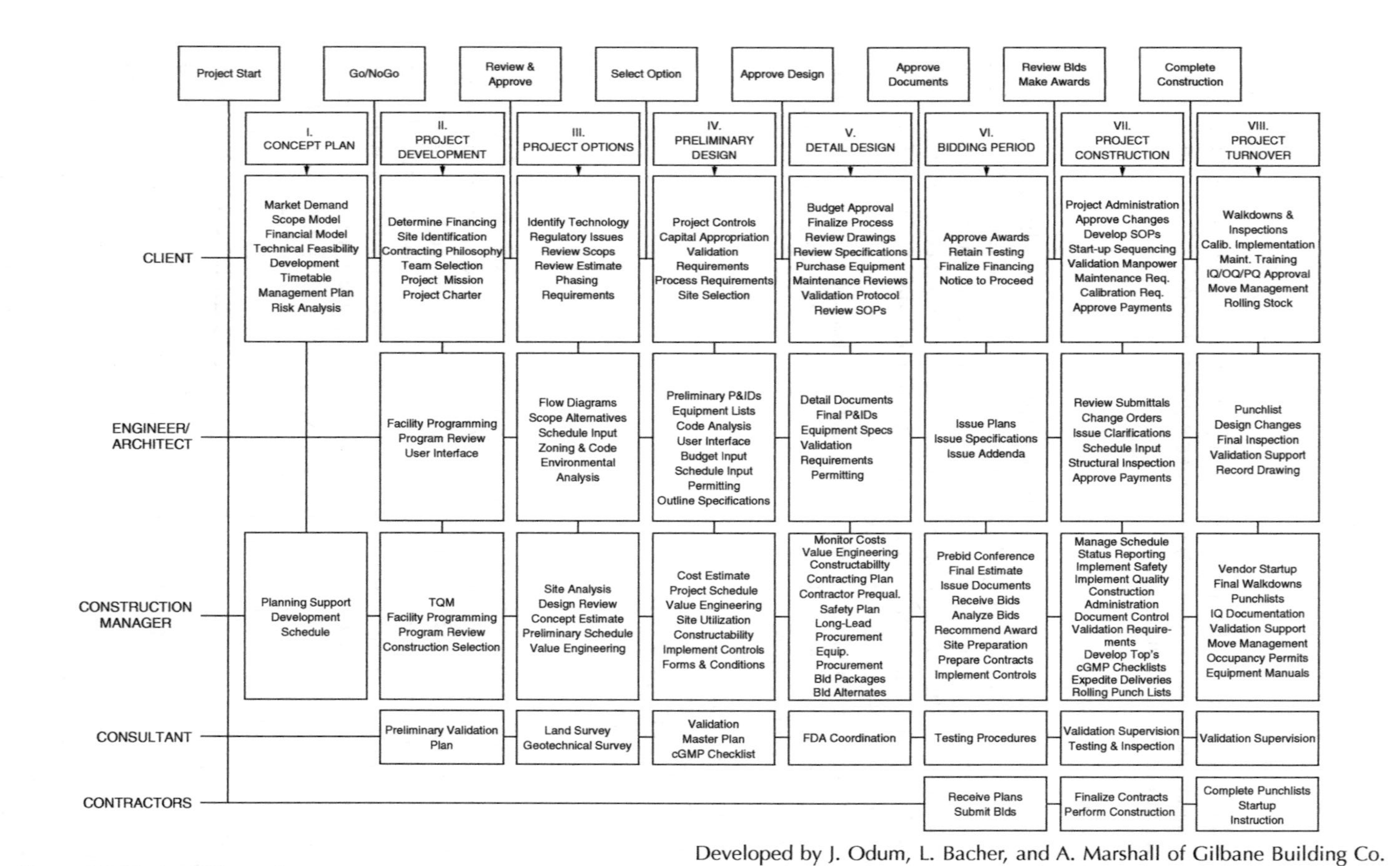

Figure 1.2 Project life-cycle program.

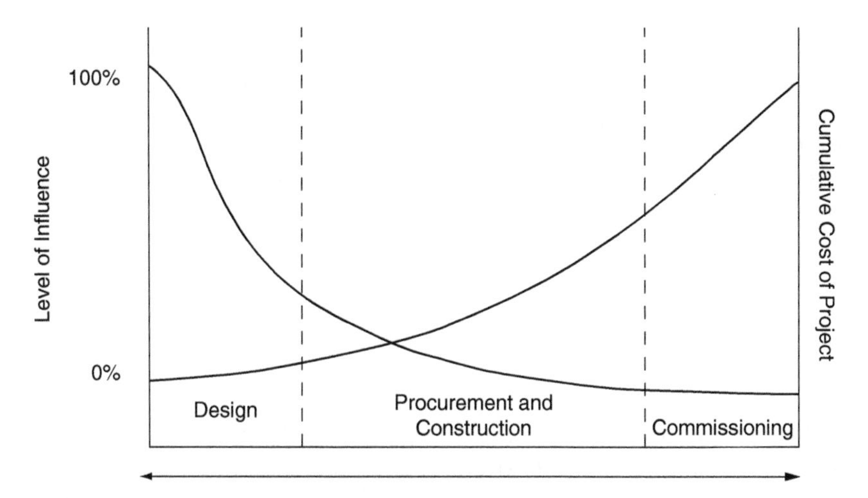

Figure 1.3 Level of influence on project costs.

schedule by as much as 39 percent on average versus the project authorization schedule. [1] While every project may not realize these levels of savings, it does provide a clear incentive to spend the necessary time to plan the project properly.

People

People execute projects. The development of the project organization structure and the selection of the right project team members, which involves obtaining their commitment to the project and managing their activities, are two of the most critical tasks facing the project manager. Building a broad-based team involves taking advantage of the various experiences of both individuals and organizations in the pharmaceutical industry.

Defining an organization is essential so that a project gains recognition within the corporate structure. People must understand who is on the team and what roles they perform. The organization will also define who the formal leader of the project is (leader being singular). Project team members cannot be expected to divide their loyalty between two or more captains. Formal means that the single leader, the project manager, has been given the responsibility and authority by corporate management to run the project.

Team members must have an undying need to succeed personally. If you, as project manager, become apathetic, so will your team. Team members must be committed to a project and motivated to achieve the project goals in a timely and cost-effective manner.

Communication is essential in effective project management. Nothing is more frustrating to team members and clients than to have the game plan changed without their knowledge. The timely and accurate exchange of information between all parties is the key to an effective communication plan. Formalize it, follow it, and continually review it during the life of the project.

Control

"Plan the work, then work the plan." Successful project execution requires not only good planning, but also the ability to follow the plan and control activities so that the defined goals of the plan are met. The project manager and the project team must be able to answer the following questions concerning the project:

- Where are we?
  Where should we be at this point in time?
  How do we get to where we are going?
- Are we getting there?

Without knowing where you are makes it difficult to know where you are going. The high costs and extremely complex schedules common to most parenteral manufacturing projects require diligence in monitoring progress and controlling change. Proper documentation of changes to the defined scope of work is essential for tracking cost. The project team must realize the impacts of additions, deletions, or revisions to the defined facility program. An added feature may require only a few man-hours of design to incorporate, but it does not necessarily reflect the magnitude of the change during construction. Studies have shown a correlation that for every dollar of design input that is missed during the planning of a project, it will cost on average $15 to correct during design and up to $150 on average to correct during actual construction. [2]

Schedules are only as good as the data that go into them; put garbage in, get garbage out. Even the most detailed schedule is worthless if it is not monitored and updated to reflect current project status. Using the right tools to measure quantities and predict progress are important ingredients in the recipe for success.

Contracting

A successful project will ultimately involve many companies and individuals outside the client organization. These relationships require formal contracts execution. How contracts are developed, executed, and managed becomes very important.

Contracting philosophy has many differing viewpoints. What works for one organization may not work for another. Regardless of project philosophy, the following key issues are fundamental to success:

- Do not expect something for nothing. Looking for value is very different from getting the cheapest price. Be prepared to evaluate price based on services and/or products delivered.
- Expect to get everything you ask for. Make your scope of work and commercial terms clear and well defined. Identify your expected

deliverables and the time frame they are required. Do not settle for anything less.
- Successful contracts are well managed. Involvement is essential. A "hands-off" approach to contract management usually spells disaster.
- No secrets. Both parties in the contractual arrangement must be open and honest with each other. Finger-pointing will never help. The flow of information must be two way, not only in one direction.

There have been numerous books written on the subject of project management; it is not my intent to expand on them. In this book, I do want to focus on several of these and other key project management activities and issues as they relate to these types of projects. Chapter 2 will discuss the critical questions that must be answered in order to define a sterile product manufacturing project. An important part of this is defining the project team. Chapter 3 describes some of the methods used for assembling the proper team and how the organization can be as strong and successful as possible.

The biggest problem that project managers face is defining the scope of the project. What does your client want? But what is wanted is not always what is actually needed. We will address facility programming in detail. This process forms the foundation of a project, what we will build, how much it will cost, and when it will be completed.

The successful delivery of complex parenteral projects is a critical engineering endeavor. There are several project delivery methods that have been used successfully. Choosing the right contracting philosophy for a project is important.

The project management environment is very dynamic; managing change requires control over cost, schedule, and resource utilization. We will look at specific case examples of how project control techniques can be applied to these projects.

Because biopharmaceuticals are produced from living organisms and are regulated by the federal government, biotech projects have their own set of criteria that make them technically challenging and difficult to deliver. Chapters 11 through 13 look at the basics of the issues that are common in the industry.

It all begins with the current Good Manufacturing Practice (GMP) and its impact on every phase of the project. The uniqueness of the architecture, the stringent requirements for hygienic piping systems, and the criteria for cleanroom design and construction will be discussed.

Commissioning is an often overlooked subject that has, undoubtedly, been a thorn in every project manager's side at one time. How commissioning should be planned and integrated into the validation program will be reviewed. Validation, from a project perspective, will be described so that an appreciation of the planning and resourcing required is found.

And what about the future? The industry is changing as you read this book. Some of the current trends will be discussed, with a focus on how they will impact project execution and the overall thinking of the industry.

References

1. Construction Industry Institute. 1994. Pre-project planning: Beginning a project the right way. No. 39-1 (May).
2. Skibo, A. 1989. Project management and contracting issues, Part 1: Project team development. Biopharm (April).

chapter 2

Project formation

The decision by any pharmaceutical or biopharmaceutical company to embark on a capital building program is one that is made in an environment of continuing change—change in the regulatory environment, in the global market, and in the financial needs and expectations of the company. The technology may be promising but, nevertheless, may still be under development. Where the penalty for exceeding project budgets in a large pharmaceutical company may often only require a shifting of capital funds from one project to another, the same overrun in a start-up company can result in bankruptcy or the need for premature equity offerings that do not meet corporate expectations.

The formation process sets the stage for the project and establishes the cornerstone for success. While many individuals acknowledge this, actual practice continues to be a rush to proceed, which forces compromises in what is considered good, commonsense formation practice. Issues impacting project cost, schedule, regulatory compliance, technology, quality, and safety are causalities when decisions are made in haste.

Where do we start? Project formation begins by asking questions. These questions are intended to define a project's needs, scope, and overall program. Knowing which questions to ask is as important as obtaining the answer. If you do not understand the question, you cannot be expected to find the answer. If you cannot find answers, you will not be able to manage the project.

Conceptual development

A major cause of a project's difficulty lies in the lack of clear objective definition, an inconsistency of objectives, and a lack of clearly communicated project objectives. Therefore, the first step in the project formation process and organizing for success is developing a clear, detailed definition of project objectives. These objectives must be clearly communicated to all project participants. They must also be effectively understood by any organization involved in the project to ensure alignment of thinking with the client's business objectives.

Project objectives must be measurable and not contain vague or misleading language. The objectives should be defined in terms of measurable results, specific in terms of time periods for accomplishment, and flexible to accommodate change as warranted through the project life cycle. Objectives such as "world-class facility," "state-of-the-art manufacturing facility," or "leading-edge technology" are too vague and open to conjecture.

Project objectives can be distinguished by the fact that they deal with function, time, and cost. There must be a balance between these three factors. Participants must realize that completing a project quickly may not be cheap. Having high functional standards may not be cheap and may take a longer period of time. Being cheap may compromise function. Examples of specific objectives include "output of 5 million finished doses annually," "produce clinical materials by third quarter 1999," or "total project cost not to exceed $75 million."

Functional objectives

Before you design and build, you must think and plan. What is the facility supposed to provide? How much material is to be made? What is the material quality? What is the physical form necessary to meet market demands? These questions relate to functional objectives.

Companies must solve complex problems that evolve from process and system design, GMP requirements, personnel and environmental protection, and regulatory compliance. They must address scale-up issues, economies of scale, and life-cycle durations of equipment. People from various organizational and technical backgrounds will need to be brought into a new facility to run and manage the operations. To do all of this, you must first know what you need.

The function of the facility is defined by the application of product, the manufacturing definition, the stages of product development to be served, and the range of activities to be performed in the facility. Is the facility going to provide primary production capacity for marketable product or is it going to be a pilot plant to produce clinical materials? Will the facility be dedicated to a single product or have multiproduct flexibility? Will the facility have multiple functions such as administration, warehousing and distribution, or research support? Do we build a new facility or renovate existing facility space?

Functional objectives require engineering information related to process and manufacturing operations, relevant regulatory data, state and local ordinance data, and site planning. To compile this information may require the development of some preliminary engineering documents that would include site plans, conceptual layouts, and process descriptions.

The development of a preliminary scope of work document will have a significant impact on design and construction costs. This document

Example 2.1 Bio CorporationBiotech Manufacturing Preliminary Scope of Work

Section I: Project Overview

a. General Project Description
b. Project Approach
c. Schedule
d. Project Funding

Section II: Technical Description

a. Process Description

Section III: Discipline Design Criteria

a. Civil
b. Architectural
c. Structural
d. Fire Protection
e. Mechanical
f. Piping
g. Electrical
h. Utilities
i. HVAC
j. Environmental
k. Process
l. Equipment
m. Instrumentation

Section IV: Execution

a. Organization
b. Reports/Communication/General
c. Project Security
d. Regulatory
e. Validation
f. CAD
g. Specifications
h. Numbering Systems
i. Document Distribution
j. Tracing/Specification Release

Section V: Scope of Architect/Engineer Services

a. Phase I—Preliminary Design
b. Phase II—Detailed Design
c. Phase III—Construction Liaison

(Example 2.1) will force the project team to define known requirements and address areas of uncertainty, as well as walk through the project's technical requirements, building design requirements, approach, funding, schedule, and validation.

Schedule objectives

Every project has a schedule, a time frame in which the scope of activities must be completed. Every schedule has milestone dates, critical reference points in time that must be met. The schedule objectives of a project may depend on a number of issues that relate to market penetration, capacity expansion, funding availability, or product development.

During conceptual development, it is important that there is agreement on the nature of the schedule goals for the project. Different functional groups will look at schedule goals and milestones from different points of view. Marketing will focus on product launch timing. Manufacturing will focus on facility start-up and commissioning. Regulatory Affairs will look at issues such as the Establishment License Application (ELA) submission and Food and Drug Administration (FDA) approval. All are correct in their thinking; they must have a coordinated effort to reach their objective. At the same time, it is important to ensure that these objectives are realistic, both from a cost and execution perspective. Obtaining something quickly rarely means getting it cheap.

A typical conceptual project milestone schedule might look similar to Figure 2.1. Project milestones are focused on the launch of a product within a predetermined market window. The design, construction, and validation of the facility all must support the production of consistency lots to support product launch.

A different schedule objective is shown in Figure 2.2, where activities center around meeting a predetermined plant shutdown in order to make necessary equipment upgrades. This represents a more focused view of project planning, but still shows the importance of identifying goals, constraints, and milestone activities.

Budget objectives

Every capital project requires the commitment of funds. While the specific procedures for submittal and authorization may vary between organizations, the process is fundamental. It is very important for all members of the project team to understand this process and realize that it is intimately tied to the capital allocation program of the company.

Budgets have a common goal—to maximize the benefits from a commitment of resources. The trick lies in the fact that budgets are often in tension with technical quality and innovation. It may sound trivial, but everything has a price. Flexible manufacturing methods, isolator technology, expert system-based process controls, and robotics may be the desire of manufacturing groups that support the project, but corporate funding constraints may dictate otherwise. A keen awareness of benefit/cost issues becomes fundamental in budget development.

The development of a project budget during the planning phase is a dynamic process. It is often based on a broad definition of project scope, quantified by the experience of individuals in defining allowances for project components, square foot costs from "similar" projects, duration and costs for facility validation, and so on. Thus, the definition of project scope is probably the single most important issue to be addressed in the development of the project costs. Going back to the functional objectives of the project, you have to know what you want. You also must be able to separate what you want from what you actually need; want and need may be two entirely different things.

Sterile product manufacturing facilities are not cheap to build (Table 2.1). A look at these costs clearly indicates a need to carefully look at design alternatives, project delivery methods, and operational philosophy. Carefully

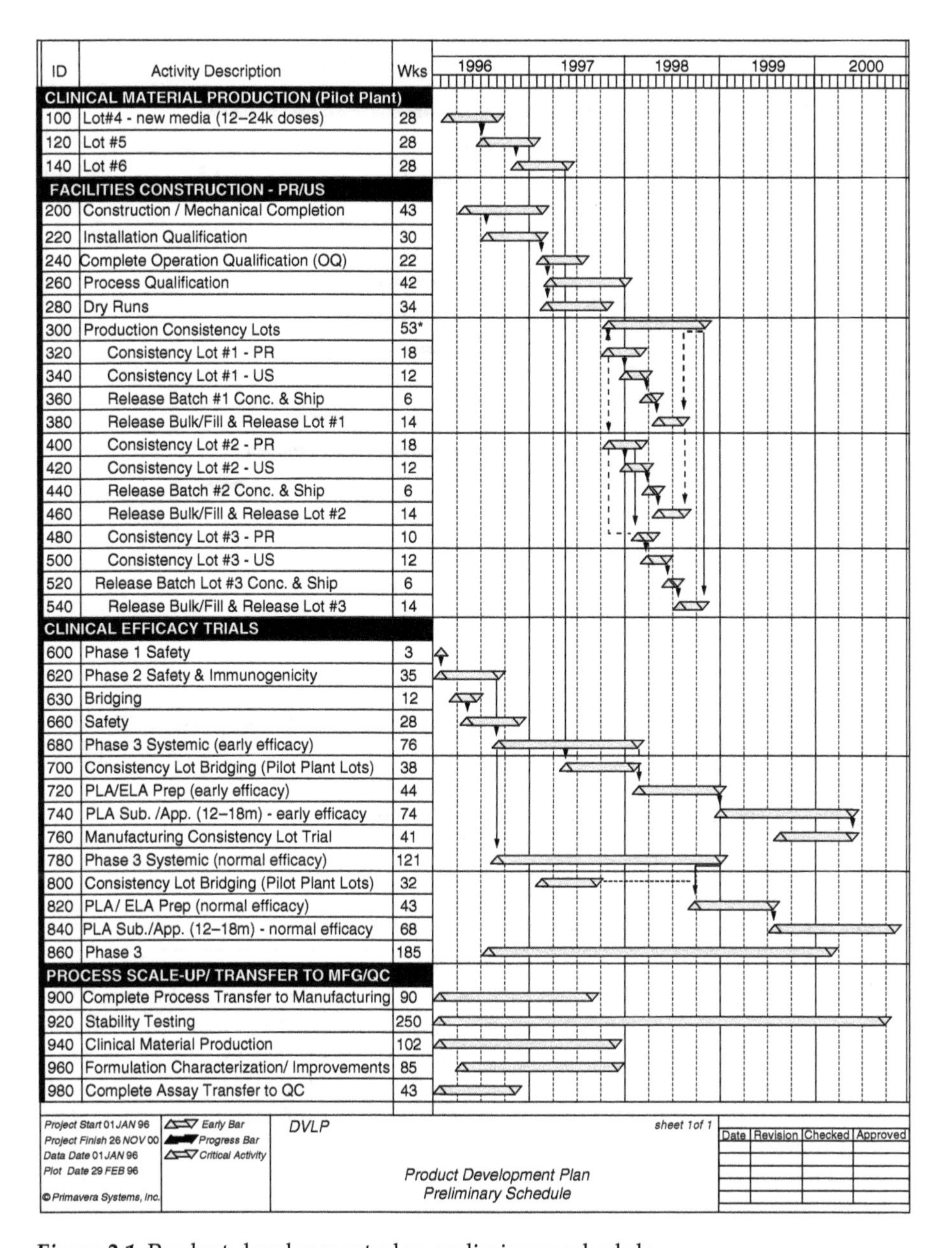

Figure 2.1 Product development plan preliminary schedule.

document and review the assumptions that go into the development of project scope (Example 2.2).

Once a conceptual budget has been established (Figure 2.3), monitor it with diligence. As the design progresses, assumptions and allowances in the conceptual estimate must be refined and revised to reflect the actual requirements of the project. Variances from the conceptual project budget should

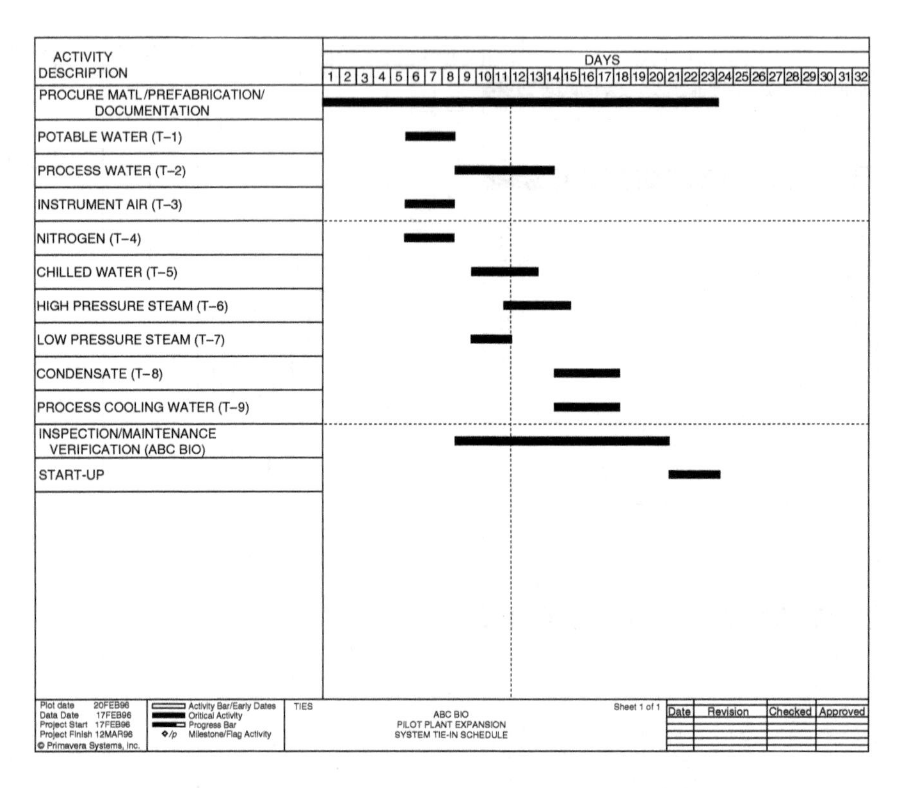

Figure **2.2** ABC Bio pilot plant expansion system tie-in schedule.

Table **2.1** Biotech Facility Costs

Project Type	Year	Size (Square Feet)	Installed Cost	Cost per Square Foot
Pilot Plant	1993	12,500	$13,000,000	$1,040
Microbial	1994	80,000	$45,000,000	$560
Cell Culture	1994	38,000	$29,000,000	$763
Pilot Plant	1995	320,000	$192,000,000	$600
Cell Culture	1995	180,000	$125,000,000	$690
Microbial	1995	37,000	$25,000,000	$675
Pilot Plant	1995	13,000	$9,000,000	$690
Vaccine Conjugation	1996	25,000	$22,000,000	$880

be identified when they occur. Regardless of whether the change that caused the variance is resolved by considering a design alternative or by revising the budget, the decision must be made in a timely manner without any surprises. Many careers have suffered permanent damage due to the compounding impact "surprises" have on project cost.

Example 2.2. Sample Project Objectives and Assumptions

Budget

- Cost range for the project is $80 to $100 million. This total installed cost is assumed to include all costs associated with the project—internal engineering costs, consultant fees, validation, moving and relocation, miscellaneous equipment, and furnishings.

Schedule

- Board approval of funding is assumed for first quarter 1993. Construction to commence first quarter 1994. Project completion date must converge on the finalization of clinical trials, providing adequate bridging and product launch capabilities. Product launch is assumed to start first quarter 1998.
- Project will be executed in a fast-track mode for design and construction.
- Conceptual design should be completed in a time frame to support early regulatory review by FDA, anticipated in early fourth quarter 1993.

Materials and Systems

- The materials of construction, finishes, and systems of the facility will be described as "above average" in overall quality. This describes the desired level of maintainability and regulatory image required of this type of facility. It also describes the corporate position in the industry.
- The useful life of the facility is seen as at least 20 years.

Controls/Automation

- There is no desire to control individual manufacturing operations from a central point, rather a desire for a system monitoring critical variables as identified, equipped with alarms and annunciation capabilities for any process deviations or discrepancies.

Capacity

- Built-in capacity for expansion and/or capacity increase, which requires preinvestment, will be considered for all main processes and critical utilities. Although a level of redundancy will be expected for equipment reliability, this redundancy will not be relied on for capacity increases.
- The assumed operating mode for the facility will consist of three shifts, with none of the three shifts operating at full capacity.
- The design will identify opportunities for the use of surge space to provide for expanded capacity in the future.

Equipment

- Dedicated major equipment trains will be utilized to afford flexibility and regulatory reliability, dedicated to the production of one product at a time on a campaigned basis.
- Due to the aggressive nature of the schedule, the use of modular skid equipment will be deemed essential.
- Multipurpose equipment will not be utilized for the project. All spare and change parts used will be consistent with existing operations where applicable.

ESTIMATE SUMMARY

CLIENT:	WORK ORDER:
PROJECT:	CONTRACT:
LOCATION:	PREPARED BY:
DESCRIPTION:	DATE:
	TIME:

A/C NO.	ITEM & DESCRIPTION	MANHOURS		COSTS ( )			
		DIRECT	CONT-RACT	LABOR	CONT-RACT	MATERIAL	TOTAL
00-00	EARTHWORK & CIVIL		4,156	115,000	35,000	19,000	169,000
10-00	CONCRETE		8,667	297,000	3,000	95,000	395,000
20-00	STRUCTURAL STEEL		6,368	238,000	0	449,000	637,000
30-00	BUILDINGS (INCL. ALL HVAC)		13,176	486,000	1,456,000	642,000	2,584,000
40-00	MACHINERY & EQUIPMENT		5,776	217,000	0	5,012,000	5,229,000
50-00	PIPING		45,712	1,632,000	10,000	1,076,000	2,718,000
60-00	ELECTRICAL		12,761	464,000	0	433,000	897,000
70-00	CONTROL SYSTEMS		4,748	76,000	0	503,000	579,000
81-00	COATINGS		0	0	0	0	
85-00	INSULATION		2,173	76,000	0	147,000	223,000
89-00							
	DIRECT FIELD COSTS		103,537	$3,601,000	$1,504,000	$8,376,000	$13,481,000
			AVG. RATE	$34.78			
90-00	FREIGHT					192,000	192,000
92-00	CONSTR. SERVICES, SUPPLIES & EXPENSES					943,000	943,000
93-00	FIELD STAFF					1,195,000	1,195,000
	FIELD DISTRIBUTABLES					$2,330,000	$2,330,000
	TOTAL FIELD COSTS		103,537	$3,601,000	$1,504,000	$10,706,000	$15,811,000
90-20	ENGINEERING					2,600,000	2,600,000
	TOTAL H.O. COSTS					$2,600,000	$2,600,000
	TOTAL FIELD & H.O. COSTS		103,537	$3,601,000	$1,504,000	$13,306,000	$18,411,000
99-10	ESCALATION					$368,000	$368,000
	SUBTOTAL						
99-20	CONTINGENCY					$2,025,000	$2,025,000
	SUBTOTAL						
99-30	TAXES					$285,000	$285,000
	SUBTOTAL						
99-60	FEE					$310,000	$310,000
	PROJECT T.I.C.						$21,399,000
	VALIDATION AND EXPENSES					$1,217,000	$1,217,000
	START-UP AND COMMISSIONING SUPPORT					$300,000	$300,000
	GRAND TOTAL PROJECT						$22,916,000
	SPARES ALLOWANCE						$200,000
	EQUIPMENT INSPECTION ALLOWANCE						$150,000

Figure 2.3 Conceptual budget.

Organizing

Organizing a project requires the identification of project participants and their relationships, responsibilities, and authorities. The project manager is responsible for defining the organizational structure of the project, ensuring the development of the project mission statement, and developing the planning tools that will be used to manage and control the project.

Team selection

How do you determine who should be on the project team? One of my favorite answers from an experienced colleague was, "Anybody who could eventually screw up the project should be included up front." The point is that building a project team will require evaluating project requirements and determining the right balance of technical, managerial, and human factors in order to select the proper representation of the core team.

Team makeup is critical. Every group that will eventually have a stake in the project should have representation on the team. Project Management, Manufacturing, Engineering and Maintenance, Quality Assurance, Regulatory Affairs, and Validation must all participate in the decision-making process.

Key team members must be selected for their experience, skills, and ability to support and contribute to the project and commit the necessary time and energy to the project. Don't expect a single individual to be a "jack-of-all-trades." When process engineering skills are required, find a process engineer who has the necessary experience (e.g., fermentation or purification system design for cell culture processes). If difficult architectural issues related to facility layout to meet GMP may be involved, find a process architect.

Realize also that some (or many) of the particular individuals may not be found in your company. This is especially true in start-up companies that lack in-house expertise. If this is the case, go out and get them, either through contract hiring, temporary service, or through a third party organization. Build a broad-based team by choosing the best people.

Every team must have a formal leader and a structure. The key word here is an adjective—singular—only one. Team members cannot divide their loyalties between different leaders. Being the formal leader also means that you manage the project, control its destiny, and influence decision making in terms of the project charter.

There is no perfect organizational structure. How you define structure depends on the type of project, the skills of the team members, the needs of the project, and the comfort of the team leader in performing his or her role. For many projects, the integration of outside resources, such as architectural and engineering (A/E) services, will become a significant aspect of organizing the project.

Organization structure can be seen as anything from the traditional hierarchical "pyramid" to a matrix type of structure that combines both vertical and horizontal integration of team members. Regardless of the form, the fundamental mission must be to organize the project along clear lines of communication and responsibility, and to formalize policies and procedures that describe the overall guidelines for the administration and management of the project.

Every project should utilize a Project Procedure Manual (PPM) to describe these guidelines. A typical PPM (Example 2.3) includes an organizational chart, project communication guidelines, design engineering and construction objectives, cost control and financial administration, scheduling, objectives, and a quality plan.

Example 2.3 Sample Contents of a Project Procedure Manual

Section No.	Description
1.0	**Organization**
1.1	Project Organization and Management
1.2	Written Project Communications
1.3	Monthly Project Status Report
1.4	Employee Confidentiality Agreements
1.5	Project File Index
1.6	Project Directory
2.0	**Design Engineering**
2.1	Introduction
2.2	Design Engineering Organization and Responsibilities
2.3	Security and Proprietary Information
2.4	Engineering Equipment Identification
2.5	Drawing Format Issue and Revision
2.6	Engineering Calculations
2.7	Engineering Specifications
2.8	Engineering Equipment Requisitioning
2.9	Vendor Drawings and Control
2.10	Information Needs List
2.11	Estimate Deviation Notices
2.12	Checking Procedure—Drawings
3.0	**Procurement**
3.1	Introduction to Purchasing—Responsibilities and Procedures
3.2	Development of Bidders List
3.3	Bid Requests
3.4	Receiving and Processing Bid Proposals
3.5	Issuing Purchasing Orders and Change Orders
3.6	Expediting and Traffic
4.0	**Controls**
4.1	Project Controls—General
4.2	Estimate Preparation
4.3	Initiating and Processing Vendor/Subcontractor Backcharges
4.4	Estimate Deviation Notices (Total Project)
4.5	Cost Reporting
4.6	Planning and Scheduling
5.0	**Construction Operations**
5.1	Construction Operations—General
5.2	Contracting and Procurement
5.3	Safety and Security
5.4	Material Control

— continued

Example* 2.3 *(continued) Sample Contents of a Project Procedure Manual

5.5	Quality Control, Inspection, and Monitoring of Subcontractors
5.6	Controls
5.7	Construction Engineering—General/Scope
5.8	Document Control and Distribution
5.9	As-Built Drawings and Sketches
5.10	Final Checking, Testing, Inspection, and Acceptance
6.0	**Accounting**
6.1	General Accounting
6.2	Project Accounting, Cash Management, and Funding
7.0	**Project Closeout**

Project charter

Corporate management, project sponsors, and project team members must have a clear understanding of the mission of a project in order to fully understand the implications of what is developed for the project. The best way to accomplish this is to develop what I will refer to as a project charter, which is sometimes called a project mission statement or a project initiation document.

The project charter begins with a statement of the business need — what is the problem or opportunity that the project addresses? This definition must clearly define the need in terms that will not prematurely focus the project in a specific direction. For example, if a business need is defined as "build a new manufacturing facility," rather than "provide additional manufacturing capacity," you may miss the alternative of renovating old or existing facility space as an acceptable alternative that meets the need.

The charter must define the project objectives — what are the end results to be achieved? End results can be defined as a product, a process, or a service. Along with the objectives must come the criteria for determining completion — how will you know when you are finished?

It is also good practice to include in the charter any assumptions that may have been made in defining the project needs and objectives. Assumptions become important because without agreement by all project participants, variations in expectations will undoubtedly occur.

The charter may also include a statement concerning the risks associated with the project. Any analysis of risk should also include constraints imposed on the project that could potentially impact the meeting of project objectives. Knowing what the constraints are early allows sufficient time to pursue alternative solutions that may become necessary in order to reduce risk.

Finally, the charter should define anticipated resource requirements (financial, time, technical, etc.). Do not make specific definitions of resources at this time; enough detailed information is not yet available. This will be a "qualified estimate" of anticipated needs.

Roles and responsibilities

In forming any project, it is important to not only know what everyone is supposed to do, but also to know who is supposed to do what. Roles and responsibilities of personnel must be well defined in order to foster communication and to define tasks.

Figure 2.4 provides a generic example of a responsibility matrix, a tool used to define responsibilities of positions and the type of responses involved in the decision-making process. In Figure 2.5, this same approach is taken in more micro terms, specifically the review and approval process for project documents.

Alignment

It is very important that all members of the project team understand the overall goals and objectives of the project, the roles and responsibilities of their respective team members, the project milestones and measurements for success, and the approach for project execution. One technique used by many companies to define this "common purpose" is to conduct project alignment sessions, sometimes referred to as kickoff meetings or goal-setting sessions.

There are many different approaches and methodologies in conducting this type of meeting. However, the following key points will ensure success:

- Be sure that all the right people attend. The important role of this meeting is to formalize a common set of goals and objectives with the project team, in particular, the project client. User group involvement is imperative. They must understand what the project is and the parameters within which it will be executed. This will be the first opportunity to obtain a close look at the client's needs. Typically, representation from Manufacturing, Regulatory Affairs, Validation, Engineering, Quality Assurance, and Business/Financial groups will be required. Depending on the nature of the project, Scale-Up and Research may also be involved.
- Do not omit outside contractors. If your team involves third party organizations to perform design, construction, and/or validation activities, include them.
- Keep the forum open and highly interactive. Communication is the key. Expectations, conflicts, and agreements must be discussed and documented in order to establish a project basis.
- Use an "unbiased" third party or individual to facilitate the session and document the issues that are addressed. It is more effective to let the project team communicate rather than worrying about managing the meeting or keeping records.

Roles and Responsibilities Matrix

Note: Use individual names rather than job titles

	Project Executive	Finance	Engineering	Compliance	Construction	Operations	Vendors
Establish Production Requirements	C					R	C
Define Production Process	C					R	C
Define Production Control System			R			C	
Identify Process Equipment			R				
Prepare Validation Master Plan			I	R		I	
Select Site	C		R	C	I	I	
Conduct Environmental Study & EIR	I		I	R			
Review Applicable FDA Regulations	I		I	R			
Geotechnical Evaluation			R				
Prepare Schematic Drawings			R			I	C
Prepare Outline Specifications			R			I	C
List Technical Standards			R			I	C
Approve Design Criteria			R	C	C	C	
Perform Constructibility Review			C		R	I	
Develop Summary Schedule	C		R			C	
Prepare Budget Estimate	C		R				
Perform Investment Analysis	C	R					
Identify Long Lead Purchases			R				C
Initiate Government Reviews	C		I	R			

R RESPONSIBLE for making the function or decision happen. Accountability and initiative are here.
A Must APPROVE, including the obligation to penetrate, question, understand, and concur.
C Must be CONSULTED by R prior to decision.
I Must be INFORMED of decision by the R person.

Figure 2.4 Roles and responsibility matrix.

- Do not be constrained by time. The time spent in this session will be some of the most effective spent in the entire life of the project. Do not constrain it to a few hours.
- Document the results. A formal report should be issued upon completion of the session to document agreements, open issues, assumptions, milestones, and so on. This document will become very important in time, providing a basis for reviewing changes to scope.

	Engineering			MFG			Scale-Up			QA/TS		
	A	A/N	R	A	R	I	A	R	I	A	R	I
Process Flow Diagrams (PFDs)		X		X			X				X	
Process Description		X		X			X				X	
Mass Balance		X		X			X				X	
Yield Balance		X		X			X				X	
Architectural Layouts	X								X			
P&IDs		X		X					X		X	
Room Criteria Sheets	X				X						X	
Detailed Engineering												
Specifications												
Process Equipment		X		X					X			X
Nonprocess Equipment	X											
Mechanical	X											
Architectural	X					X						
Electrical	X											
Civil/Structural	X											
Detail Drawings												
Process Equipment		X		X					X			X
Facility Layout	X				X				X		X	
Other	X											
Equipment List	X				X				X		X	
Bid Packages	X				X						X	
Contracts	X				X							X
Vendor Drawings												
Process Equipment		X		X					X			X
Other												
Vendor Submittals												
Process Equipment		X		X					X			X
Other												
Schedules												
Engineering	X					X			X			X
Construction	X					X			X			X
Project Start-Up			X	X					X	X		
Project Validation			X	X					X	X		
Validation Documentation												
Master Plan			X	X					X	X		
Turnover Packages			X		X					X		
SOPs				X						X		
Protocols			X	X						X		
GMP Checklist			X		X					X		

Figure 2.5 Approval matrix.

Another advantage to this type of meeting is from the team building that will occur. Team members learn more about their peers and their roles and expectations for the project. This will also begin to establish the continuity on the project for later activities that will occur during project execution.

Project alternatives

In the execution of any project, there is always more than "one way to skin a cat." Alternatives must be reviewed in relation to project needs and constraints in order to arrive at the best "fit." Alternatives will have different impacts on project cost, schedule, technology implementation, regulatory compliance, and risk. Project managers must remain aware that there are alternatives in not only what they deliver, but also in how they deliver it.

Analyzing technology

There must be an organized approach to assessing available technologies in relation to project needs and constraints in order to arrive at the best practical solution. It is common practice within most engineering and manufacturing organizations to look at technology from the standpoint of minimizing risk—in the delivery of finished product, in meeting regulatory requirements, and in the selection of materials of construction. The traditional result has been spending more time and money and not receiving commensurate value in return.

Technology assessment must begin by asking some basic questions:

- Is the proposed technology new or can existing technology be used to meet project objectives? One example here could be the implementation of a barrier technology microenvironment for filling operations instead of the more traditional approach of a class 100 filling room.
- For scale-up operations, what are the problems identified? Anyone who has been through scale-up of any new process knows that it is not as easy as it looks. Issues related to product purification, chromatography operations, and lyophilization can create major engineering issues.
- How much experience is there with the process? At what scale have you worked with the process? Is this similar to other products, does it have similar unit operations, or does it use similar raw materials?
- What is the risk philosophy of the corporation? How much are they willing to trade higher risk for lower cost?

Once all of the relevant technology information has been gathered, it then becomes important to develop a set of criteria that takes in all of the objectives, constraints, and guidelines of the project. Criteria may include the following:

- Process flexibility
- Operating costs
- Environmental considerations
- Site compatibility
- Equipment availability
- Regulatory compliance

Technology selection criteria usually fall into the following three general categories:

1. **Go or no go** — These criteria require compliance or the technology is dropped from consideration as being successful. If a technology would not pass regulatory requirements, it would obviously be dropped as an alternative. Another case might be if there are legal issues related to patent clearance.
2. **Needs** — These items are contained in the stated goals and objectives of the project. Meeting dosage capacity would be a critical need.
3. **Wants** — These are criteria that reflect the user group's satisfaction level. They are issues that would make them happy and that can add value to the project, but are not categorized as absolute needs. Process control philosophy, preferred equipment vendors, or level of automation could be considered technology wants.

Project management

Alternatives to the overall management philosophy of the project should also be explored for ways of achieving cost savings or improving the overall execution process. One possible alternative that could be explored is negotiating critical trade contracts early in the design stage in order to obtain value-added services instead of the traditional approach of bidding all subcontract work and taking the low bid. Many firms are finding value in this approach for hygienic piping and HVAC systems.

Another alternative could involve maximizing the use of preengineered equipment and systems that utilize proven components to meet process parameters instead of custom-designing systems to meet the process. This could enhance the ability to gain performance warranties from equipment vendors.

Site evaluation

The process of selecting a site for any manufacturing facility involves the evaluation of many different considerations and capabilities. There are both near-term and long-term issues to be addressed, as well as capabilities that may be unique to a biotech-oriented utilization.

Site evaluation is based on the assessment of relative strengths and weaknesses of alternative locations in order to meet project requirements. Simply put, the location must maximize the benefits for the owner—whether in existing space, on existing property, or on a new "green" site.

Overall business objectives are very important in the selection process. The final decision should be based on long-term needs, not just short-term considerations. This is especially true for reviewing the option of building on a new site, where the cost of the property will add a significant amount to the overall project cost. The considerations that should be reviewed in this case include the following:

- **Business climate** The business environment in a state (or country) should be very important in the site selection process. Slow business growth at the state (or federal) level is a matter of increasing concern to many owners, because they fear that a large proportion of the tax burden will be borne by businesses in order to help remedy an increasing budget deficit. The financial ramifications of short- and long-term problems of any region (e.g., high personal, business, and corporate tax structures or an aging infrastructure), must be evaluated and factored into the final decision-making process.

Many states (and countries) are currently considering ways in which to bolster their declining manufacturing base. Biotechnology is one of the industries targeted for new job creation. In many cases, firms are aggressively pursued by providing incentives such as job training, financing, and tax breaks.

- **Utilities** A new site must have adequate utilities to support the manufacturing operation. Availability, adequacy, and cost are the key factors, especially for water, sewer, and electricity.
- **Zoning** The regulatory environment at the federal, state, and local level must be factored into the evaluation process. While many areas are targeting biotechnology as a growth industry, few have encouraged or facilitated the creation affirmative zoning regulations. This will often become a critical issue as states or municipalities struggle with economic interests versus social goals and environmental concerns. A careful review of zoning requirements is recommended. Like it or not, visions of mutant bacterial strains causing a plague still exist.
- **Demographics** Information concerning the overall population, the labor force, and wage levels is available from many sources. The type of information that should be considered includes the following:
  - Historical and projected population growth
  - Historical and projected employment growth
  - Composition of the labor force
  - Median years of education
  - Percentage of college graduates
  - Median household income
  - Percentage of women and minorities in labor force
  - Labor force participation rate
- **Housing** An analysis of the housing market on a local and regional level is important. Look at individual towns and communities relative to cost, stability, and availability.
- **Quality of life** Different people have different viewpoints on quality of life. Assets such as beaches, mountains, recreational facilities, cultural events, educational institutions, and accessibility to transportation should always be considered in site selection.

It will be necessary to develop a large amount of information in order to conduct a site evaluation. One proven method of assembling this information is to develop a Request for Proposal, similar to Example 2.4, that requests the type of information necessary to start the evaluation process. This information will be developed for a number of site alternatives.

Example 2.4 Request for Proposal for ABC Biopharmaceutical Corporation

ABC Biopharmaceutical Corporation has a requirement for office, research and development, QA/QC, and manufacturing space in the Raleigh/Durham area. CenterWest is invited to submit a proposal based on the following criteria. The proposal format has been standardized in order to provide a fair comparison and to permit full consideration to each proposal submitted; we ask that you present your proposal in this format to facilitate the subsequent analysis of all proposals.

ABC Biopharmaceutical Corporation will be reviewing the proposal primarily from an economic perspective. In as much as some proposals may be dropped from consideration after an initial economic review, we strongly encourage you to respond with your most aggressive terms.

The specific requirements for the ABC Biopharmaceutical Corporation are as follows:

1. Location:	Land located along Evans Road. Please describe the previous uses of the land.
2. Initial Premises:	Approximately 150,000RSF
3. Design/Protective Covenants:	As part of this proposal, please provide a list of protective covenants relative to construction on this site and any other restrictions relative to design and construction.
4. Infrastructure/ Predevelopment Costs:	Please specify the cost of estimated infrastructure improvements to provide roads and utilities to the site. In addition, please describe in detail the nature and timing of any future infrastructure improvements that will impact the ABC Biopharmaceutical Corporation site.
5. Subsoil Conditions:	Please describe any unusual subsoil conditions that would impact the development of the building.
6. Environmental Status:	Please provide an estimate of any costs associated with hazardous waste remediation. Specify what, if any, environmental testing has been done on the site.
7. Land:	Please specify the size of the land area to be acquired by the ABC Biopharmaceutical Corporation and outline if the parcel of land can be subdivided.
8. Purchase Price:	Please quote a purchase price based on a price per acre.
9. Parking:	Please specify what minimum parking ratio is required under zoning for the development on this parcel of land. If there is a different ratio for different uses, please specify.

— continued

Example 2.4 (continued) Request for Proposal for ABC Biopharmaceutical Corporation

10. Permits and Approvals:	Please provide a schedule that outlines the details on the necessary permits and approvals required to commence and complete construction. Outline various local and state agencies and boards involved in the permitting and approval process.
11. Utilities:	Please specify the availability, cost, and entity that provides the following utilities: water, sewer, gas, electricity, and steam. In addition, indicate if any of these entities will offer any rate incentives to the ABC Biopharmaceutical Corporation.
12. Real Estate Taxes:	Please provide an estimate of real estate taxes and comment on the likelihood, timing, and valuation of any real estate tax incentives available to the ABC Biopharmaceutical Corporation. Specify the valuation method used to determine real estate taxes.
13. Use:	ABC Biopharmaceutical Corporation will be using the site for office, research and development, laboratory, and manufacturing purposes. Provide a copy of those provisions of the zoning regulations concerning permitted uses for the site.
14. Architectural, Enginering Construction	ABC Biopharmaceutical Corporation will be fully responsible for the design and construction of initial and future buildings, including the selection of the architect, engineers, and general contractor.
15. Area Amenities:	Please provide a list of amenities in the area, including transportation, retail, and hotel.
16. Other Companies:	Please provide a list of other owners and major tenants at CenterWest, indicating their usage and approximate square footage. Specify who owns the abutting parcels of land to the land along Evans Road and what the intended uses for the parcel is.

Conceptual scope and estimate

One of the most difficult tasks of any project manager is to develop a project scope definition that accurately reflects the requirements of the project based on early development information. This scope of work must be developed using input from a number of varied sources, it must define the boundaries of the project, and must reduce to the extent possible the uncertainties of the project in order to define budget and schedule parameters.

From the conceptual scope will come the first look at the anticipated cost of the project. Budget estimates that are developed at this stage are based on the experience of the estimators, available data on past projects that are similar in nature, and the agreements that are reached by the project team as to the provisions to be included in the project.

Scope definition

Scope definition refers to an idea or understanding; that is, an idea of what the project will be. Information gained from the project team must be analyzed to develop a technical description of the project and the engineering and maintenance design criteria.

The technical description of the project will include a process description that reflects the process operations as they are known to this point. This narrative will define process steps, raw materials, clean and plant utility requirements, and equipment identification. Included will be some preliminary information regarding operating temperatures, process times, and volumes.

From this information will be developed the first Process Flow Diagrams (Figure 2.6), the initial project equipment list and descriptions, mass balance information, energy balances, and a general facility layout that defines area classifications and preliminary material, personnel, product, and waste flows.

Conceptual estimate

The conceptual estimate is the first look at the cost of the project. This estimate is known by many names: factored, order-of-magnitude, design basis, and so on. Estimates done at this point usually involve a parametric or factoring technique to produce an expected cost.

Many A/E firms use the method of factoring installed costs from major equipment costs. The equipment costs are based on the preliminary information found in the technical documents that are used to rough size the equipment and provide assumptions on materials of construction. In many cases, budget-type quotations are received from equipment vendors as a cross-check to the costs developed. A preliminary project schedule is also developed to provide a basis for the estimate.

These type estimates have an accuracy range of 25–30 percent at best. The accuracy is a function of the information available. It is also a function of the experience of the estimator, particularly with newer biotech facilities. Table 2.2 gives an example of a factored estimate summary.

Estimates at this stage must necessarily carry a large contingency factor. Special provisions, escalation, and risk provisions must also be factored to produce a total cost for the alternatives given. The most important point to remember is that conceptual estimates should be used as a means to evaluate alternatives and make decisions on whether to proceed with a project as it

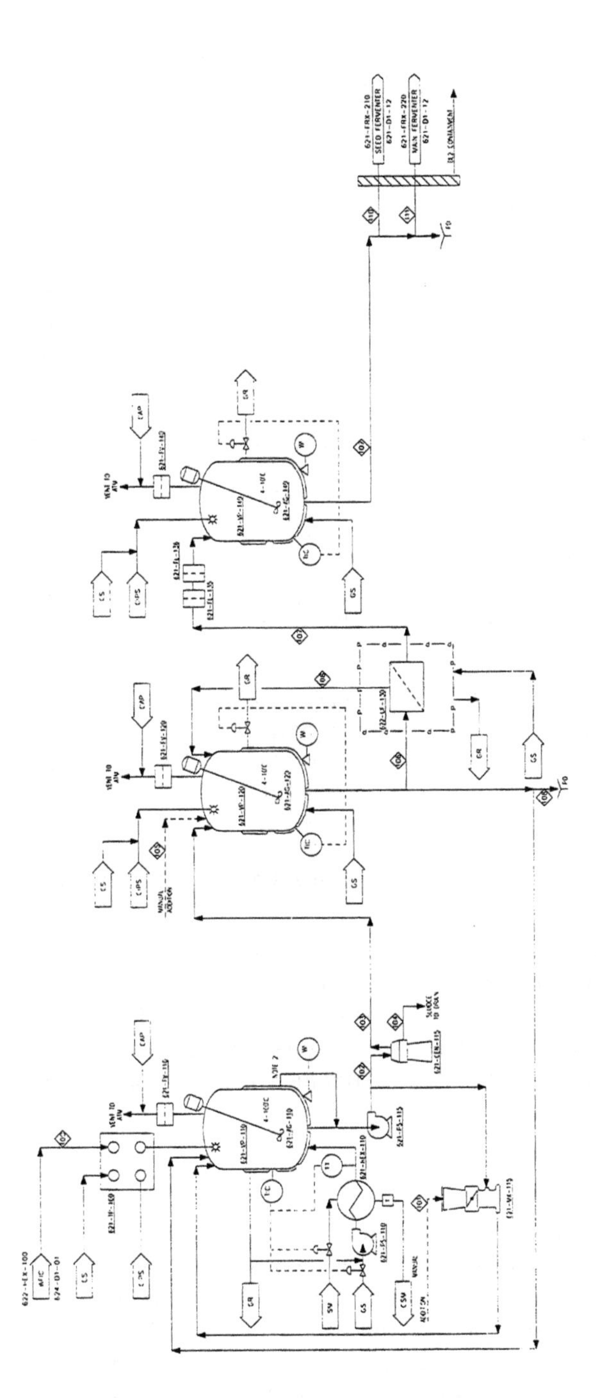

Figure 2.6 Process flow diagram for a fermentation process.

Table 2.2 Option #6B: Factored Estimate Discipline Breakdown for 31,744 ft² Facility

Item No.	Description	Item Cost	Cost per Square Foot
1	Site Work (incl. Site Electrical of $120,000)	$320,000	$10.07
2	Building Shell	1,077,330	33.91
3	Architectural Treatments	1,198,920	37.73
4	Building HVAC	1,471,290	46.30
5	Plumbing and Drainage	196,856	6.20
6	Fire Protection	83,331	2.62
7	Electrical	911,612	28.69
8	Process and Utility Piping	2,754,045	86.68
9	Instrumentation	1,292,305	40.67
10	Insulation/Coatings	284,775	8.96
11	Equipment (incl. Freight & Taxes)	5,965,480	187.75
12	Equipment Support	390,595	12.29
13	Equipment Installation, Handling/Erection	476,203	14.99
14	Permits	39,000	1.23
	Subtotal Direct Field Cost	**$16,461,742**	**$518.09**
	Construction Management	1,481,556	46.63
	Start-up and Commissioning	312,798	9.84
	Subtotal Field Cost	**$18,256,096**	**$574.56**
	Engineering Services	2,633,879	82.89
	Validation	1,037,261	32.64
	Fee	395,017	12.43
	Escalation	397,123	12.50
	Contingency	3,289,085	103.51
	Total Installed Cost	**$26,008,461**	**$818.55**

is defined. Many companies make the mistake of basing a final decision on estimates generated during conceptual development; it is the first number that anyone sees or hears and it is assumed to become the "gospel." By doing this, you run the risk of having uncertainty in scope definition translate into added costs to meet project needs as more detailed information becomes available.

Evaluation

The evaluation of project alternatives can only be accomplished by establishing a consistent basis for selection; you must compare apples to apples. In most cases, economics becomes the primary factor. Often, there will be no clear economic advantage between alternatives. When additional criteria must be considered (I believe there is always more to a decision than money), each criterion should be evaluated using tools that address risk, project needs criteria, and the potential range of outcomes.

We have found that using identified criteria developed by the project team and assigning weights to each criterion provides a systematic approach. Every criterion will not carry the same level of importance or confidence. Some of the information being evaluated will not have the same level of accuracy. This methodology will level the field with regard to these types of inconsistencies. It also ensures that individual agendas or predetermined perceptions do not have a disproportionate influence on the final decision.

Example 2.5 provides a sample matrix where this methodology was used to evaluate two hygienic piping contractors selected for a biopharmaceutical manufacturing plant upgrade. The criteria shown were developed in meetings involving the client user group representatives, the A/E firm, and the construction manager. Final selection was not based on price.

Project definition

Defining a project is more than looking at technical, operational, and regulatory issues. Current business conditions in the pharmaceutical industry have produced increased pressures on profit margins. Project managers are asked to treat projects as strategic tools that contribute to that bottom-line profit. To do so requires identifying, measuring, and managing risk both qualitatively and quantitatively.

Risk analysis

Risk management deals with established thresholds of cost, schedule, and technical performance. To manage risk effectively, you must first identify where the risk lies and to what extent. Risk identification can be difficult since it will rely heavily on the experience and insight of key project team members. In defining a project, the constant flow of information and feedback from individuals plays an important role in defining areas of risk and the ways in which it can be reduced.

The first area of risk analysis should be for the business elements of the project. These elements might include the following:

- Capital costs
- Operation/maintenance costs
- Start-up/commissioning costs
- Market considerations (size, share, life span)
- Process/technology issues
- Regulatory issues
- Financing
- Raw material availability

There are also a number of risks associated with the actual construction of the facility including the following:

Example 2.5 Final Evaluation Criteria Comparison Form: Process Piping

By: Subcontractor No. 1
Date: Subcontractor No. 2

Criteria	**Weighted**	**Eval. 1**		**Eval . 2**		**Eval. 3**		**Eval. 4**		**Eval. 5**		**Eval. 6**		**Eval. 7**		**Eval. 8**	
Description	**Value**	**Sub 1**	**Sub 2**	**Sub 1**	**Sub 2**	**Sub 1**	**Sub 2**	**Sub 1**	**Sub 2**	**Sub 1**	**Sub 2**	**Sub 1**	**Sub 2**	**Sub 1**	**Sub 2**	**Sub 1**	**Sub 2**
Validation Document	5																
Support Capability																	
Schedule Capability	5																
Quality of Trades/Subs	5																
Quantity of Trades	4																
Quality Assurance	5																
On-Site Project																	
Manager	5																
On-Site Staff																	
Capabiliity	4																
Total GMP	5																
Fee	4																
General Conditions																	
Costs	4																
Material Costs	2																

— *continued*

Example 2.5 (continued) Final Evaluation Criteria Comparison Form: Process Piping

Criteria	**Weighted**	**Eval. 1**		**Eva1 . 2**		**Eval. 3**		**Eval. 4**		**Eval. 5**		**Eval. 6**		**Eval. 7**		**Eval. 8**	
Description	**Value**	**Sub 1**	**Sub 2**	**Sub 1**	**Sub 2**	**Sub 1**	**Sub 2**	**Sub 1**	**Sub 2**	**Sub 1**	**Sub 2**	**Sub 1**	**Sub 2**	**Sub 1**	**Sub 2**	**Sub 1**	**Sub 2**
Labor Costs	3																
Stop Work Potential	1																
Criteria																	
Description																	
Safety	5																
Shop Fabrication	3																
Field Fabrication	3																
Working Relationship	3																
Understanding of Scope	5																
Ability to Manage																	
Changes	4																
	Totals																

Legend:

Most Proficient in Category = 2 3 Weighted Value (WV)

Least Proficient in Category = 1 3 Weighted Value (WV)

Both Bidders Evenly Proficient = Each Bidder 1 3 WV

Example: Quality of Trades = 5; Sub 1—Most Proficient = 2; Sub 1—Most Proficient = 2; Sub 2—Least Proficient = 1 (5 3 2 = Sub 1—10; 5 3 1 = Sub 2—5)

Evaluators:

- Site conditions
- Labor skill and availability
- Contractor/supplier performance
- Labor posture (strikes, work stoppages)
- Material shortages
- Constructability of design
- Weather

It is also important that every member of the project team understand that risk can increase or decrease due to

- Inaccurate estimates or schedules that define unrealistic goals
- The human factor—errors of omission, poor judgment, lack of understanding
- The failure to make timely decisions (time is money)

The final analysis of risk will usually come down to an analysis of business risk associated with profitability, risk due to cost variability in estimates, or risk due to uncertainty in the project schedule. Many tools are used for analyzing the available project data to assess risk, including the Monte Carlo method and PERT (performance evaluation and review technique) and CPM (critical path method) scheduling. Their accuracy depends on the quality of the information being evaluated.

Execution approach

Sterile manufacturing facility project execution requires the performance of numerous tasks by a variety of individuals, many times representing different groups and organizations both inside and outside the company. A formal execution approach is necessary to ensure that all of these tasks are identified and carried out in a timely manner by the appropriate members of the project team. This approach is defined in the project execution plan.

The plan must be specific and as detailed as possible in terms of roles and responsibilities. It should address every key project element relative to "how" and "when" the project will be executed as well as the procedures to be used. It must also identify the resources that will execute the project — job functions and their location relative to organization.

Project team members must recognize that the plan will be a dynamic document during the conceptual formation of the project. Plans and assumptions may change and the plan must have the flexibility to change as needed. Effectively managing revisions to the project is important. Example 2.6 provides an execution plan table of contents.

Project control guidelines

Effective project control is based on the following basic guidelines:

Example 2.6 Contents for a Summary Project Execution Plan

1.0	Introduction
2.0	Project Management Plan
3.0	Personnel/Staffing Plan
4.0	Project Controls Plan
5.0	Design/Engineering Plan
6.0	Validation
7.0	Information Management Systems
8.0	Procurement Plan
	8.1 General
	8.2 Contracting Plan
	8.3 Logistics Plan
9.0	Finance Plan
10.0	Construction Plan
11.0	Quality Plan
12.0	Safety Plan
13.0	Environmental Plan
14.0	Project Turnover, Commissioning, & Acceptance Plan
15.0	Project Closeout Plan
16.0	Management Participation Plan

- The scope must be adequately defined early in the project.
- The control budget and project schedule must be defined early in the project.
- A contracting strategy must be developed that reflects the consensus of the project team.
- Engineering design activities must support procurement and construction need dates.
- Procurement activities must be planned to support construction needs.
- A control system must be implemented that provides management with the information necessary to access project status from a cost and schedule basis and forecast progress against defined project goals.

Good planning and control require good communication. The project team must have all the necessary information to make prudent decisions related to project cost and schedule and be empowered to take corrective action when necessary. You must plan the work, then work the plan.

Project control involves the control of work and the control of cost. Control of work is often based on a work breakdown structure (WBS). Physical work activities are categorized by area and discipline in order to collect components of cost — labor man-hours and materials. The Construction Standards Institute (CSI) and the American Institute of Architects (AIA) have a 16-division standard WBS that is utilized by many projects. It is important that the project WBS reflect the project scope of work and facilities.

The WBS is then used to divide the project work elements into a finer level of detail for control. A work package matrix can then be developed for each WBS area. These packages will be used to schedule engineering design work to support construction contract releases. The packages are based on the timing of design release, contracting philosophy, cost efficiencies, and so on.

chapter 3

Defining the project team

We have previously stated that it takes people to execute projects. No surprise. The number of people, their backgrounds and experience level, their availability, the organization structure they work under, and their roles and responsibilities play a very important part in the execution process and whether the project is a success or a failure.

For many small or start-up companies, the development of a sound project team is a major undertaking; small companies with limited personnel resources may have little or no experience with major building projects. These teams are being asked to develop, execute, and control projects in an environment where few team members have any depth of related experience from which to draw. They are, in essence, being thrown to the lions in a situation where the financial risk is tremendous.

The owner's team

During our discussion of the formation process, we stated that it was important for every user group within the organization to have some form of representation on the team. We also said that the team must have a central point of contact—a leader that is empowered to make the necessary decisions to run the project.

Looking closer at the makeup of the Owner's team, you must recognize that input into important project decisions, whether they relate to the process, quality control issues, or validation, must be reviewed by more than just the contact person. In many cases, companies will have department managers or directors identified as the representative for a particular user group. This is done because these individuals already have some form of approval authority from management for their operations.

Assuming that these individuals also have detailed knowledge of day-to-day operations may be a mistake. In many cases, you will find that the detailed questions that must be answered in the development of a project can only be correctly answered by the operators that work in the plant, by the maintenance mechanics that routinely service equipment, and by the quality control inspectors that take daily samples and perform the necessary analysis. These individuals provide a wealth of knowledge during the conceptual development of a project. Don't leave them out.

Approach

The first step of the Owner's project team is to assess their own in-house capabilities by asking three basic questions:

1. How much expertise do you have in the areas necessary to execute the project (e.g., process design, HVAC design, architectural design, and construction administration)?
2. How much time can individuals devote to executing all of the necessary elements of the project?
3. Are you willing to accept the risks associated with having responsibility for the design and construction of the facility?

While the answers to the first two questions are easy to develop, the third raises many other questions that normally lead to the decision to hire outside contractors to handle the design and construction aspects of the project. Few small or young companies have the resources, both from a personnel and financial perspective, to tackle a major capital project.

Having made the decision to include outside contractors as a part of the project team, the next question becomes how do you choose the right firms for the job?

Selecting the architectural and engineering firm

Architectural and Engineering (A/E) firms can offer a variety of services for a project. These services may include engineering design, both conceptual and detailed; estimating, again both conceptual and detailed; architectural design, both of the facility and the process architecture to meet GMP; procurement of equipment and materials; validation support or actual validation execution; and, in some cases, construction services. These services can be packaged together or separated to meet specific objectives. And of course, this is all done for a fee.

When

Contrary to what many people believe, the first task of the Owner's team members is not to immediately rush out and hire an A/E firm. As previously stated, the Owner's team members must first ask questions pertaining to the needs, scope, and overall program. Knowing the objectives of the project are a must.

Without a preliminary scope document, an outside firm cannot provide efficient and cost-effective support to your company. If you do not know what you want, how do you expect someone from another firm to know? The premature hiring of an A/E firm is a costly mistake that still occurs today. These firms are expensive; rates for an average size conceptual design team can easily run $20,000–$30,000 per week or more. That rate and two months of unproductive time can easily produce design fees that well exceed $150,000.

There is also the problem of maintaining focus and momentum. The early days of a project are euphoric; everyone is excited and eager to start, focusing on that all important groundbreaking. But time spent getting bogged-down in trivial questions and the exchange of data seen as obvious by the Owner's team members can lead to bitter feelings and a loss of confidence in the A/E firm. Remember that A/E firms function best when you tell them what it is you want, when you want it, and how it is to be delivered.

How

Selecting the right firm to design your facility need not be a time-consuming, painful, unpleasant experience. The reasons why a particular firm is selected are as numerous as the sales personnel who work for them. The selection process, like the formation of the project, begins by asking the right questions.

The first step will be to develop a list of firms that are considered candidates based on the type of facility to be built. Some firms have better reputations as process engineers. Others are more experienced in fill-finish applications. Some firms specialize in small capital projects, say under $10 million. Other firms can provide the entire scope of services necessary for large "mega" projects, whether in the United States or overseas.

With this list in hand, develop a qualification questionnaire to be sent to each firm. Besides the basic general information relating to name, address, and telephone number of a contact individual, questions that might be asked include the following:

- Where are your office locations?
- Are you listed in Dunn & Bradstreet? If so, what is your current rating?
- What services do you provide?
- What is the number of current employees?
- What is the annual dollar volume of your professional design services?
- What is your current backlog?
- How many similar projects have you completed over the past three to five years?
- Has your firm been dismissed from a project or had a project canceled in the last five years?
- Have you ever failed to complete a contract?
- Does your firm have any pending legal action, suits, or claims? If yes, please explain.

All of these questions, and many more, are designed to provide some basic, yet important, information about candidate firms. In addition to this information, talk with your peers in the industry. Ask them about their experiences with firms, whether good or bad. Also meet with some of the representatives of the A/E firm for informal discussions on capabilities and services. This entire process should take no more than two weeks.

When this information is returned, review the questionnaire carefully. Which companies offer the services that you need? Which ones have the experience that fits your needs? If location is an issue, do they have an office close enough to respond rapidly?

Now you should narrow the list of candidates down to no more than three. These will be the firms that will receive the detailed qualification package and Request for Proposal (RFP). Using more than three firms means more work. Every additional firm means another proposal to review, another interview to conduct, and another evaluation to develop. This takes time!

Preparing the request for proposal

The solicitation of proposals from A/E firms is an art. You must ask the right questions, make your intentions clear, ensure that the scope basis for the bid is well defined to allow for uniform pricing, and make sure you get all the information you need. Clearly define responsibilities and deliverables if possible. Separate the project into the appropriate phases—conceptual design, preliminary engineering, detailed engineering—and give estimated time frames for the overall project.

The RFP should ask specific questions related to the organization of the A/E firm. Who are the proposed team members? What is their background? What will your organization look like in the various phases of the project (preconstruction, detailed design, and construction support)? What are the roles and responsibilities of each individual in each of the project phases?

Look carefully at past project references and ask pointed questions as to the person's involvement and whether he or she was with the current company when doing that work. Ask for personnel references of the key team members—the design manager, process lead, GMP coordinator, piping lead, project control manager. Check not only those references, but others from peers in the industry.

Also look closely at past project references of the company as a whole. Many times, companies will include outdated project references. Many of the personnel who worked on those projects may have died, retired, or left the company, which is of no use to you.

Ask for information concerning the execution philosophy and the ways in which the project will be managed and controlled. What systems do they use and how are they implemented? If information management is an issue, ask specific questions concerning the use of computer aided design systems, local area networks, and "paperless" document flow. Request examples of the various examples and reports that you feel are important.

The RFP should be very specific in defining the basis for the bid. Clearly define the services required—design, procurement, and/or validation. Based on your defined project schedule, request that the A/E firm provide schedule durations for the various phases of design. These durations should include estimated man-hours required to perform the scope of work.

It is also good practice to request a listing of deliverables that the A/E firm anticipates. The estimated number of drawings and specifications that will be generated will give insight into the overall understanding of your project's scope and the services necessary to develop a complete design package. The list of drawings can be developed on a discipline basis (civil, structural, piping, etc.). Specifications might normally be defined for equipment items and standard American Institute of Architects divisions (concrete, masonry, HVAC ductwork, etc.).

The RFP should also include a nondisclosure agreement for services and questions pertaining to preferred contracting methods. In general, most firms will be receptive to any contract format. Including a sample document can eliminate any confusion or incorrect assumptions early in the evaluation process.

Be specific in the method of pricing that is to be used. Several methods exist:

- **Reimbursable** The A/E firm will bill for the time its personnel spend on the project and the expenses they incur according to a predetermined rate. You should ask for an estimate of the man-hours and expenses the A/E firm believes will be necessary to complete the scope of services.
- **Lump sum** The A/E firm will quote a fixed price for the project scope of services. This price will include all costs, including any contingency that the A/E firms feels necessary to cover risk. With this method, deviations from scope that require extra man-hour expenditure will require Owner approval before implementation. It is good to ask for the rates of individual personnel as a cross-check against the overall price and contingency.
- **Guaranteed maximum (GMAX)** The A/E firm will guarantee a not-to-exceed price. The books are open so that if the project comes in under the GMAX price, the savings are either returned to the Owner or shared with the A/E firm based on a predetermined formula. Changes in scope will increase the GMAX amount. Any costs above the GMAX are incurred by the A/E firm.

Example 3.1 gives an RFP from a case project that involves the design of a pilot facility.

Allow sufficient time for the RFP to be completed by the A/E firm. In most cases, two weeks is an adequate amount of time for this information to be collected and put into an acceptable format. Much of the information you request will be readily available through previous proposal documents

Example 3.1 Request for Proposal for Engineering Services ABC Biotech, Inc. Pilot Plant Expansion Project

---

Your firm is invited to submit a proposal to provide design and engineering services to ABC Biotech and to assist in the construction and commissioning of a pilot plant expansion project, to be located in Anywhere, USA. Your firm, along with others, has been selected for this invitation because of your reputation, experience, and qualifications in the biotech industry.

The following information is being provided as the basis for the preparation of your proposal:

1. Proposal and Execution Requirements
2. Project Scope
3. Engineering Standards and Specifications for A/E Contractors
4. Standard A/E Contract

Please provide, as a minimum, the following information in your proposal:

1. A detailed explanation of the services you see as necessary to satisfy the Execution Requirements, including a detailed description of your project approach and timeline.
2. A comprehensive drawing list defining the deliverables necessary to construct the facility.
3. Total man-hours required to accomplish the proposed services. Include a breakdown of man-hours for each major phase of your project approach, and by discipline.
4. The proposed project organization for each phase of your project approach. Include detailed resumes of the key project personnel who will contribute to the project. Also include the office location for each of these key personnel and two references that can be contacted.
5. Specify the billing rate for each person and whether it is based on a salary multiplier or a fixed rate. Indicate the multiplier rate for this project if a salary multiplier will be utilized.
6. A detailed description of anticipated expenses, including a breakdown for each major phase of your project approach. Indicate how these expenses will be billed.
7. Time and material cost to complete the preliminary design and finalize the project scope. In addition, provide a time and material cost summary based on the total engineering services in your proposal.

Please review the enclosed sample contract and respond, in writing, to your willingness to work under this contract format or any particular items that you may have questions on.

Any questions should be directed to the names indicated in the Request for Proposal. Five copies of your formal proposal must be delivered to the Director of Engineering by noon on September 1, 1996.

---

and in-house information maintained by the A/E firm. The organization structure, defining personnel, and establishing the price will be the time-consuming activities.

Proposal review

When all of the proposal information is received, you should establish selection criteria that reflect the issues of importance. These will usually involve an understanding of the scope of work, the deliverables that will be provided, the estimated amount of labor man-hours that will be expended, the experience of the project team relevant to your type of project, the overall execution methodology that the A/E firm will use, the schedule in which the scope of work will be completed, and the estimated cost of providing all of the above services. Putting this information into some type of matrix form will help in comparing the scope and price of proposals (Example 3.2).

As your team reviews the proposals, you will find that each one will differ in some form. Execution philosophy may vary. The schedules will be different, with some A/E firms committing to a tighter schedule than others. Deliverables is another area where differences will occur. Be aware of the type and number of items, say PFDs, that an A/E firm defines as being required. The man-hours required to complete the scope will differ, possibly reflecting a more efficient method of execution or a lack of understanding of your scope. And it is a given that the prices will vary, sometimes by great margins.

During this internal review, questions will arise that can only be answered by the A/E firm. These answers will come during the A/E firm's interview. This interview will be a formal presentation by the A/E firm of their proposal, including the key project team members, their understanding of their roles and responsibilities, execution philosophy, control mechanisms, quality assurance, interpretation of the scope of work, familiarity of process requirements, and any other issues that might be deemed important. The important point to remember is don't hold back.

The interview location can be at either your office or the office of the A/E firm. Travel budgets and time will be the key factors. Allow a full day for the interview process. This may include a tour of your facilities or a tour of the A/E firm's offices, depending on the location.

The agenda for the interview should be flexible. Feel free to ask questions at any point. Do not feel obligated to follow a predetermined agenda by the A/E firm; seeing how flexible and fast thinking a proposed project team can be in an interview situation may give you some interesting insights into how they will perform on your project. Do not leave the interview until all of your issues have been addressed and all of your questions satisfactorily answered.

Conducting all of your A/E firm interviews in the same week is best if it can be accomplished. Information will be fresh in your mind. Issues from one interview may have to be addressed in the others. And it completes the interview process quickly, allowing you to proceed with the final evaluation and selection.

Based on the final analysis of proposal information and information gathered from the interviews, select the firm that most meets your needs

Example 3.2 Engineering Proposal Summary (Bids from Eight Different Companies)

Preliminary Design (weeks)	10	15	7	12	6	16	8	14
Detailed Engineering (weeks)	16	16	14	28	24	28	28	18
Design Fees								
Preliminary Design Phase								
Hours	2,478	3,076	4,439	4,349	2,400	6,360	2,720	6,533
Fee	$177,000	$199,940	$287,252	$256,500	$160,000	$347,200	$153,500	$374,348
Average Hourly Rate	$71.43	$65.00	$64.71	$58.98	$66.67	$54.59	$56.43	$57.30
Engineering Phase								
Hours	6,080	9,669	9,193	13,580	12,400	15,575	24,210	20,186
Fees	$392,000	$507,622	$566,158	$846,000	$740,000	$820,000	$1,270,900	$1,230,489
Average Hourly Rate	$64.47	$52.50	$61.59	$62.30	$59.68	$52.65	$52.49	$60.96
Cost Estimates	included	included	included	$24,800	included	$65,000	included	included
Haz-Op Review	??	included	included	$7,900	included	??	included	included
Design Cost	$569,000	$707,562	$853,410	$1,135,200	$900,000	$1,232,200	$1,424,400	$1,604,837
Total Design Hours	8,558	12,745	13,632	17,929	14,800	21,935	26,930	26,719
Design Cost per Hour	$66.49	$55.52	$62.60	$63.32	$60.81	$56.18	$52.89	$60.06
Design Reimbursables								
Preliminary Design	$4,500	$14,000	$27,280	$14,200	$27,000	$76,000	$37,200	$30,051
Engineering	$15,000	$50,760	$40,380	$45,300	$104,000	$166,100	$207,900	$92,856
Total Design Reimbursables	$19,500	$64,760	$67,660	$59,500	$131,000	$242,100	$245,100	$122,907

Design & Reimbursable Total	$588,500	$772,322	$921,070	$1,194,700	$1,031,000	$1,474,300	$1,669,500	$1,727,744
Construction Phase								
Hours	834	1,725	3,348	3,576	4,800	1,446	1,180	3,852
Fee	$72,000	$100,912	$243,026	$159,000	$290,000	$87,500	$66,800	$188,024
Average Hourly Rate	$86.33	$58.50	$72.59	$44.46	$60.42	$60.51	$56.61	$48.81
Construction Visits	1X/month	16 trips	2X/month	2X/month	2X/month	1X/month	2X/month	2X/month
Reimbursables	$15,000	$11,000	$40,260	$9,100	$44,000	??	$6,500	$17,719
Total Construction Phase	$87,000	$111,912	$283,286	$168,100	$334,000	$87,500	$73,300	$205,741
Total Engineering Cost	$675,500	$884,234	$1,204,356	$1,362,800	$1,365,000	$1,561,800	$1,742,800	$1,933,487

from a technical perspective. This will include their abilities and skills, their execution and control philosophy, their understanding of the scope of work, their deliverables, and the time frame in which they will complete the work.

Where many firms fall into a trap is first looking at the cost of services. Until you know what that cost represents in terms of deliverables, schedule, and execution, avoid concentrating on the numbers. Only after you have resolved the technical aspects of the proposals and have ensured that the comparison will be on an "apples-to-apples" basis should you worry about the price.

Have your legal department review liability issues, contracting philosophy, and reimbursement schedules. Once a selection is made and contract formation begins, keep legal representatives in the loop.

Having made your final selection and your first alternate, have the winning firm's representatives come in for any final negotiations that may occur. Stipulate that an agreement must be reached before the project begins, even if you are in a fast-track execution mode. It is extremely dangerous to enter into a project and have work begin without some legal agreement in place to cover costs, responsibilities, risk, and payment.

Selection criteria

The criteria used to select the right A/E firm will be based on a number of factors. Some of these will be related to technical issues and abilities. Some will be based on commercial terms and conditions. Some will be based on the comfort level that you have for the individuals and the firm. All are important.

Technical qualifications

We stated earlier that you should find a firm that has relevant experience with facilities similar to yours. If you plan to build a cell culture facility, experience in mammalian processes will be a factor. If your plans are for expanding clinical manufacturing facilities or building a pilot plant, look for that experience.

In evaluating technical qualifications, be sure to cover the following issues:

- **Process requirements** Look carefully at the experience of the project team related to process development and your needs. Traditional fermentation/purification process development capabilities are easily found. If your facility will utilize special conjugation processes for vaccines, deal with hazardous materials requiring special handling, require the design of a BL–3LS containment, or involve the implementation of barrier/isolation technology for final sterile filling, you must know that the people proposed for your project have that type of experience. Having someone learn on your project is not the best way to ensure success.

- **Validation** Also look at the facilities that the firm has completed and whether they have been successfully validated and put into commercial operation. I know of a situation where a client was looking at a firm to possibly perform design on a unique facility for human therapeutics based on a similar facility done overseas. After investigating the company's performance, it was found that the original facility took over two years to validate due to problems in the design of HVAC and Water for Injection (WFI) systems. Other firms were brought in to correct mistakes made by the original designer.
- **Site location** Depending on the chosen site location for your project, the A/E firm will encounter differing rules and regulations relating to environmental and permit issues. Selecting a firm that has done work successfully in your chosen state or country will be very beneficial.
- **Type of project** There are three main types of projects: grass-roots, expansions, and renovations. Grass-roots projects involve the design of a facility on a new, empty site. There are no constraints except the boundaries of the property. Expansions involve working on an existing site where ongoing operations and site constraints will come into play. Renovation involves taking an existing structure and converting it into something different.

  Of all these types, renovations represent the biggest challenge. Trying to fit a process to an existing shell, ensuring minimal disruption of ongoing operations, conforming to existing site conditions, and ensuring the safety of plant personnel is a monumental task. Be sure that the firm you select has a project team with a proven background in the type of project you have defined.
- **Project management** Executing a parenteral facility design program is not like executing a similar program for microelectronics or power facilities. While they all involve the management of technical professionals and the resources required to execute the project, they differ greatly in approach, compliance to regulatory requirements, and technical characteristics. Be sure that the proposed managers for the key project functions have pertinent project experience related to the size and type of facility.

Management qualifications

Firms and individuals have different management styles. This is not to say that any one style is right or wrong, but the A/E firm's style of management should fit with your style of management to ensure that the flow of information, the decision-making process, the emphasis on quality, the control of changes, and the issue and control of documentation meet your needs.

There must be an understanding of the approval process and the expectations that both sides will have related to the timely flow of information. Roles and responsibilities must be clear in order for these decisions to be made. If controlling access to certain individuals for information is an issue, be sure to identify it and obtain concurrence with the A/E firm.

Cooperation and a good rapport between the client and A/E firm is an absolute key to a project's success. Both parties must understand the requirements of the project and how the management structure will function to execute the project within the guidelines and objectives of the project.

The subtleties

In choosing the A/E firm, there will be those issues that I call the "warm fuzzies"; the subtleties that define the comfort level between individuals go a long way in determining the success of a project. Unless your team has previously worked with the A/E firm, the interview may provide the only chance for one-on-one interaction with the overall team.

It is true what they say about first impressions—you normally only get one chance. The comfort level that a group of individuals has toward one another is important. In many cases, careers will be on the line in the execution of a major facility project. There will undoubtedly be pressure to keep cost and schedule goals in line with project objectives. How the team members interact will be a key to success.

How do you address these "warm fuzzies"? There is not a concrete answer. Go with your intuition. Ask a lot of questions. If necessary, spend some private one-on-one time with an individual during the interview. Be comfortable with your final decision.

Selecting the construction contractor

The selection of a construction organization to build your facility must first begin with a decision on the delivery method that you wish to use for the execution of construction activities. The delivery method will define the execution philosophy that the construction firm will employ to manage and deliver the finished facility. There are several approaches that are used in the construction industry that must be reviewed before a decision is made. These approaches can be very confusing, especially to those who are not very familiar with the construction industry and its nomenclature. The important point is to choose a delivery approach that best fits your company's management style and philosophy, that works well in the local construction environment, and that will meet the project demands for cost control and schedule adherence.

There are many different ways to execute a project. Even to those in the industry, the boundaries between definitions of contractor philosophy can sometimes be unclear. The following is a discussion on the prevailing delivery philosophies that are in use today, based on the experiences of the author in each situation.

Engineer/procure/construct (EPC)

EPC execution normally refers to a situation where the engineering design firm also performs the procurement of equipment and materials and acts as the constructor for the project. In some cases, EPC execution is performed by two separate companies, one focusing on the design and the other on the construction, that enter into some type of joint venture or alliance for the purpose of executing a particular project.

The advantages of the EPC approach to executing a project are that you have only one firm to deal with. Coordination between the design, procurement, and construction efforts falls under a single point of authority, usually a project director or executive. The personnel involved with the construction team will be very familiar with the design approach, the management philosophy, and the methods of estimating and scheduling used by the design team since they often work for the same company or have worked together on previous projects.

EPC execution is also referred to as design/build by many people. This again emphasizes the single point of responsibility for the design and construction of the facility. The development of detailed design drawings and specifications, the procurement of equipment and materials, the construction of the building, the installation and testing of the systems, and, in some cases, the execution of some or all of the validation would be the contractual responsibility of a single firm.

This approach finds favor with many clients that are taking a "fast track" approach to their project execution; the philosophy being one of design and construction occurring concurrently instead of sequentially in a more traditional design-bid-build approach. Many large EPC firms also have the ability to perform some or all of the actual construction work through their own craft labor force. By doing so saves time in the procurement process due to the elimination of bid cycles for a majority of the work activities.

General contractor

General contractors are builders; they perform no design services. They take the full responsibility of construction based on the design provided by others. A general contractor will take design documents and either self-perform or bid out blocks of work. In most cases, the general contractor will have relationships (e.g., an alliance) with certain local specialty contractors for providing work in return for a discount on pricing. The general contractor will go to these preferred contractors in order to get the best price to develop a lump sum bid for the scope of work.

In this scenario, there is very little input from the builder into the design process. Design documents are completed, issued for bids, and the design is built by the general contractor. Errors in the design documents due to omissions or discrepancies are handled through change orders to the general contractor's original contract.

The traditional general contractor will have limited technical expertise on staff—technical referring to engineers, quality control specialists, or validation support personnel. The objective is to build what is designed.

Construction management

A construction manager is defined as an organization that acts in the role of Owner's representative for the management and execution of construction activities. A construction manager only manages; it will not perform any actual construction activities. The construction manager will take the design documents from a third party and package them for bid in a manner that will give the biggest advantage to the client. The construction manager will then manage the various specialty contractors on the project to ensure compliance to cost, schedule, safety, and quality requirements.

The construction manager will also challenge design issues through programs of value engineering and constructability analysis to ensure that design documents are in the best "buildable" state at the time of issue for bids. This is usually done during a period known as preconstruction, when engineers from the construction manager will review design documents as they are being developed by the A/E firm.

Construction managers can hold trade contracts directly with subcontractors, taking on some of the project risk from the client, or merely act as an agent of the client, allowing the client to hold contracts with the various subcontractors.

Selection process

The selection process for the construction organization should follow the same rationale as that for the engineering design firm. Asking the right questions will again be the key in developing the RFP.

Organization structure and the experience of individuals is again important. The roles and responsibilities of particular job functions will differ between delivery methods. Be sure that you (and the proposed construction team!) understand what role each individual has in managing and/or executing the work in the field. The roles of individuals will clearly differ based on the functionality of the organization.

Be sure that the individuals proposed have relevant construction experience, preferably in biotech or pharmaceutical projects where parenteral products were to be manufactured. Some people will argue that building is building; if you can manage one type of project, you can manage any kind of project. Not true! The issues related to regulatory compliance and the sanitary nature of the processes involved make sterile product facility construction one of the most challenging areas of the construction industry. That, coupled with the equipment-intensive nature of the processes, creates a difficult coordination challenge for meeting schedules that always have too little time for too much work. Having individuals learn on a project in this environment is not the ideal situation.

Remember that all of the individuals identified in a proposal have a strong likelihood of being on a current project assignment. It is rare that good, capable people are sitting around an office waiting! Be sure to question the availability of personnel and whether the construction company will be able to deliver them as promised. Many times, project schedule delays, client intervention, or personal issues will render proposed staffing assignments useless. While some switching of personnel will occur, be sure that the key team members that have influenced your selection have a firm corporate and personal commitment to your project. Beware of the "bait-and-switch" routine, where people are proposed for the purpose of getting a project when it is never the actual intent to deliver them.

Look for similar past projects as references. Contact these clients for information related to cost, safety, and schedule performance. Most construction firms will not give you a project reference on a project that went sour, but this type of information can be found through industry contacts with relative ease. If your project will be located in an area where labor posture is a concern, be sure that the RFP requests information pertaining to closed, open, or merit shop construction experience, and working with contractors in the local area. It is also important that the company be familiar with the construction industry in your project's region or state. Experience and a knowledge of the local contractor's labor pool experience and prevailing wages is very important.

Safety performance is another critical area that must not be overlooked in the evaluation process. As Occupational Safety and Health Administration (OSHA) regulations become stricter and insurance costs become higher, the ability to manage a construction site safely is essential. Companies are required to provide a vast array of safety statistics to state and federal OSHA offices. Asking for these statistics as part of the RFP is a common practice that will shed light on a company's record.

In the situation where an EPC approach to execution is preferred, treat the engineering and construction aspects of the project separately. The success attributes that are applicable to one may not be the same for the other. For example, in construction, safety performance in terms of lost-time accidents will be a definite factor; for engineering this is not a factor.

All nationally recognized construction firms have systems in place for tracking progress from a cost and schedule basis. Today, most are utilizing personal computer–based systems that allow greater flexibility at job-site locations. These systems have the ability to schedule activities, track materials and equipment, track progress, and provide a multitude of reports. But the issue should not be in the system that is used; it should be a focus on the ability of the personnel managing the construction to develop the correct data, interpret physical progress against predetermined goals, and manage construction activities. There is a disturbing trend in the industry of becoming dependent on what a schedule analysis performed by a computer tells you; if the machine says it, it must be correct. The danger in this approach can best be summed in a simple phrase: "garbage in, garbage out."

A more detailed discussion on cost evaluation and scheduling techniques is found in chapter 5.

chapter 4

Facility programming

Exceptional facilities do not just happen. They are planned to be functional, to be efficient, and to meet regulatory requirements. They are planned to be cost-effective. They are planned to meet market demands for product. They are planned to be environmentally pleasing to those who must work in them on a daily basis. They are planned to be safe, protecting the workers as well as the outside environment.

To properly plan, there is a need for information—information related to known facts, such as production volumes, site constraints, or regulatory requirements; information on project goals, such as schedule time frames, budget targets, and process flexibility; and information on needs, such as cleanliness requirements, utility requirements, and safety requirements. There must be information related to risk and how it can be minimized. There must be information concerning ideas that will improve productivity through improved technology.

Where will all of this information come from? How will it be collected, reviewed, and evaluated? And most important, how will it be implemented into the design of the facility? The answer to these questions is from programming—the process leading to the statement of a facility design problem and of the requirements to be met in offering a solution. [1]

First step

The longest journeys in life begin with the first step. In developing a program for a biopharmaceutical facility design, that first step comes in the recognition that successful projects, regardless of size or type, have two common attributes that can best be summed up from responses to a CII survey. Respondents were asked the question, "What makes a project a success?" The two almost unanimous answers were "a well-defined scope" and "early extensive planning."[2] Nowhere is it more important to recognize these attributes than in the planning for a biotech or pharmaceutical project where tighter product schedule demands and high facility costs create the potential for unacceptable risk exposure.

Although many other activities will have an impact on the overall success of a project, the emphasis must be on making certain that all members of the project team—the Owner, the A/E firm, and the constructor—all understand and agree on the project needs and how these needs will be met. Realizing that the ability to influence project cost and schedule is greatest in the early stages of planning is the most important point in defining project scope. To do this successfully requires input from numerous individuals dedicated to finding the facts that will clarify and state the problems facing the facility's A/E firm and constructor. The important first step is in facility programming.

The process

Programming is a process—a search for information to clarify, to understand, to clearly define the scope of the facility. Programming is known by many names, such as goal setting, alignment, or problem seeking. Each may employ a different methodology to collect, record, and review data that are generated. Their common attribute is that each begins by asking questions—the right questions. The method utilized should emphasize the goals of the project and the means of meeting those goals, whether they are driven by cost, schedule, or quality.

It is important to distinguish the difference between facility programming and facility design. Programming can be viewed as problem seeking (looking for the right information); design is traditionally seen as problem solving (finding the solution). These are two distinct processes that require different attitudes and different capabilities; yet they are both essential to responding to the needs of a client in defining the scope of a facility.

The theoretical design process involves two stages: analysis and synthesis. In analysis, the parts of a design problem are separated and identified. In synthesis, the parts are put together to form a design solution. The difference between programming and design is the difference between analysis and synthesis. A simple example will help to clarify this. The identified need may be for containment of a process step such as seed preparation; the design solution for that need may involve the use of a class 3 biological safety cabinet.

It should now be apparent that programming must precede design—coming before either preliminary or detailed design. Only after a thorough search of all the pertinent data regarding the needs of the facility and the concepts to meet these needs can the design problems be defined and resolved in an efficient and effective manner. But how are data collected?

Gathering data

The goal of programming is to begin the process of scope development and cost control; separating project needs, items that are essential to the manufacture of a sterile product or to meet some regulatory requirement, from wants—the inclusion of a new automated control process, the implementation of a new technology, or the desire for upgraded architectural finish materials. But before looking at individual design questions, you must have an understanding of the entire design philosophy—the facility design basis.

Scope development requires information and ideas. They both must come from many different groups within the client organization—Manufacturing, Engineering, Quality Control, Validation, Regulatory Affairs, Safety, Materials Management, and Maintenance. Any group that has the potential to directly or indirectly impact the final scope of a project should be required to provide information. And be aware that each will bring to the table a different set of needs versus wants. Example 4.1 shows the difference between wants and needs from the viewpoint of a manufacturing group for their process expansion.

The most common and effective way to collect data is to conduct programming sessions with designated representatives from these groups (Example 4.2) in an interactive forum that allows for the free flow of information that can be instantly reviewed. Usually led by an experienced programming facilitator, these sessions have as their main goal to build a consensus among the team members as to the scope of the facility.

The information exchange will normally begin with the use of questionnaires. Forms and instructions for their use will be sent to designated team members requesting basic information related to the facility (Example 4.3) — information that is factual and quantitative. This information will be compiled and a set of qualitative issues and questions developed for each discipline that will form the basis for the interview sessions with the team members (Example 4.4).

During this type of interactive session, the orderly display of information on a wall is one method of providing a good visual tool for keeping score. It becomes very easy to generate dialogue between the team members as information and ideas are graphically depicted and it allows individuals to easily spot information that is missing or interpreted incorrectly.

Steps

For this discussion on programming, five distinct steps of the process are defined:

1. Establish goals
2. Collect facts
3. Review concepts
4. Determine needs
5. State the path forward

Again, different approaches and methodologies may define these by other means. But these steps are essential in any systematic approach.

Example 4.1 Needs vs. Wants

"Must Have" List

1. All surfaces highly cleanable, including doors, window frames, floors, walls, and ceilings without causing "rust" or "peeling" to occur from disinfectants, detergents, hot WFI, and high moisture exposures. Typically, this is solved by PVC wall and ceiling covers, stainless steel window frames, stainless steel doors, stainless steel vent covers, stainless steel wall brackets, epoxy troweled floors, etc. Other engineering alternatives are acceptable but we must have surfaces that can tolerate the above mentioned agents and not result in peeling or rusting.
2. All critical outlets moisture sealed and covered to protect from electric shock. Outlets distributed to allow agitators and process equipment multiple-simultaneous use.
3. All sprinkler heads designed to create a sealed off area from the manufacturing suites to the interstitial spaces.
4. Emergency steam shut off valves located at the exit or outside the rooms where steam drop exists.
5. Predetermined routes to remove and/or install permanent tanks without tearing down existing walls. Typically, popout panels are used for this purpose.
6. All electrical lights sealed and recessed. Approximately 70–80 candle foot of light at the work surface is standard criteria for this type of manufacturing.
7. Differential air pressure gauges located outside manufacturing rooms.
8. All platforms used in product should be stainless steel, including decking support, legs, stairs, etc. This eliminates point chipping and rusting. Must have toe boards and nonskid surfaces.
9. A three (3) foot clearance around all equipment without interference from valves, piping, or other obstructions. This is for both cleaning access and safety from steam burns, etc. Also the process is not 100% defined or scaled to production volumes. Therefore, additional equipment, such as columns, filters, portable tanks, etc., may be required in the future.
10. Sufficient floor space within the manufacturing suite to allow for additional pieces of equipment (e.g., portable tanks, filters, and columns) that may be required during start-up of production.
11. All tanks should be 316L stainless steel with all supporting equipment constructed for triclover style fittings.
12. Bumper guards on all doors.
13. Cable channel for conduit runs for future MES (manufacturing execution system) wiring.
14. All portable tanks stainless steel with white wheels to prevent skid marks on the floor and bumper guards.
15. Reserve rooms for tables (desks) and file cabinets.
16. Reserve room for chart recorders.
17. Designated first aid kits in area.
18. Designated space for fire alarm.

"Strongly Desire" List

1. Stainless steel bumper guards on all doors must be replaceable.
2. Stainless steel benches to include stainless steel legs or supports in gowning and degowning rooms. Necessary to prevent corrosion from disinfectants.

Example 4.1 (continued) Needs vs. Wants

3. PVC covers on all walls in the processing, filling, glassware rooms (repairable and replaceable).
4. Recessed fire extinguishers throughout manufacturing.
5. All Edge Gard hoods of stainless steel construction. Light diffuser panels secured to allow cleaning without moving (flat opaque).
6. All corners in the processing and filling rooms covered from wall to ceiling, floor, and corners to improve cleaning.
7. All HEPA filter screens stainless steel.
8. All freezers, incubators, etc., using stainless steel platforms to allow cleaning under equipment.
9. All access panels to be stainless steel.
10. All utility piping, such as glycol and CIP, recessed within the wall within manufacturing, processing, and filling rooms.
11. All pressure relief valves stainless steel on all tanks.
12. Stainless steel drain covers throughout areas. Rupture discs for jackets are stainless steel.
13. Automatic valves on the bottom of permanent tanks to minimize operator crawling under tanks.
14. All chart recorders recessed into walls using same chart size.
15. Close in all exposed piping and wiring around autoclaves, oven, and bottle washers to prevent dust buildup.
16. Recess autoclave and over control panel including printer, etc., in wall recess.
17. Remote particulate monitoring throughout area.
18. Walkable metal deck above ceiling.
19. Stainless steel retracting hose reel on clean air and nitrogen systems.

The first three steps are based on the search for pertinent information related to the project:

- **Goals** What does the client want to achieve and why? Goals are seen as the ends to a project—the final results of your efforts. Goals should be specific (e.g., a licensed facility to meet CBER [Center for Biologics Evaluation and Research] guidelines, final production output of 50 million doses per year, total project cost of under $50 million, or project completion to meet a specific product launch date from marketing). Vague or "motherhood" goals should be avoided; examples include to provide a pleasant working environment, implement state-of-the-art technology, or provide a world-class facility.
- **Facts** These are the "knowns." Facts are used to describe existing conditions, whether related to process steps, economic data, site constraints, or user characteristics. It is important to note that facts are only important if they are appropriate and if they have a bearing on the design of the facility. The number of employees working in the facility, the existence of wetlands on the proposed site, the containment classification of the process, or any hazardous raw materials used in the process are examples of important facts.

Example 4.2 Facility Programming Session

ABC Biopharmaceutical, Inc. List of Attendees

ABC Biopharmaceutical:

1. Plant Manager
2. Director of Manufacturing
3. Lead Manufacturing Supervisors
4. Director of Facilities/Engineering
5. Maintenance Supervisors
6. Regulatory Affairs
7. Manager of QA Services
8. Manager of Quality Control
9. Materials Manager

DEF Engineering:

1. Project Manager
2. Director of Engineering
3. Process Architect
4. Lead Process Engineer
5. Lead HVAC Engineer
6. Project Architect
7. Procurement Manager
8. QA Manager
9. Validation Support
10. Lead Mechanical Engineer

GB Construction:

1. Project Manager
2. Construction Engineer
3. Project Services Manager

- **Concepts** How does the client want to achieve the goals of the project? There will always be more than one way to solve a problem. Concepts provide the means to an end—in this case the project goals. It is important to understand that programming concepts differ from design concepts. Programming concepts are ideas that can be seen as functional solutions to a client's goals and needs. Design concepts refer to concrete solutions to design problems. Flexibility is a programming concept; providing multiple processing suites or designing for over capacity could be a design solution for the concept of flexibility. Expandability is another programming concept; building in shell space could be the design solution. Example 4.5 provides a list of some of the basic concepts that are seen in many biotech facilities.

The next step of the programming process, focusing on the determination of project needs versus wants, will be the most difficult to complete. Determining needs inevitably becomes an economic feasibility step to see if a budget can be developed or met. It is at this step that the process of cost

Example 4.3 Example of Facility Programming Questions

1. What is the basic philosophy for conducting research in the pilot facility: team-oriented or individual researchers working with their assistants?
2. Will utilities be provided from a central source or from individual support locations in some proximity to the manufacturing operations?
3. Will glassware and equipment be washed in a central facility or within the manufacturing suites?
4. Will there be a central storage area for materials or will there be some satellite storage?
5. What do you see as the need for interaction between groups of people from different groups?
6. Are there any special containment issues related to the process; to hazardous materials?
7. What is the philosophy regarding redundancy of utilities and support systems?
8. What is the expansion philosophy for the facility?
9. Will the facility manufacture a single or multiple products? What will be the basis for multiproduct manufacturing; concurrent or campaign?
10. Will there be a need for redundant capacity to support the manufacturing process?

control begins on a project. The cost estimates that come from a programming effort are conceptual in scope, but are also very predictive of the realistic ability to meet a target budget.

Needs can be defined in four basic categories:

1. Space requirements
2. Quality of construction
3. Availability of funds (budget)
4. Time (schedule)

Space requirements

The functional needs of the facility, as defined by the process, will have the most direct bearing on the space requirements of the facility. Except where renovation projects create space constraints due to existing building footprints, the building space will be defined in a manner to fit the process and the people required to support the process.

Process areas or suites must be large enough to accommodate the vessels, support equipment, and personnel that will occupy them during production. Facilities that require BL-3 containment must have designated areas for biowaste collection and a kill system. Material storage areas should have the proper cold room space for the storage of intermediates that require special temperatures during designated QC hold points. These are examples of functional needs dictating space requirements.

Example 4.4 Issues and Questions

Existing facility expansion requirements:
- Raw materials
- Intermediate product
- Final product
- Utilities
- Personnel

Completion schedule for each individual building.
Expansion philosophy:
- New space
- Cannibalize and build new
- Shell space
- Surge space

Discuss central vs. local services:
- Lockers
- Glass wash
- Cafeteria
- Office/Administrative
- Maintenance
- Material staging
- Quality control
- Plant utilities
- Clean utilities

Discuss approach to FDA interface.
Discuss gowning philosophy.
Discuss preferred vendors.
Define cleaning and sanitization procedures and identify type of cleaning solutions used.
Review site material and people flow.
Review security requirements:
- Fencing
- Computer-controlled access gates/doors
- Closed circuit TV
- Flood lights
- Guard house
- Patrol roads

Discuss shop inspection philosophy.
Is office size criteria in study applicable?
Process related items:
- Define the packaging and shipping requirements for the product.
- Any existing equipment to be relocated to the new facility?
- Do you plan on manufacturing or recycling any of your raw materials?
- Define storage requirements (quantity, conditions, quality control) for both raw materials and finished product.
- Define the number of people required to perform the various production operations.
- Discuss implications of barrier technology.

Example 4.5 Basic Programming Concepts for Biotech Facilities

Priority	The concept of priority is related to the order of importance. The goal is to establish a priority based on the ranking of values. For example, emphasis on product flow may be more important than personnel flow when looking at contamination concerns, and, therefore, becomes the primary concern when developing flow diagrams.
Hierarchy	The concept of hierarchy is based on the need to express symbols of authority. This may be shown in the size and location of offices for personnel based on their "rank" within the organization.
Character	The corporate image that is to be projected is the basis of character. This image can be projected to potential customers as well as to the FDA—the perception that "we know what we are doing."
Accessibility	This will involve controlling the entrance to clean rooms through proper gowning areas, establishing designated routes for personnel and material flow, and meeting building code requirements for emergency and handicap access.
Separated Flow	Separate traffic patterns, both within the facility (operations personnel in clean areas versus lab personnel) and external (pedestrian traffic and automobile/truck traffic) are both important.
Mixed Flow	For common areas or "social spaces" only (break rooms, cafeterias).
Sequential Flow	The progression of people (personnel flow). This must address contamination issues, access to clean rooms, and maintaining product integrity. Flow diagrams are essential.
Orientation	Provide proper direction or reference, to prevent the feeling of "being lost," and avoid wrongful entry into restricted areas, clean rooms, or processing areas. Ease of entry and exiting.
Flexibility	Versatility, multiproduct, multifunction, expandability for future growth or new products. Campaign manufacturing, dedicated vs. nondedicated equipment. You must consider services and utilities as well as the architectural space and how it relates to the process.
Tolerance	Is a particular space tailored precisely for a static activity or is it provided with a "loose fit" allowing for change? Addition of a laminar flow workstation, a new fermentation train, or new automated warehousing equipment could be options.
Safety	Worker safety and environmental safety issues, such as OSHA standards and containment criteria to NIH standards.
Security Controls	You must be concerned with protection of property as well as control of personnel movement.
Density	A goal for efficient space use will result in the appropriate degree of density, for example, storage space for lab equipment, glassware, or utility support items. This concept can also be applicable to warehousing and storage issues.

— *continued*

Example 4.5 (continued) Basic Programming Concepts for Biotech Facilities

Service Grouping	Should services be centralized or decentralized? Not only the obvious utility services such as WFI, DI, or clean steam, but also lab gasses, storage areas, quality control support areas, or product testing.
Activity Grouping	Should activities be integrated or compartmentalized? Contamination concerns will be a factor as will containment issues.
People Grouping	Functional organization of people and the tasks they perform can dictate access into certain areas, gowning activities, or access/security control measures.
Home Base	Home base is related to the idea of territoriality—easily defined space for individual activities such as sampling, product testing, packaging areas, etc.
Relationships	The correct interrelation of space promotes efficiencies and effectiveness of people and their activities. Laboratory results will promote efficient testing procedures. Correct sampling locations must be effective. Material flow through the facility must be efficient and effective in preventing cross-contamination concerns
Communications	To promote the effective exchange of information, viewing areas or special intercom systems may be required, especially for cleaning and sterilization operations. The use of laboratory information management systems (LIMS) or MES are examples of communication issues.
Neighbors	The fit within the community, whether it is an established business park or downtown center. Compliance to zoning restrictions and/or community action plans, as well as building "covenants" are issues. To "be a good neighbor" is the goal.
Energy Conservation	Efficient operation of building systems for personnel comfort and cost control, as well as designing for an energy efficient building by keeping heated areas to a minimum and keeping heat flow to a minimum.
Environmental Controls	Controlling variables of the environment to meet clean classification areas as well as industry standards for the proposed working environment.
Phasing	Will phasing of construction be required to complete the project on a definite time and cost schedule? Is there a definite "window" for market entry? Are there urgent needs for validation timing to meet production schedules?
Cost Control	Search for economy ideas that will lead to a realistic preview of costs that will be in line with available funds. Cost control begins with the programming phase of any project.

Source: Odum, J. 1994. "Developing a Program for Biopharmaceutical Facility Design." Biopharm (May): 36–39.

Quality of construction

Quality of construction has a direct impact on the cost of a facility, usually expressed in the final cost per square foot that has become a familiar benchmark for discussion when addressing overall facility costs. When speaking about the quality of construction, we are referring to the alternative means that can be taken to achieve the end result required by the client. In parenteral facilities, there is an obvious quality issue related to compliance with FDA guidelines. This is coupled with corporate standards that will dictate the approach that is taken in meeting a set of predetermined minimum standards of construction.

A good example of quality of construction lies in the architectural finishes that are specified for a facility to meet regulatory guidelines. GMP requires that "Floors, walls, and ceilings be of smooth, hard surfaces that are easily cleanable." [3] Process requirements, such as materials used and cleaning procedures, will have an impact on the finishes that are selected. But in cases where all process issues are equal, some companies will choose a wall finish of epoxy paint over gypsum wallboard or concrete block walls with a plaster coat and epoxy finish, while another may elect to use a PVC wall covering over the wallboard instead of the epoxy paint. All are acceptable approaches in the mind of the FDA; the differences are shown below in the cost of materials and installation and a comparison of the life cycle economies of each alternative.

- Alternate A: Gypsum wallboard with epoxy paint
- Alternate B: Concrete block wall with plaster coat and epoxy finish
- Alternate C: Gypsum wallboard with vinyl sheet surface

Assume a 2000 ft^2 total wall area based on 10 ft ceiling height.

	Materials	Labor	Total
Alternate A:	$1,920	$2,900	$4,820
Alternate B:	8,440	12,040	20,480
Alternate C:	12,320	5,500	17,820

This same type of philosophy can be expanded into other architectural finish areas, such as floors and ceilings, the level of automation that is designed into the process, the finish requirements for hygienic equipment and sanitary process piping materials, the environmental monitoring methods that are employed, or the use of advanced systems such as barrier systems or computer integrated manufacturing. Each area will have alternatives that will meet GMP, while having different costs associated with their level of quality.

Budget

No company has unlimited funds to spend on every project. That is why there are budgets. A budget depends on realistic predictions of facility size, equipment requirements, type of construction, and other expenditures as percentages of building cost.

All of the decisions made during programming will have a dramatic impact on the budget. The selection of building exterior finish, the decision to utilize automated process controls, or the determination of required room classifications will determine the range of the facility budget. A simple example of how programming decisions impact the budget can be taken from a previous project.

The decision was made during the initial programming sessions to include in the scope of the project automated particulate monitoring of the process rooms, which were classified as class 100,000 areas. This area totaled approximately 8100 ft^2 in the preliminary layout drawings that were developed. The basis for the decision was a desire to reduce QC man-hours required to perform particle counts per current SOP manually.

Based on Federal Standard 209E, paragraph 5.1.3.2, there must be a sensor per every 100 ft^2 of area to be measured. Assuming particulate sizes of 0.5–10 mm, nonunidirectional airflow, and a laser type system, the estimated cost came to $137,000, including all material, labor, and engineering. Subsequent reviews of the project budget to reduce costs eliminated this system from the project scope, determining this was not a project need, but more of a want since the current method of manual monitoring was functional and acceptable to the FDA.

Time

Most projects will have time constraints associated with their required completion. This may be to meet market demand, to support research and development of clinical materials, or to replace existing facilities earmarked for renovation or demolition. Whatever the case, the project scope as defined from the programming sessions will give a clear indication as to the project team's ability to meet the schedule working under the defined project budget.

Some defined project needs will have major schedule impacts on a project. Process equipment requiring long lead times for design and fabrication is one example. Lyophilization operations, fermentation, and filling operations require specialized equipment that, in many cases, is manufactured overseas. The lead times for these types of equipment can reach 48–52 weeks.

Permitting is another area that can have a major impact on the timing of a project. If a project need defined in the process requires special permitting applications, for example, the disturbing of wetlands or the storage of materials classified as hazardous or toxic, the time required by many states and/or municipalities for review and approval is considerably longer. This time must be factored into the schedule during programming.

A different timing need may be associated with coordinating the completion and validation of a facility in order to meet a predetermined window of time for market launch or product transfer from another facility. In these instances, issues related to overall project duration would be defined as needs and budgets may reflect increases due to decisions made to meet the schedule through extended work hours, fabrication premiums for equipment, or accelerated engineering completion.

Path forward

The final step, a statement of the path forward, is really a statement of the premise for design of the facility—a package that will define the design criteria for the facility. Again, this document will not provide the detailed design solutions for the designer. It will identify the criteria that will establish the direction the design will take in meeting the needs of the client and the goals of the project.

This package will take all the information from the programming sessions and provide examples of the architectural flow diagrams (Figure 4.1) and the first look at the anticipated architectural program (Table 4.1).

Types of information

The information required for the effective programming of a sterile facility must address the entire scope of the design. Too little information can lead to only a partial statement of the design scope. Too much information, much of which will be easily seen as frivolous, can create "data clog" and burden the programming process to the extent that team members lose interest in the task at hand. The right amount of information will focus on the scope of the design, not on some universal problem. The designer must have a clear understanding of the problem before he or she can come up with a solution.

It is also necessary to address the issues of function, form, economy, and time in order to classify information.

Function

Function relates to what will occur in the building space. It involves the activities of the facility, such as scale-up, fermentation and purification, sterile filling, or packaging; the people who will be required to carry out the operations related to the activity, and the relationships of the spaces within the facility.

Form

Form defines the physical environment; it concerns what currently exists or what will exist in the facility space. Form refers to what you will see and how the environment will "feel" to the individuals who will work within it.

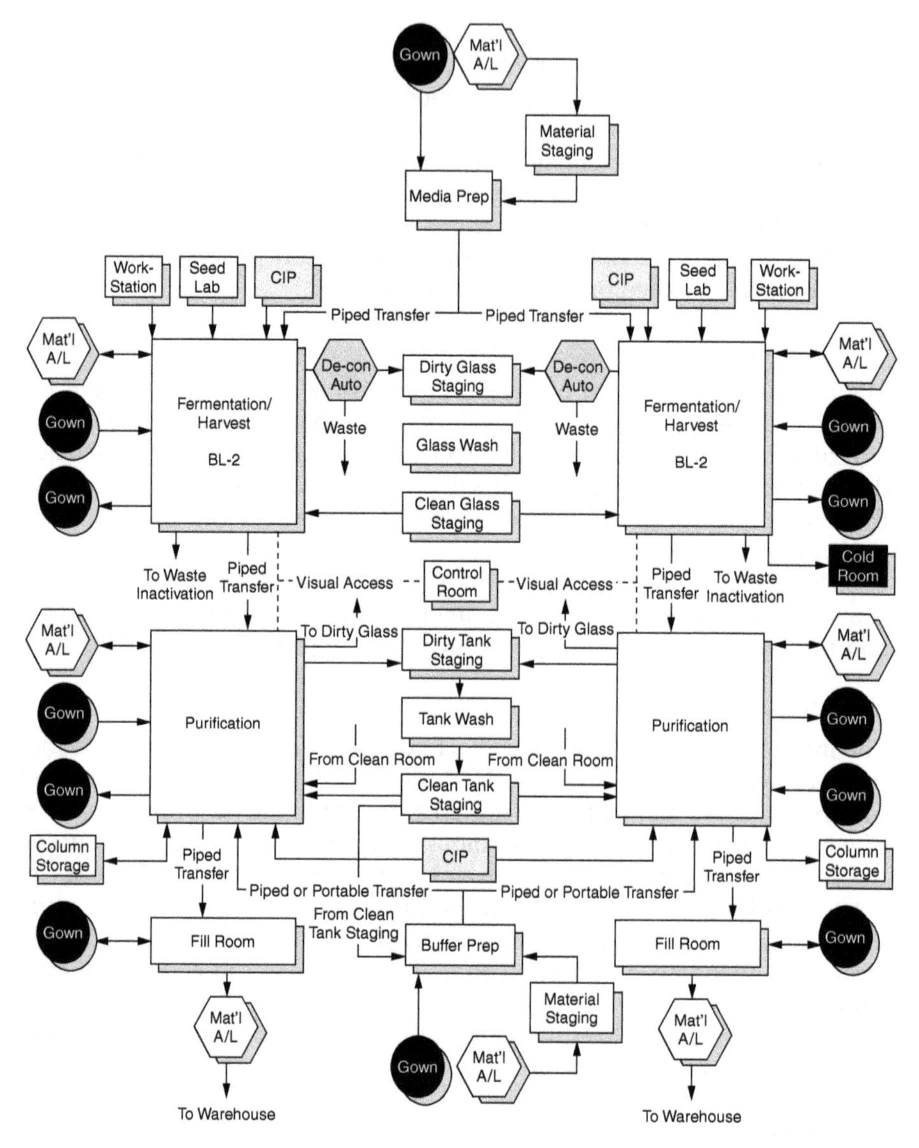

Figure 4.1 Architectural flow diagram.

Personnel and material flow through the facility and efficient space utilization are form issues. Bubble diagrams (Figure 4.2) are effective tools to show these relationships.

Economy

Economy concerns the cost of the facility and the budget for the facility. Quality of construction, operating cycles, and facility and equipment life cycle costs will all impact the economies and return on investment.

Table 4.1 Sample Architectural Program

Area	Function	Remarks	1997 ft^2 (net)	Area Status	HVAC	Finishes
Production						
Administration						
General	Managers	5@120 ft^2 Closed Office	600		95% ASHRAE	A
	Clerks	4@40 ft^2 Open Workstations	160		95% ASHRAE	A
Ferm. (Train 1)	Supervisor (1st Shift)	2@100 ft^2 Closed Office	200	Verify 1/2	95% ASHRAE	A
	Supervisor (2nd shift)	2@100 ft^2 Closed Office	200	Verify 1/2	95% ASHRAE	A
	Supervisor (3rd shift)	2@100 ft^2 Closed Office	200	Verify 1/2	95% ASHRAE	A
	Operators (1st shift)	5@50 ft^2 Open Workstations	200	Verify 50	95% ASHRAE	A
	Operators (2nd shift)	5@50 ft^2 Open Workstations	200	Verify 50	95% ASHRAE	A
	Operators (3rd shift)	5@50 ft^2 Open Workstations	200	Verify 50	95% ASHRAE	A
Ferm. (Train 2)	Supervisor (1st shift)	2@100 ft^2 Closed Office	200	Verify 1/2	95% ASHRAE	A
	Supervisor (2nd shift)	2@100 ft^2 Closed Office	200	Verify 1/2	95% ASHRAE	A
	Supervisor (3rd Shift)	2@100 ft^2 Closed Office	200	Verify 1/2	95% ASHRAE	A
	Operators (1st Shift)	5@50 ft^2 Open Workstations	200	Verify 50	95% ASHRAE	A
	Operators (2nd shift)	5@50 ft^2 Open Workstations	200	Verify 50	95% ASHRAE	A
	Operators (3rd shift)	5@50 ft^2 Open Workstations	200	Verify 50	95% ASHRAE	A
Purif. (Train 1)	Supervisor (1st shift)	1@100 ft^2 Closed Office	100		95% ASHRAE	A
	Supervisor (2nd shift)	1@100 ft^2 Closed Office	100		95% ASHRAE	A
	Supervisor (3rd shift)		0			
	Operators (1st shift)	3@40 ft^2 Open Workstations	120	Verify 40	95% ASHRAE	A
	Operators (2nd shift)	3@40 ft^2 Open Workstations	120	Verify 40	95% ASHRAE	A
	Operators (3rd shift)		0			
Purif. (Train 2)	Supervisor (1st shift)	1@100 ft^2 Closed Office	100		95% ASHRAE	A
	Supervisor (2nd shift)	1@100 ft^2 Closed Office	100		95% ASHRAE	A
	Supervisor (3rd shift)		0			
	Operators (1st shift)	3@40 ft^2 Open Workstations	120	Verify 40	95% ASHRAE	A
	Operators (2nd shift)	3@40 ft^2 Open Workstations	120	Verify 40	95% ASHRAE	A
	Operators (3rd shift)		0			

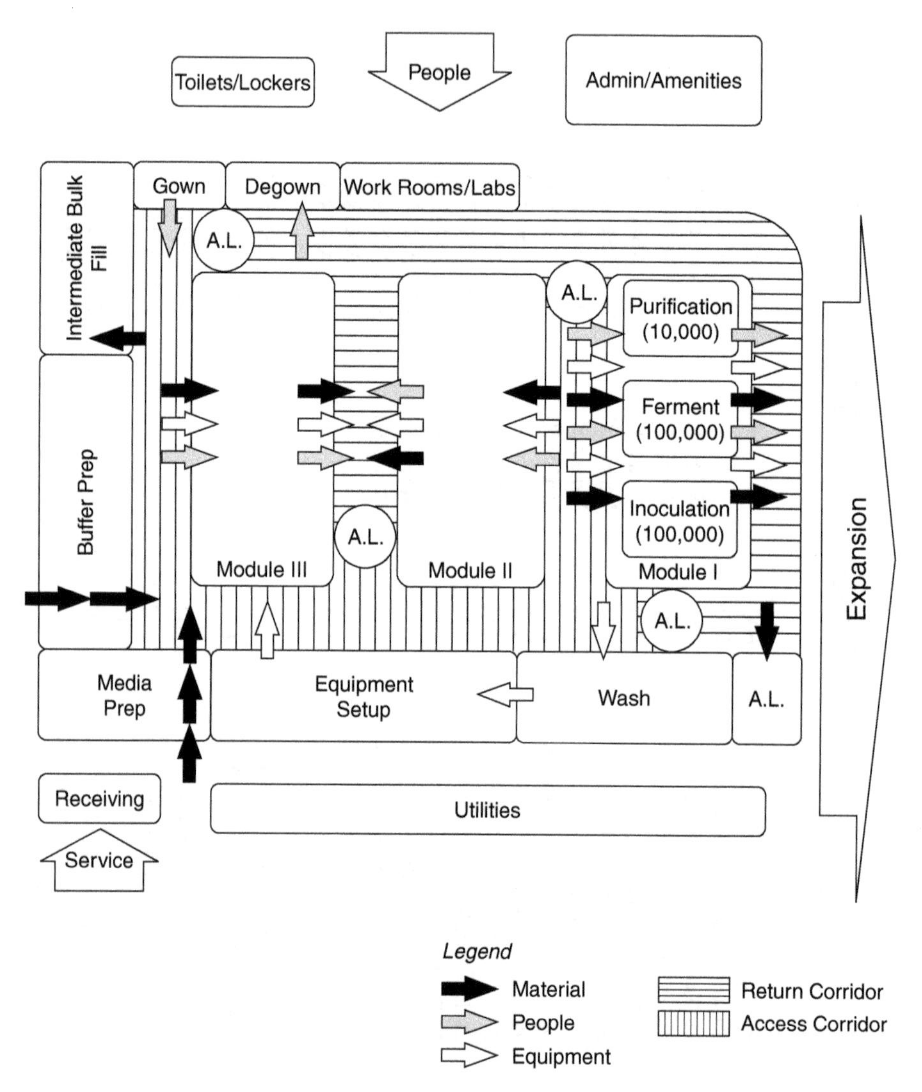

Figure 4.2 Bubble diagram of personnel flow. Source: Odum, J. 1994. Developing a program for biopharmaceutical facility design. Biopharm (May): 36–39.

Time

Time relates to more than an anticipated project schedule for completion. Time issues — past, present, and future — concern considerations such as the influence of previous project experiences, what worked and what did not work; the impact of current regulatory decisions, the change in the USP monograph for WFI; and the impact of future business growth, designing flexibility for increased production capacity.

Communication

Any team effort requires open communication — the need to communicate thoughts, ideas, and facts. Programming is no different. It is important for the participants to understand the need to make decisions during programming. Every decision that is made during programming will simplify the design effort. It is also important to remember that you are not trying to design the facility; you are only trying to define the characteristics of the facility.

An effective programming session is usually led by a facilitator or programmer, someone who will have responsibility for directing the programming effort and communicating the data from the session to the team members. Programmers do not have to know everything that the client knows, but they should know enough about the client's operations, needs, current facilities, and ideas to process efficiently and discard information that will allow for effective organization of the pertinent data. Anyone can add data; it takes an experienced individual to know how to discard.

There are several proven ways of communicating the data from interactive programming sessions — analysis cards, flip charts, brown sheets, or, most recently, computer graphic displays. The importance is in providing the programmer the ability to instantly give images of the ideas being expressed to speed up comprehension by the team members over a written description in a document. In Figure 4.3, analysis cards clearly depict the ideas and facts that were documented during an actual programming session.

Results

What do you get from programming? The results from this type of scope-defining session will be based on a consensus of the ideas and issues defined by the entire project team, focusing on separating needs from wants in order to begin controlling project cost while defining project scope.

The actual deliverables from this effort will come in many different forms, depending on the needs of the client. A conceptual layout of the facility will be developed based on process flows, space requirements, and spatial relationships reviewed during the programming sessions. This will give a first look at the anticipated footprint size, the arrangement of rooms, and the sizing of each room.

From this conceptual layout will also come the first material, personnel, product, equipment, and waste flows. Reviews for GMP against process operations will give early indications of whether these flows make sense based on operational philosophy and SOP. Contamination issues can also be addressed as a review of crossing of flow paths occurs.

Order-of-magnitude costs based on the simple square footage of the layout can be developed. The deliverables will also include a preliminary equipment list based on the process description, process flows, and layout.

Figure 4.3 Analysis cards from a programming session.

Some programming reports might also include site layouts, basic utility descriptions, and general HVAC criteria and classifications.

The programming report will identify the agreed-upon goals of the project, prioritized by the team. The goals will be specific in nature and will give management assurances that the project team understands the goals based on overall corporate objectives.

The programming report must also provide concepts that address the resolution of problems or questions that remained open at the end of the sessions. Optional facility layouts or site locations may be provided. Evaluations of new technologies, such as the implementation of barrier technology or computer integrated manufacturing systems, may also be provided to allow for additional review by management. Issues related to environmental regulations or building code requirements, such as waste disposal or fire code compliance, may also be included in summary form.

References

1. Pena, W. 1987. Problem seeking. Washington, DC: AIA Press.
2. Diekmann, J., and K. B. Thrush. 1990. A report to the Construction Industry Institute: Cost and schedule controls. Boulder: University of Colorado.
3. Code of Federal Regulations, Title 21, Part 211.42. 1990. Washington, DC: U.S. Government Printing Office.

chapter 5

Project control

One of the most important challenges facing management today is controlling the all-too-frequent cost and schedule overruns that impact pharmaceutical and biotech manufacturing projects, where such overruns can have a disastrous impact on corporate stability. Design and construction projects for validated facilities, with their associated long durations, complex systems, and large amounts of materials and manpower, present management with a formidable task of tracking resources and developing timely and efficient schedules.

A project control system must provide a framework for management to develop a project execution plan and to establish a system by which to control the scope of work. This system must provide the information needed by the project team to identify and correct problem areas and, ultimately, keep project cost and schedule under control.

Basic elements

There are two basic elements when dealing with project control: resources and time. It is one of the primary functions of the project manager to see that the required resources are available to execute the project and that they are expended in as effective a manner as possible.

All resources, whether they are labor hours, materials, process or utility equipment, or project indirects, can be converted to a dollar amount. The control of these resources involves three primary tasks:

1. A budget must be constructed for each resource.
2. The performance of each resource must be measured and evaluated to determine whether the resource is expended efficiently.
3. A forecast of resource needs must be made against present performance. The fundamental goal of project control is resource productivity. Are project resources being expended in the most efficient manner?

The second element of project control is time. Resources must be expended to meet a specified time frame for the project. Controlling time involves three primary activities:

1. The project must be planned: What are the required activities and in what sequence must they occur?
2. The project must be scheduled: How long will each activity take and when should it take place?
3. Progress must be monitored to ensure that activities are performed on time. Time control is concerned with production.

Project execution begins with a complete definition of the scope of work. This is more than just a good, sound technical design that meets all of the process and regulatory needs of the project. The scope must also provide the cost basis for the project and the time frame for completion.

The project budget must identify and quantify all the components that make up the total estimated cost for the project — design costs, material and labor costs, equipment procurement, general conditions, taxes, insurance, bonds, validation expenses, and start-up and testing costs. The budget must also define the contingency funds for the project and the parameters under which that contingency can and will dispersed. Finally, the budget must be tied to an anticipated cash flow that will allow for sound financial planning during project execution.

The project schedule will define the execution plan for the construction of the facility. It must detail the activities required and their most efficient sequence of execution. It should also reflect a sound utilization of resources that is based on a logical and achievable approach, not some "pie in the sky" approach of "more means quicker." The project schedule will be a living, dynamic document that is developed with sound logic and a clear understanding of the scope of work activities and the effort required to complete them.

The scope will define responsibilities — what is to be provided by the A/E firm, the construction organization, and the client in regard to specific project needs, such as purchasing of equipment or obtaining the necessary permits. It also relates to the providing of services — equipment inspections at vendor facilities, expediting, or the coordination of vendor services for start-up.

All of this information comes from an extensive planning effort that begins well before the actual construction of the facility. Once the project has been properly defined, it then becomes the responsibility of project management to answer some key questions: Where are we? Where do we want to be? How do we get there? Are we reaching our goals? The answers to these questions come through project control.

Plan the work, work the plan

Project managers in any industry know that a good plan is worthless if you do not know where you are, where you are going, and if you are getting

there. Project control, as practiced through a formal project program, provides the answers to these questions, allowing the project manager to adapt to changes in a manner that maintains a focus on the needs and goals of the project.

The tracking of progress and the managing of resources on a biotech project requires taking steps to ensure that actual performance conforms to the project plan. The project plan will act as the overall map for the project team's efforts, by establishing the project expectations as defined by the project goals. To see that those expectations are met, the project manager must

- Continually update the status.
- Analyze the impact of change.
- Act on any variances to the plan.
- Publish any revisions to the plan.

Let us look at the basic tools for controlling a project and how these basic steps for project control are implemented.

Work breakdown structure

Planning a sterile manufacturing facility project will require that the project be divided into identifiable parts in order to make it more manageable. As projects become larger, this breakdown becomes more critical. A work breakdown structure (WBS) creates these divisions, providing a framework for integrated cost and schedule planning. It also allows for easier monitoring and control by establishing the format in which estimates are assigned and costs are accumulated.

The WBS (Figure 5.1) should be established early. As the scope of work is defined and estimates and bid documents are prepared, the WBS can be adjusted to accommodate the data. Only after contracts have been awarded and work activities stated is the final WBS form set.

The smallest unit in the WBS is defined as a work package, or contract package, when the work is contracted out. Each package is a well-defined scope of work that is measurable and controllable. Packages are identifiable in a numeric system in order to capture both budgeted and actual performance data. Example 5.1 shows a work packaging listing from the WBS in Figure 5.1.

The size of each work package may vary, depending on the breakdown of work defined in the scope. Since the objective is to divide the work into the most manageable parts, all packages will not be the same. Smaller or less complex activities, such as electrical grounding, would have fewer levels of detail than a hygienic process piping package.

Each work package and its components will have an identification number that will be used by the accounting and project control systems. Each segment of the identification code represents a level of the WBS. The code

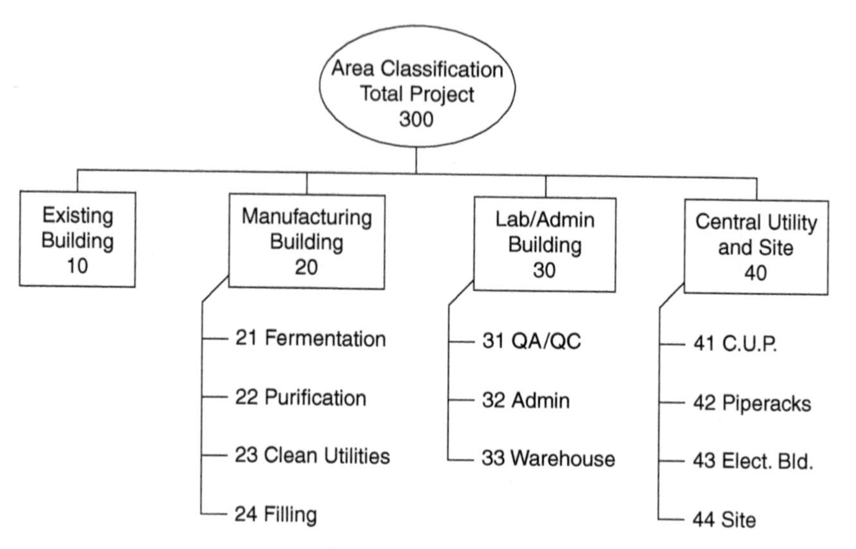

Figure 5.1 Work breakdown structure.

is nothing more than a group of digits that represents a building to be built (refer to Figure 5.1). The next level below the building represents the elements or contracts that comprise the building scope.

Estimating

Estimating is an integral part of project planning and, therefore, is an important tool in controlling a project's costs. Most biotech/pharmaceutical projects involve estimating: ROM (rough order-of-magnitude) estimates, trends, appropriation grade estimates, detailed estimates, and so on. A great deal of time is spent in collecting data, analyzing data, and refining data to come up with a "magic" number that will be held up for all the world to see.

Unfortunately, regardless of how good a job is done, there is one safe bet—the actual cost of the project will differ from the estimate. Since the first number that is seen by management tends to take on a somewhat biblical importance, the danger of under/overestimating a project becomes significant. Especially for large cost overruns, a search to pinpoint the reasons will often include a search for the guilty.

Estimating can be defined as the method used to forecast the bottom line cost of a project, based on a forecast of the value of each element that plays a role in determining that bottom cost. In this discussion, we will consider estimates to fall into two basic categories: preliminary, which would include conceptual, factored, and order-of-magnitude estimates, and detailed estimates.

Example 5.1 Work Packaging

Revision 0

WORK PACKAGING

Work Item / WBS Area →	Exist. Bldg Area 10	Bldg Area 20	QA/QC Bldg Area 31	Admin Bldg Area 32	Whse Bldg Area 33	C.U.P. Bldg Area 41	Site Work Area 42,43, 44,45	Total Project Area 90
00 Site Items								
01 – Site Prep,Grading,U/G Util.,etc.							X	
02 – Permanent Fencing							X	
03 – Fire Loop							X	
04 – Asphalt Paving							X	
05 – Finish Grading/Landscaping							X	
06 – Temporary Facilities							X	
10 Concrete								
11 – Building Foundations		X	X	X	X	X		
12 – Equipment Foundations		X	X		X	X	X	
13 – Slabs		X	X	X	X	X	X	
14 – Misc. Concrete							X	
15 – Fireproofing								X
20 Steel								
21 – Main Steel		X	X	X	X	X	X	
22 – Misc Steel		X	X	X	X	X	X	
23 – Pre-Engrd Bldg								
30 Architectural								
31 – Roofing		X	X	X	X	X		
32 – Siding		X	X	X	X	X		
33 – Masonry		X	X	X	X	X		
34 – Glass & Glazing								X
35 – Toilet Partn & Access.								X
36 – Carpeting								X
37 – Specialty Items								X
38 – Arch. Int. Finishes								X
3A – Demolition	X							
3B – Retrofit Existing	X							
40 Equipment								
41 – Process Equipment		X	X					
42 – Utility Equipment		X	X		X	X	X	
43 – Lab Equipment		X	X					
44 – Cooling Tower							X	
45 – Field Erected tanks							X	
46 H.V.A.C								
47 – Ductwork (incl. Insulation)		X	X	X	X	X		
48 – HVAC Equip/Controls		X	X	X	X	X		
50 Piping								
51 – Underslab Process Pipe		X	X					
52 – Underslab Util. Pipe		X	X	X	X	X		
53 – A/G Process Pipe		X	X					
54 – A/G Utility Pipe		X	X		X	X	X	
55 – Sprinkler System								X
56 – Plumbing		X	X	X	X	X		
60 Electrical								
61 – Grounding/U/G Elect.								X
62 – Lighting		X	X	X	X	X	X	
63 – Primary Power Dist.							X	
64 – Secondary Power Dist.		X	X	X	X	X	X	
65 – Telephone/Security/Communication								X
66 – Fire Alarm								X
70 Instrumentation								
71 – Instruments		X	X		X	X	X	
72 – Controls Wiring		X	X		X	X	X	
80 Insulation & Coatings								
81 – Pipe Insulation								X
82 – Equipment Insulation								X
83 – Painting								X

NOTES:

1. X denotes a work package
2. A contract will be made up of one or more work packages. These packages will be issued simultaneously or in some cases, at different times as addendum to the original contract. This will be a factor of the schedule.

Preliminary estimates

The accuracy of any estimate is a function of the information available used to develop the estimate. In preliminary estimates, the information that is used for development purposes is, you guessed it, preliminary in nature. The scope of the project is in its early development stages. Information may include a process description, preliminary PFDs, a conceptual layout of the facility that shows general room orientations and the overall facility footprint, facility flows that will begin early definition of room classifications, a preliminary list of process and utility equipment, and a preliminary site plan.

The factored estimate is one type of preliminary estimate that takes this type of information and, utilizing historical data for similar projects, estimates overall project cost as a factor of process equipment cost and adds building costs estimated for square footages of facilities based on preliminary layouts. Installation factors are applied against equipment categories within an area or building (Table 5.1). Indirect costs for items such as engineering, procurement, validation, contingency, and fees are estimated as a percentage of the total field cost and are summed to give a total project cost (Table 5.2).

Factored estimates have an accuracy range of 30–40 percent based on the detail of information and accuracy of historical data used. These types of estimates can be produced early in the project life cycle for minimal cost. However, because of their potential for deviation, users of this information must realize the potential for change that exists as more detailed engineering information is developed.

Another type of preliminary estimate is known as an appropriation or authorization grade estimate. This estimate is normally produced at the end of the preliminary engineering phase of the project. At this point, layouts and equipment arrangements have been established; the equipment list is more detailed since the P&IDs (piping and instrumentation diagrams) for all process and utility systems are complete and some preliminary pricing of items has occurred, there is more definition in the instrumentation and control philosophy for the facility, and building details related to room finishes, exterior treatments, and overall structure are complete. With this added detail, the accuracy improves to around 20 percent.

Detailed estimates

Detailed estimates contain very detailed information related to the design of systems and building facilities. Actual take-offs of materials for process and utility systems are made from orthographic and isometric drawings. Building costs are also developed from actual take-offs of materials. Equipment costs are actual prices from awarded contracts. Quantities of valves, instruments, specialties, and control/automation devices are defined. Unit rates for labor and contract costs will be included.

The detailed estimate has an accuracy of 5–10 percent. It will be based on substantially complete (70–85 percent) design documents. Allowances may still be made for items such as validation costs or Owner's cost related to items such as consumables or spare parts.

Table 5.1 Third Factored Estimate—Base Case

Item Description	QTY	U/M	Unit Rate	Cost
Central Utility Plant (Area 41) Summary				
Building Shell				$1,363,250
Interior Build-Out				$1,458,550
Subtotal Bldg. Cost				**$2,821,800**
Capital Equipment (including frt & tax)				$1,882,925
Equipment Installation (w/pipe, elect., instr., insul.)				$2,113,060
Subtotal Equipment & Install				**$3,995,985**
BMS/PLC/PC Interface (including Software)				$200,000
Direct Field Cost (DFC)				**$7,017,785**
Construction Management, Indirects, Construction Support	9%	DFC		$631,601
Start-Up/ Commissioning Assistance	2%	DFC		$140,356
Total Field Cost				**$7,789,741**
Engineering, Procurement & Support	16%	DFC		$1,122,846
Validation	7%	DFC		$491,245
Central Utility Plant Subtotal				**$9,403,832**
Escalation	4%	CUP		$376,153
Contingency (Design Development, Subpricing, etc.)	12%	CUP		$1,128,460
Fee	2%	DFC+Contingency		$162,925
Central Utility Plant Total Installed Cost (TIC)				**$10,532,292**

Estimate development

The development of the cost estimate for a sterile product manufacturing project is a challenge—feared by some, mastered by others, but appreciated by all when completed in a manner that is accurate within the established goals of the project. The technologies, industry standards, and regulatory guidelines that are intertwined into the fabric of a parenteral facility create very expensive pitfalls when not properly accounted for in the estimate.

Table 5.2 Equipment Factors for Factored Estimate

Facility Location	Equipment Category						
	1	2	3	4	5	6	7
Site Work	3.03	0.0	0.0	0.0	0.0	0.0	0.0
Production Area	0.0	2.63	2.07	1.50	1.15	1.08	1.31
Warehouse Area	0.0	2.68	2.09	1.13	1.20	1.13	0.0
QA/QC Area	0.0	2.68	2.12	1.50	1.20	1.13	0.0
CUP Area	0.0	2.58	1.57	1.13	1.20	1.13	0.0

Notes:

1. The above factors include freight and sales tax in the estimate.
2. The above factors are direct costs only and do not include engineering, construction management, or other fees.

Factored estimates

Developing a preliminary project estimate begins with an understanding of the information that is available. The PFDs and the preliminary equipment list must be cross-checked to be sure that all items are taken into account. The preliminary layouts must provide for all of the unit operational "areas" that the PFDs define, such as fermentation, glass wash, tank wash, and purification. The layouts must also reflect the material, personnel, and process flow philosophy that is dictated by the PFDs and the process description.

The estimate format will follow the WBS for the project, defined for specific elements on a cost-per-square foot basis for building elements, an installation factor based on equipment costs for process and utility systems, and some defined allowances for elements such as validation costs or spare parts based on historical data.

Table 5.3 shows a factored estimate format. In this example, capital equipment and installation costs are calculated as percentages of equipment costs. These equipment costs are based on a rough sizing of equipment and the anticipated materials of construction. Preliminary pricing information may be obtained from vendors to provide greater accuracy to the anticipated costs.

Different companies will use different estimating factors. In Table 5.3, the factor used for instrumentation is shown at 20 percent of equipment cost. Factors for instrumentation materials can be as diverse as

- Up to 75 percent of equipment costs for "heavy" instrumentation.
- 30–45 percent of equipment cost for "medium" instrumentation.
- 15–20 percent of equipment cost for "light" instrumentation.

Labor factors would be estimated within a range of 20–30 percent of material costs.

Table 5.3 Factored Estimate

Item Description	QTY	U/M	Unit Rate	Cost
Capital Equipment & Installation				
Category 2				
Standard process/ utility equipment	1	PCS	4,444,900.00	$4,444,000
Foundations and support steel	5%	EQ$	4,444,900.00	$222,245
Equipment installation	8%	EQ$	4,444,900.00	$355,592
Process & utility piping	75%	EQ$	4,444,900.00	$3,333,675
Instrumentation	45%	EQ$	4,444,900.00	$2,000,205
Process electrical	20%	EQ$	4,444,900.00	$888,980
Insulation/coating	5%	EQ$	4,444,900.00	$222,245
	Subtotal			**$11,467,842**
Category 3				
Skid-mounted equipment	1	PCS	6,164,300.00	$6,164,300
Foundations and support steel	5%	EQ$	6,164,300.00	$308,215
Equipment installation	12%	EQ$	6,164,300.00	$739,716
Process & utility piping	50%	EQ$	6,164,300.00	$3,082,150
Instrumentation	20%	EQ$	6,164,300.00	$1,232,860
Process electrical	10%	EQ$	6,164,300.00	$616,430
Insulation/coatings	5%	EQ$	6,164,300.00	$308,215
	Subtotal			**$12,451,886**
Category 4				
Portable equipment	1	PCS	1,547,400.00	$1,547,400
Equipment installation	5%	EQ$	1,547,400.00	$77,370
Piping	25%	EQ$	1,547,400.00	$386,850
Electrical	5%	EQ$	1,547,400.00	$77,370
Insulation/coating	5%	EQ$	1,547,400.00	$77,370
	Subtotal			**$2,166,360**
Category 5				
Benchtop equipment	1	PCS	75,500.00	$75,500
Handle/install equipment	10%	EQ$	75,500.00	$7,550
	Subtotal			**$83,050**
Category 6				
Equipment requiring no installation	1	PCS	283,500.00	$283,500
Handle equipment requiring no installation	3%	EQ$	283,500.00	$8,505
	Subtotal			**$292,005**

— *continued*

Table 5.3 (continued) Factored Estimate

Item Description	QTY	U/M	Unit Rate	Cost
Category 7				
Lyophilizers equipment cost	1	PCS	1,650,000.00	$1,650,000
Foundations and support steel	5%	EQ$	1,650,000.00	$82,500
Process equipment installation	2%	EQ$	1,650,000.00	$33,000
Process piping	10%	EQ$	1,650,000.00	$165,000
Instrumentation	5%	EQ$	1,650,000.00	$82,500
Process electrical	3%	EQ$	1,650,000.00	$49,500
Insulation/coatings	1%	EQ$	1,650,000.00	$16,500
	Subtotal			**$2,079,000**
Total Equipment Cost Only				$14,165,600
Freight on Equipment	4%	EQ$		$566,624
State Sales Tax	1%	EQ$		$141,656
Total Equipment (incl. Frt & Tax)				**$14,873,880**
Total Equipment Installation				**$14,374,543**

Equipment must also be categorized based on its configuration—skid-mounted equipment, portable equipment, benchtop equipment, or field assembly (standard installation and hookup of components). Each will have a different installation factor relating to the labor requirements for installation.

Estimating the "soft costs" in a project, those costs that do not contribute to the actual physical construction of the building, is where many projects run into trouble. Estimated costs for engineering services, procurement services, fees, general conditions, start-up assistance, construction management services, and contingencies in a factored estimate is an area to scrutinized carefully, especially when the final estimated cost exceeds your expectations. The summary will give the breakdown of these costs as percentages of either the direct field cost (DFC) or the total field cost (TFC).

Analysis of factored estimates — Rules of thumb

Once the estimate is complete, how do you know that the numbers reflected are good? There are some areas to analyze carefully that will give you a higher level of confidence that the projected cost of your project is within a reliable range. These "rules of thumb" are general in nature, based on experience, not science. Specific projects may vary.

- **Rule of Thumb 1** One of the critical areas of the estimate will be in equipment cost. Parenteral manufacturing facilities are generally equipment intensive, with high standards for fabrication, materials of construction, material finish, and process control. For this type of manufacturing facility, the total project cost should generally equal

Table 5.4 Total Cost to Equipment Cost Ratio for Sample Biotech Projects

Project	Total $/Equipment $ (in Millions @ Yr. Expended)	Ratio
Primary Manufacturing 1996	75.9/14.2	5.3
Primary Manufacturing 1996	22.9/5.2	4.4
Primary Manufacturing 1991	19.6/4.6	4.9
Clinical Manufacturing 1991	8.5/2.9	2.9
Multiproduct Manufacturing 1998 est.	110.0/23.0	4.8
Primary Manufacturing 1992	48.7/8.4	5.8
Pilot Plant 1995	9.8/2.5	3.9
Manufacturing Renovation 1993	29.5/6.0	4.9

four to six times the cost of equipment. Table 5.4 gives examples of this rule applied to actual project cost/equipment cost ratios for several biotech facilities constructed since 1990.

- **Rule of Thumb 2** Biotech and pharmaceutical manufacturing facilities are very mechanical system-intensive projects. The estimate should reflect the direct cost of mechanical systems (process piping, HVAC, utility piping) in a 30–40 percent range of total project direct costs.
- **Rule of Thumb 3** The estimated value of total "soft costs" for this type of manufacturing project should be in a range of 25–35 percent of total project cost. Although these costs vary, depending on policy and procedure between companies, identifying the cost exposure related to them early is important for controlling them as the project scope becomes better defined. If "soft costs" exceed this range, look carefully at each of the components to see if there is some valid reason for this high level. It may be that high engineering fees due to schedule constraints, high project contingency levels, or high insurance premiums (as examples) do in fact drive the cost to unacceptable levels.
- **Rule of Thumb 4** The typical ratio of direct manufacturing space to other support spaces within the facility is normally in a range of 35–50 percent of total gross square footage of the facility. If you find estimated areas that are grossly above or below this range, a review of the facility programming documents may be a good idea. You may find too much space allotted for wants versus needs, or you may find too little manufacturing space to support the process needs.
- **Rule of Thumb 5** For renovations involving fit-out of existing shell space, fit-out costs for full equipment installation, piping and HVAC, electrical/instrumentation, and finishes should run in a range of $1,500 to $2,500 per square foot of net manufacturing space. This incorporates all the necessary process and support systems. Numbers outside this range could indicate problems in scope definition that should be investigated.

Estimate trending

Project scenario You have issued the preliminary estimate for a project at a cost level that meets current projections for funding approval. Management is pleased with the cost information provided. As the design effort moves forward, more details of the process, utilities, and the needs of the user groups become integrated into design documents based on appropriate approvals. It is now three months later and the funding approval is coming before the board; it is time for the appropriation grade estimate. Upon completion of the estimate, you find that the project cost has risen 10 percent. What happened?

There is no greater importance than that of trending costs on a project during the development of the design. It is important to be able to identify quickly cost variances once preliminary estimates are developed, at a time when critical engineering decisions are being made.

Trend estimates are intended to give the project manager a quick order-of-magnitude cost of proposed changes in design—changes that can either add to or delete from the initial project cost estimate. Trending focuses on time; a timely identification of potential deviations from the project budget allows the project manager time to react in a manner that is in the best interest of the project. For projects that are schedule driven, this is very important since the design is evolving rapidly and many engineering decisions are made at a rapid pace.

The execution of a trending program is simple and can be implemented in a number of ways. In some instances, meetings are held on a regular basis to identify the proposed changes that will impact cost. Another method is to issue formal notices of potential changes for review and approval. Regardless of the method, the importance is to identify the change and its potential impact to the cost of the project.

Realize that the change is not only something new added to the scope, but in many instances the change comes from a more defined definition of the engineering design. For example, the original estimate could have factored the need for clean utility equipment; as the design develops for the process and the factors of capacity and system redundancy are calculated, it becomes apparent that additional and/or larger equipment for generating clean utilities is required. Another example is the amount of emergency generator capacity required for the backup of critical systems exceeds what was in the original estimate, requiring the addition of another emergency generator. In both examples, the basis of design has not really changed, only the details of the engineering solution.

The objectives of trending are as follows:

- Provide early indications of unfavorable cost trends. No one likes surprises when it comes to the cost of a project. Minimizing the impacts of potential surprises is a key function of trending.

- Identify potential schedule delays. Many times, cost impacts will also include potential schedule impacts due to increased design effort, procurement delays, or increased construction durations.
- Allow the most time for corrective action. Having the necessary information early can expedite decisions before detail design and/or procurement activities begin.
- Heighten cost awareness. Making everyone conscious of the cost impacts of engineering decisions is another key to successful cost management on a project. People will begin to think before they change the scope.

In Example 5.2, a sample cost trending report illustrates the documentation of cost trends that could occur during design development.

Contingency development and management

Contingency allowances in a budget are intended to protect against cost increases related to items that cannot be quantified and estimated, yet are known to exist in the project. These cost increases may occur due to variances in estimated quantities, the pricing of purchase orders and contracts, labor performance, or escalation beyond that assumed for equipment or materials. Other contingency items could involve delays in construction due to inclement weather, unknown site conditions, or schedule recovery due to delays from late equipment deliveries.

In many cases, a design contingency will also be added to a preliminary estimate due to continuing changes in process development or the need to hedge against potential operational philosophy brought about from pending regulatory concerns.

It is important to recognize that neither a design nor a project contingency should be viewed as a reserve fund to account for cost increases due to a change in project scope, changes in project schedule due to scope growth or ineffective project control, additional equipment or buildings not in the original scope, or extensions in the work week due to overtime. It is equally important to recognize that when the

engineering design is complete, design contingency should be eliminated and the project contingency used to provide funds for any expenditures meeting the contingency litmus test.

How much contingency is enough? The level of contingency funds carried in a preliminary estimate will depend on the type of project and the level of uncertainty associated with the design. For a traditional fermentation-driven process involving classical purification technology, the level would be lower than that of a new fill facility incorporating the latest in barrier technology. Renovation projects will traditionally carry higher levels of contingency funding than grassroots projects on "green" sites due to the numerous pitfalls that are inherent with renovations in operating facilities.

Example 5.2 Cost Trend Report

CLIENT: ______________
PROJECT: ______________
JOB NO.: ______________

BY : ______________
DATE: ______________
PAGE NO.: ________ of ________

COST TREND REPORT

<u>NOTES</u>: 1. Trend changes are based on the Preliminary Estimate issued 02–01–94.
2. Indicated trend value includes base item change plus all associated costs, such as foundations, piping, electrical, engineering, general conditions, and escalaction.

DATE	TREND NO.	DESCRIPTION	TREND ORDER OF MAGNITUDE	CUMULATIVE TRENDS VALUE	LATEST TOTAL ESTIMATE
02/01/94		Preliminary estimate issue – accuracy + 25%.			55,540,000
03/01/94	1	Additional demolition of Building 'A'.	450,000	450,000	55,590,000
03/15/94	2	Add spare plant boiler – BL – 1001. Increase capacity on reactor – R 1000.	250,000 400,000	1,100,000	56,640,000
04/01/94	3	Delete pumps 101,102,103.	(200,000)	900,000	56,440,000
04/15/94	4	Add CIP system to project scope.	1,400,000	2,300,000	57,840,000
04/30/94	5	Add motor control center MCC–5. Increase capacity of aeration tank. Change piping specification P–5000 from SS316L to C.S.–240.	150,000 200,000 (250,000)	2,400,000	57,940,000

Schedule

All sterile product facility project contracts, whether for engineering or construction, require the performance of a specific scope of work with a definite starting point and ending point. Every project manager needs a basis to evaluate a contractor's progress in meeting contractual obligations. The project manager also needs a tool for forecasting financial obligations, such as the amounts of progress payments, cash flows, and the supplying of owner-furnished equipment or services. There must also be a mechanism for coordinating the sequence of work activities and monitoring the progress made against predetermined milestones. The project schedule will provide these tools.

Schedule development and evaluation

A good schedule is really nothing more than a "paper model" of the execution of the project. Scheduling operations can be accomplished by many different methods, from simple bar charts done manually to sophisticated computer-based networks that can manipulate thousands of data points. Regardless of the method, the emphasis of the schedule must be to define the plan for executing the work and provide a means of monitoring performance against predetermined project goals.

Schedules are generally produced in a hierarchical manner, beginning with a summary project schedule in bar chart form (Figure 5.2). This summary schedule will identify major project milestones and estimated durations for major activities based on preliminary project information and previous project experience. The more detailed CPM logic networks that follow the summary schedule will detail hundreds and sometimes thousands of activities based on a detailed analysis of execution sequence.

There are numerous books addressing the basics of CPM scheduling and the process of developing networks. This discussion will only summarize by saying that there are dozens of computer software packages available to develop complex schedules from vast databases. But the real test of a schedule comes not from developing reams of data simply for the sake of the schedule, because schedules are only as good as the data that go into them and the effort made in monitoring progress against them. Even with the most advanced computer software, "garbage in" will result in "garbage out."

Summary schedules

Like preliminary budgets, the first summary schedule that is produced is the one that will be remembered by most of the management team in terms of milestones for starting and completing a project. It is, therefore, good practice to develop this schedule in an atmosphere of open communication with the project team in order to gain consensus on need dates that encompass the activities of each project participant.

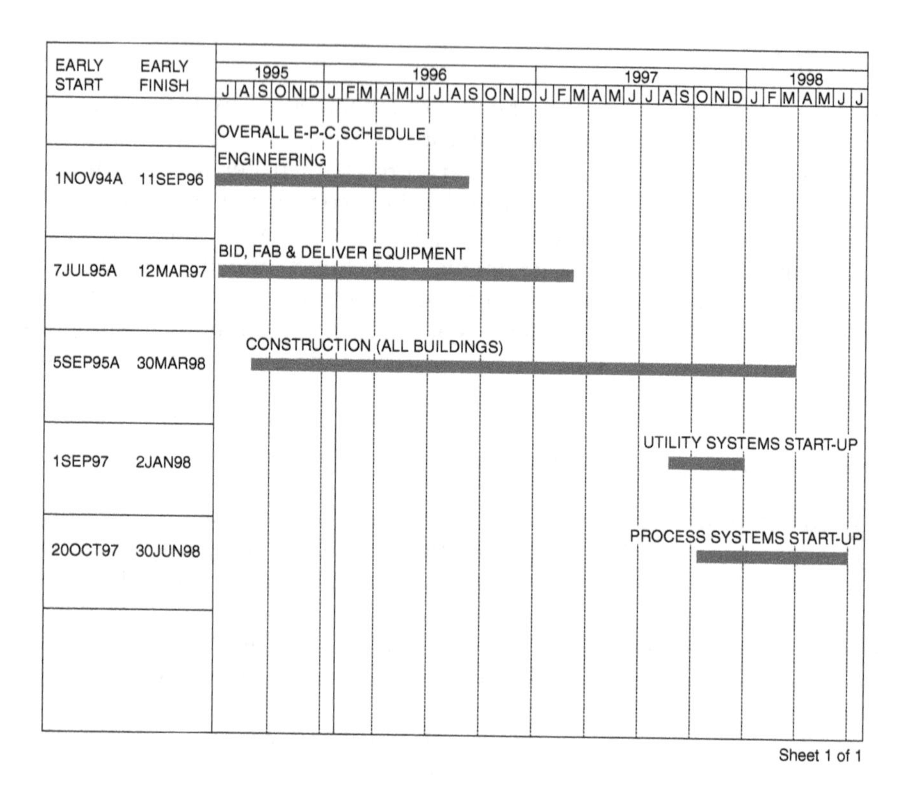

Figure 5.2 Overall EPC schedule.

One methodology that is used by many firms is an interactive schedule development approach involving the project team. In this method, each project representative organization (A/E firm, constructor, Owner) is assigned responsibility for identifying milestones that they have control over and plotting them individually on a timeline of the project. For example, the A/E firm would have the responsibility for the issue of design packages for civil work, concrete, structural steel, piping, and so on. The construction organization would identify when those design packages were needed in order to support the procurement of materials and the issue of contracts and in what sequence they must be issued. Certain approvals of design documents, the release of procurement funds, or the obtaining of some permits for locally or state-regulated functions would be the responsibility of the Owner.

Each of these milestones is placed in the appropriate timeline location, with all the members of the project team able to discuss the logic of the placement and the decision on its sequence. Cards identifying each activity and its appropriate start and end dates are color coded for each group.

Upon completion of this exercise, the resulting timeline with activities is converted to a computerized summary schedule and distributed for review and comment to the project team. After final corrections, it can then be issued to all project participants.

This interactive approach will provide many advantages over the development of a summary schedule by a single group. First, there is input from all project participants; no group is able to make commitments for another. Buy-in is immediate or the dates and/or sequence are changed. Second, it provides immediate visual recognition of where problem areas are located in terms of overlap of activities, out-of-sequence activities, and potential resource deviations such as delayed issue of design information or equipment deliveries. Third, it allows all parties to see the impact of delays and how the roles, responsibilities, and activities of each participant affect the activities of other team members. And it also starts the communication process off on the right foot; if we have made a schedule, then we need to do everything in our power to see that it works by upholding our commitments.

Activity breakdown

Unfortunately, it is a common practice for many contractors to oversimplify the analysis of their work activities. In many instances, this is due to unfamiliarity of the work, lack of experienced personnel to develop schedules, or to the fact that the work has been contracted to the lowest bidder where the preparation of detailed schedules on a regular basis is not in the budget.

Oversimplification can mask many areas of potential conflict. A simple example would be a single line identifying the work activities for interior architectural finishes over a specified duration. What this single line really represents is the coordination and sequencing of stud and drywall installation, millwork, casework, installation of terrazzo flooring, and application of epoxy surfaces on walls and ceilings, all of which must be coordinated with equipment installation and sequenced to avoid constraining other discipline work in a given room or area.

This lack of detail in a schedule is also a warning sign, indicating the possible omission of activities from the scope of work and the potential for a lack of controlling tools for work in the field. This lack of detail will not allow a project manager the ability to identify where a problem is, such as lack of productivity or insufficient manpower. By providing the necessary breakdown activities, a clear picture of the scope of work and the relationships of the activities is seen.

Long lead-time activities

There are many activities that occur away from the actual construction site that will have a tremendous impact on the actual sequencing and execution of construction work. These activities are known as long-lead items. Included in this category is the issue time for permits, procurement and delivery of specialized process equipment and components, and certain engineering design data that require certified information from equipment vendors. Schedules must take these types of extended durations into account in order to be accurate (Table 5.5).

Table 5.5 Common Long Lead-Time Procurement Items for Biotech Facilities

Item	Estimated Delivery Time (in weeks)
Lyophilizers	42–52
Fermentors	28–32
Centrifuges	32–40
Autoclaves	25–30
WFI Still	26–30
Clean steam generator	26–28
Biowaste treatment package	28–30
Stainless steel process vessels	23–26
Ultrafilters	22–26
DI water systems	24–28
CIP system	24–26
GMP glassware washer	22–24
Depyrogenation oven	22–24
Chromatography skids	24–26
Isolators	22–28
Boilers	20–24

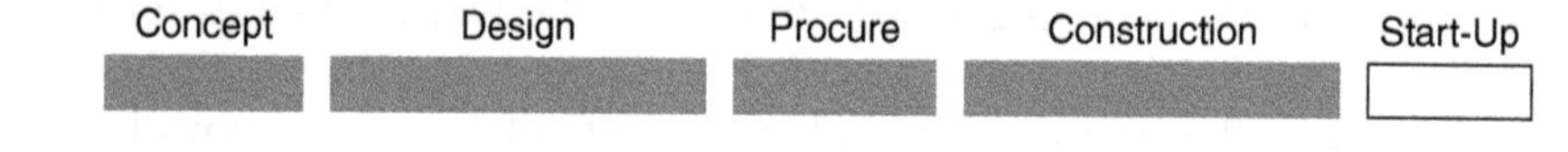

Figure 5.3 Conventional execution vs. fast-track execution.

No project can be built without design information. This simple concept is where many biotech projects run into trouble, especially in "fast-track" project schedules where the phasing of design and construction creates a heightened need for design information issued for construction (Figure 5.3). Orthographic and isometric piping drawings for sanitary process systems and HVAC duct routings must be available as early as practical to allow contractors time to evaluate material procurement and handling, access, and sequencing or work requirements.

As design continues, procurement dates will inevitably change due to extended bid dates, longer-than-anticipated contractual negotiations, or delays in final approval. These delays will result in later-than-expected delivery dates and delay in the receipt of the certified vendor data necessary to complete detailed mechanical and electrical design, such as drain locations and utility service hookup requirements.

Sequencing

Identifying all the pieces to a puzzle can be easy, as easy as turning the box over and watching them fall out. Putting them all together to complete the picture is more difficult. Doing so in a set time frame can be very frustrating, sometimes seemingly impossible. In scheduling, like the puzzle analogy, the sequence of how the pieces all fit together is very important and can represent the difference between meeting an end date or causing a costly project delay.

Many of the elements of the schedule will be obvious in terms of the proper sequence of events—foundations will follow site preparation, steel erection will follow foundation work, and underground systems installation must precede the placement of concrete floor slabs. Other elements will require closer review—which discipline leads, piping or HVAC? Does epoxy flooring precede equipment installation? When is the best time to paint?

Details become very important. Developing a good work sequence is something that must be done by the project manager and the engineers responsible for executing the work. Review dates for design availability, equipment deliveries, proposed contractor start and completion of activities, resource allocation, and accessibility issues must be done early and monitored continually. While commercial software can identify a problem in the sequence, the engineer must define how best to modify the sequence of activities to accomplish the overall schedule goal.

In areas where the sequence becomes critical, especially when dealing with multiple discipline contractors working in the same area, it is advisable to develop a specific schedule to look closely at just that small portion of the overall project. A good example of this type of special sequence is in the process of making system tie-ins to existing plant systems in operation. System downtime in an operating facility is costly, yet this must occur in some instances in order to bring new segments of a system into operation. Developing a special tie-in schedule (see Figure 2.2 on page 16) will clearly define the activities and their sequence of execution in order to minimize the impact of system downtime. This same approach should be evaluated for sequencing equipment installation and hookup, architectural finishes within congested rooms, the installation of mechanical and electrical system components in utility chases and on service racks, and in the final turnover and start-up of systems.

Activity durations

There is no standard time estimate for the performance of most construction operations. Unlike other industries where unit operation and process times can be predicted and controlled by sophisticated automation systems, sterile product facility construction is dependent on many unpredictable elements: human labor performance, weather, multiple suppliers of equipment and materials, and numerous government and regulatory agencies. Depending on the method of construction, size of the crew, work space, number of shifts, availability of materials, or use of premium time, any activity may be accomplished in different time durations.

It is important to identify the conditions that will affect the construction environment. For sterile product projects, measures to control cleanliness, such as capping open ends of cleanroom duct and sanitary process pipe, must be addressed. Environmental conditions related to the fabrication of sanitary process systems, the installation of equipment, or the installation of specialized flooring materials will impact durations. The level of quality control and inspection, the complexity of automation testing, and the complexity of facility commissioning also impact overall durations.

The analysis of any work activity duration must involve the allocation of manpower resources. By "loading" each schedule activity with the defined man-hours, any project manager can begin to determine the feasibility of the planned duration and the relationship of each activity to the overall manpower forecast.

Manpower loading is a simple exercise; simply take the estimated man-hours for a given work activity and allocate them over the planned duration of the activity. Today, this is as simple as pushing a key on your computer and letting the scheduling software produce manpower curves for you (Figure 5.4).

Manpower loading will reveal many potential problems with a schedule that would not be obvious on initial inspection. One primary concern will be the restrictions involved in limited space availability in the primary manufacturing suites and utility chases of these facilities where space is at a premium and the congestion of sanitary process piping, instrumentation, and equipment produces less-than-desirable working conditions. The physical space available to put the number of workers identified by loading the schedule activity may not be available.

Stacking of trades is a familiar term to all contractors, but its effects are especially difficult in sterile facility construction. The costs associated with "trade stacking" are magnified when dealing with restrictive environments, limited space, and the need to reassign manpower that has not been planned. Labor correction factors published by the Mechanical Contractors Association of America (MCA) reflect the impacts of "trades stacking" at 3 percent, "joint occupancy between trades" at 3 percent, and "reassignment of manpower" at 8 percent. Besides cost, these situations also impact overall worker productivity, increasing required man-hours, and further increasing overall project cost.

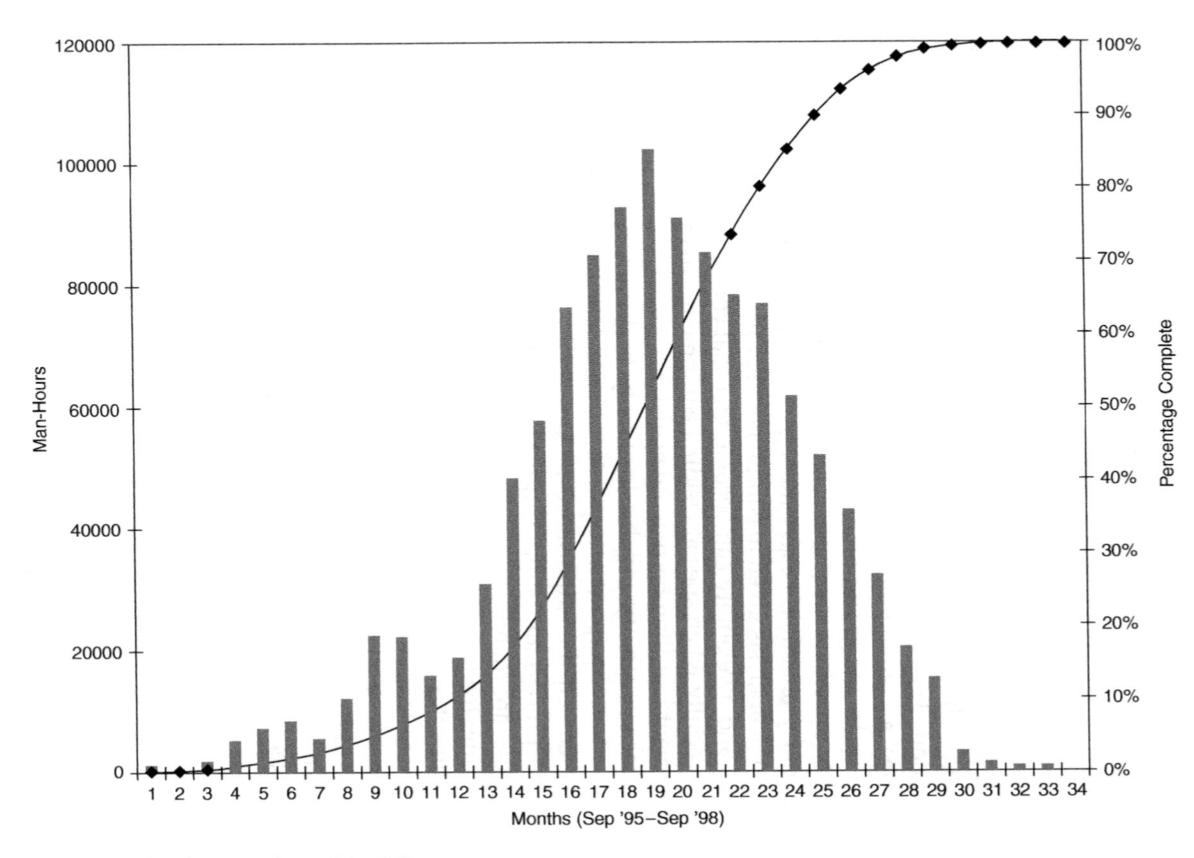

Figure 5.4 Overall man-loading totals—all buildings.

Durations of contractor schedules must also be carefully reviewed to ensure that their contract scope of work and their contract schedule represent the same scope. If your contract package identifies a schedule duration for the scope of work, many times contractors will simply agree with your schedule, hoping that they can complete the work within the allocated time or that other issues will delay the overall project duration sufficiently to allow them to complete the work. Again, take man-hour estimates from your budget estimate and compare them to those provided by the contractor in their bid and man-load their schedule in the same manner. You may find that the peak levels of manpower cannot be accommodated in the areas where the work will be performed.

The final analysis of the overall project manpower curve should show the project reaching peak manpower at 55–65 percent of the actual construction duration. This is true for most sterile product facility projects since process mechanical and electrical/instrumentation are the last activities to start but require the greatest manpower.

Activity durations must also reflect the impact of equipment allocation and material availability. Project managers should ask questions related to the amount of analysis given to equipment allocation by contractors. Equipment rates and availability can become financial constraints to any biotech project, especially when special welding equipment, environmental control, and equipment handling become unavailable.

Material availability involves not only bulk materials, but also the nonconstruction equipment required to complete the project. We have already discussed the importance of long-lead equipment identification. Since many process and support utility items for biotech and pharmaceutical facilities are manufactured overseas, schedule durations must allow sufficient time for not only actual delivery, but time to clear customs and time to allow for vendor assistance in installation if it is required. The availability of long-lead bulk material items, such as polyresin pipe, sanitary stainless tube and fittings, and sanitary process valves and assemblies, must also support the construction schedule. The availability of suppliers and their current backlog of orders will be important pieces of information for the project manager in evaluating overall activity durations if shortages become an issue.

Cash flow

Once schedule durations and resource loading have been completed, the schedule can be used to analyze the financial impacts of the project on a time scale. Cash flow is important to any project, but especially to start-up companies, pilot plants, or first-time manufacturing operations. Being able to show cumulative cash flow, cumulative commitments-to-date, and the differences between the two will give a good perspective of the financial status of the project. This will also allow cash flow projections for monthly, quarterly, and annual periods.

Project completion

When does a project end? When it is validated? When it is commissioned? When the construction work is complete? All of these may, in fact, be true, depending on the contractual definition of services to be provided by the various project participants. When developing a project schedule, it is very important to understand the definitions associated with the completion of the project, both contractually and functionally. If not, the results can lead to additional costs to the project and delays in the completion of physical work.

Many individuals in our industry have difficulty with the meaning of mechanical completion, the term most often used to describe the end point of the project's construction phase. You hear people discuss terms such as turnover, start-up, and commissioning and you begin to wonder, "Exactly where does mechanical completion fit?" The answer will depend on the project and the contractual definitions that are applied.

Mechanical completion vs. commissioning

Before any sterile product facility can be validated, it must go through a start-up phase, where systems are put through a series of functional tests and checks to see that they are ready for operation. Before this start-up work can be performed, there must be a "completion" of the systems and a program of static testing that ensures that the systems have been installed per design and that the scope of work has been completed. This is the aspect of the project referred to as mechanical completion.

Mechanical completion is usually defined in relation to the performance of work by the primary contractors on the project. The completion of their work on a typical sterile product facility project usually includes the following:

- Erection and installation of all equipment in accordance with design drawings and specifications and in accordance with vendor recommendations (Installation Qualification [IQ])
- Cold alignment of all equipment
- Mechanical, hydrostatic, and pneumatic testing of all piping systems for integrity
- Electric polarity checks of all equipment
- Loop checks of all instrumentation
- Airflow balancing of HVAC systems
- Activation of all plant utility systems, including charging of lines
- Completion of any vendor furnished start-up assistance program prior to functional checkout of systems

Commissioning can be thought of as the verification of the functional operation of the equipment and systems in the facility. Commissioning ensures that any system is operating properly and that all of the necessary procedures are in place to ensure consistent operation—the boiler system will consistently produce plant steam, the WFI system will produce the required quality of water, or the HVAC system provides class 100,000 air where it is necessary by system design.

In developing a project schedule for the construction duration of a facility, it is important to know how the contractual agreements in terms of scope of work to be performed assign responsibility for the completion of activities related to mechanical completion and commissioning. In Examples 5.3 and 5.4, two very different definitions of mechanical completion are given for the scope of work to be performed by the construction contractor(s). In Example 5.3, the owner has chosen to have the construction organization's scope of work include the system operations verification that is associated with completion of the Operational Qualification (OQ) required for validation. This will include all functional checking of equipment and systems, cleanroom certification, and complete debugging of system operational sequences for instrumentation. This is well beyond the scope of mechanical completion defined in Example 5.4, which follows the more traditional approach where the Owner's staff will have the lead role in performing OQ tests and verifications.

The development of the schedule must clearly reflect the definition of mechanical completion and the responsibilities that are associated with the various groups in performing the necessary functions. Do not lose sight of the fact that the work must still be completed, regardless of which organization does it. But clearly reflecting schedule milestones in terms of

Example 5.3 Mechanical Completion

The following activities define the requirements for contractual completion under the definition of Mechanical Completion for the ABC Biotech manufacturing project. All identified activities must be complete and approved by ABC Biotech for final project close-out of Process, Inc. contract activities.

1. Activate all plant utility systems, including plant steam, cooling tower water, potable water, compressed air, hot water, glycol, etc.
2. Activate all plant clean utility systems, including WFI, clean steam, and process air.
3. Energize the HVAC system to include final air and water balancing, installation of HEPA filters, replacement of primary and secondary filters installed for start-up, and functional checkout of the building management system. Cleanroom certification is not included in the scope.
4. Test all sequences of operation for automation systems.
5. Calibrate all instruments.
6. Systematically start up and run in all individual pieces and packaged units of new equipment installed for utility and process systems. Confirm that these pieces/units operate in accordance with design specifications, complying with their respective operational qualifications (OQ).
7. Provide all vendor training and start-up support.
8. Obtain final Certificate of Occupancy from local building officials.

Example 5.4 Definition of Mechanical Completion

The following definition of Mechanical Completion will be used to define the completion of the scope of services of Process, Inc. for the ABC Biotech manufacturing project.

- Mechanical Completion is defined to mean that all equipment has all necessary utilities hooked up, is passivated, calibrated, and all IQ documentation turned over to ABC Biotech.
- All utilities are started up.
- All laboratories and lab services are complete.
- All instrumentation loops checked, all motors checked for proper rotation.
- Rough balance of all HVAC systems.
- All building punch list items complete related to the structure and interiors.
- Final cleanup.

responsibilities will avoid any confusion late in the project when time is critical leading up to the start of validation activities to support licensing and manufacturing launch from the facility.

Schedule integration

Project schedules are developed for each of the different phases of a project's life cycle. Every organization that is providing services toward the completion of the project will develop schedules for their scope of work (the A/E firm, construction contractors, the validation contractor, etc.). The data contained in these schedules reflects the execution philosophy of each organization, focused on meeting the predetermined goals of when the project must be complete.

All of these data must be integrated into a project master schedule in order to allow the project manager to see that all of the pieces of the puzzle come together with the greatest level of resource utilization, efficiency, and effectiveness. In this day of advanced computer technology and lightning-fast software applications for scheduling, this may sound as simple as inputting data into a commercial software package and letting the computer develop the overall master schedule. The industry literature and software applications for developing a detailed schedule are endless, and will not be discussed in this text.

However, there is much more analysis that must go into the integration of a sound project schedule than a simple "data dump" into the computer. Unfortunately, many projects (and project managers) find this out too late. The relationships between engineering data and construction activities, the tie between procurement activities and construction execution, and the transition from "building the building" to "starting the systems" must be understood by the project manager if there is any hope for a successful project.

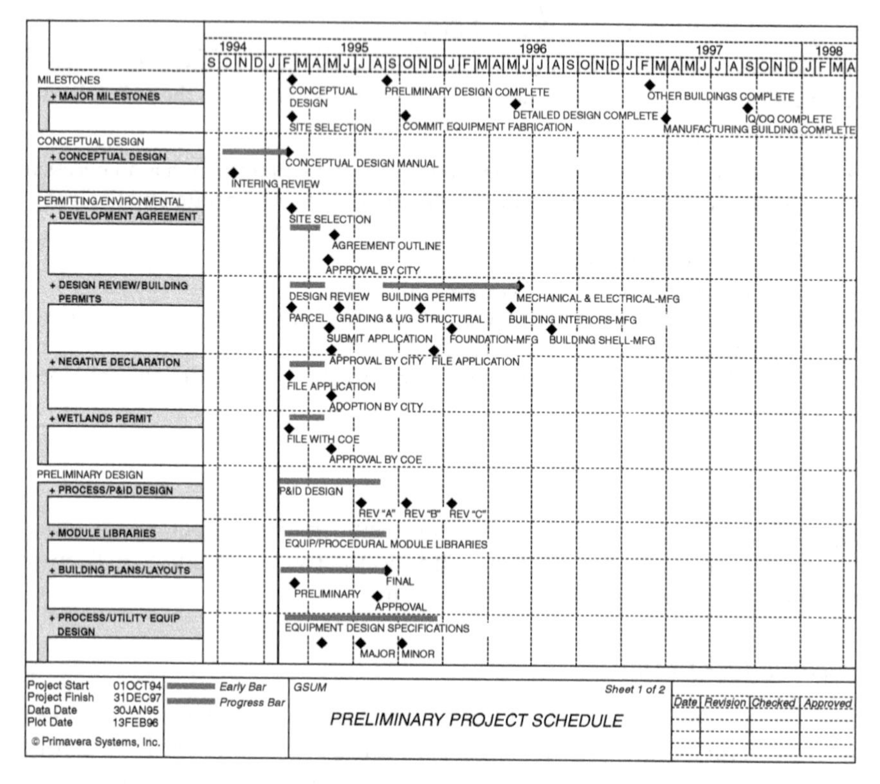

Figure 5.5a Preliminary project schedule.

Engineering and procurement

Everyone knows that it takes design information to build a facility—vast amounts of design information from the various disciplines involved with the project. Examples of engineering phase schedules (Figure 5.5) show the anticipated durations of activities based on the deliverables agreed to in the engineering contract and the estimated man-hours required to complete those deliverables.

To completely develop a facility design, the A/E firm will require information from outside sources related to equipment that will be purchased for the facility. In most cases, the A/E firm will also perform the procurement functions for the majority of process equipment and other major utility equipment, such as WFI stills and clean-in-place (CIP) systems, if requested by the owner. While this is not mandatory or always the case, many owners find it beneficial, due to the type and number of resources required to implement an effective procurement program. It is the responsibility of the A/E firm to ensure that the schedule for engineering deliverables is integrated with the procurement schedule so that purchase requisitions are issued and contracts awarded in a time frame that will allow vendors to produce the necessary design data for their equipment that is required by the A/E firm for the completion of the overall facility design.

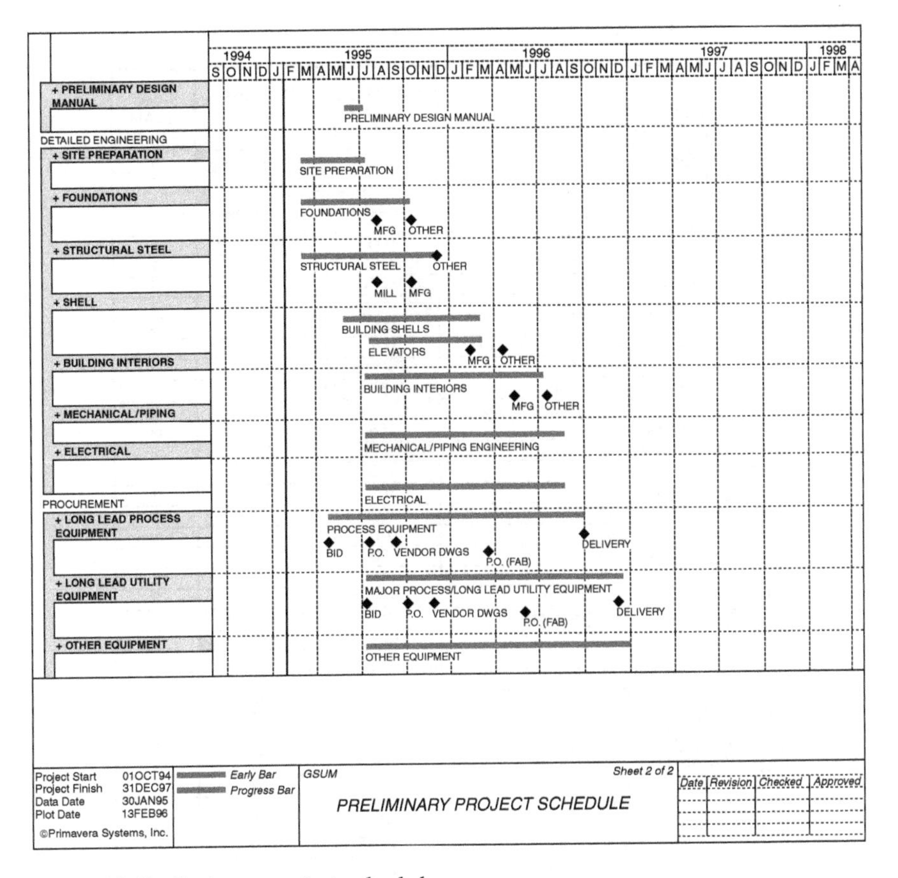

Figure 5.5b Preliminary project schedule.

Engineering and construction

Construction cannot begin until design information is issued to the field. In the current trend of "fast track" project execution, the design information is issued in phases to support expedited construction activities. Construction schedules must be based on sound access to design information that can minimize the risk associated with building to incomplete documents to the greatest extent possible.

The construction organization will determine the plan of executing the project through a contracting plan or a project execution plan. In this document, the facility is divided into work areas or packages that will define the boundaries of responsibility for contractors working on the project. By breaking a project down in this manner, the construction organization identifies the sequence in which design information is to be received and how that design information is to be packaged. Example 5.5 provides a sample of a contract plan for a manufacturing facility. The integration of this information into the engineering schedule is very important; design information received

Example 5.5 Proposed Bid Package Numbering

DESCRIPTION	LAB/WHSE FS/SP/SITE	MFG/CUP YARDS	LEAD TEAM
TRADE CONTRACTOR BID PACKAGES			
Site Preparation	001		1
Site Grading and Utilities	002		1
Site Finishes	003		1
Preliminary Site Electric	022		1
Concrete Foundations Labs/WH	005		1
Concrete Foundations Manufacturing		050	2
Concrete Foundations Cup/FS/Sp		067	2
Concrete Foundations - EW Yards		068	2
Flat Concrete and Slab on Grade	006	051	Each
Structural Steel - Primary/Secondary - LW	007		1
Structural Steel - Primary - M		052	2
Structural Steel - Secondary - M		053	2
Structural Steel - Primary/Secondary - SUF		054	2
Miscellaneous Metals	008	055	Each
Spray-on-Fireproofing		056	2
EIFS Exterior Wall Panels	100	100	1
Pre-cast Concrete	101	101	1
Metal Panels and Louvers	102	102	1
Roofing and Sheet Metal	010	057	Each
Windows and Curtain Wall	103	103	1
Exterior Caulking	115	115	1
Drywall, Acoustical,Carp, Millwork, H.M.		059	2
Drywall, Acoustical and Hollow Metal	013		1
Carpentry, Millwork & Misc. Specialties	021		1
Interior Finish Caulking	020	066	Each
Casework and Hoods	104	104	1
Finish Flooring and Tile	116	116	1
Chemical Resistant Floorings	105	105	2
Painting and Wallcoverings	015	061	Each
Environmental Rooms	106	106	1
Loading Dock Equipment	107	107	1
Food Service Equipment	016		1
Elevators	108	108	2
HVAC, Test & Balance		062	2
Mechanical & Plumbing Systems, TAB	017		1
Sanitary Piping	109	109	2
Utility Piping/Plumbing		063	2
Fire Protection	110	110	2
Building Electric	019	064	Each
Fire Alarm	111	111	2
Security	117	117	2
Communication	118	118	2
Instrumentation and Controls	112	112	2
Facility Management System	113	113	2
Field Erected Tanks		065	2
Passivation of Sanitary Process Systems	114	114	2
PURCHASE ONLY BID PACKAGES			
Finish Hardware	151	151	2
Depyrogenation Oven	152		1
Steam Sterilizers	153		1
Glass Washers	154		1
Autoclave	155		1
Floor Scales	156		1
High Density Storage Shelving	157		1
Pallet Rack Storage Shelving	158		1
Mezzanine Storage System	159		1
Bio-Safety Cabinets	160		1

BP 1 to 49 = Lab/Admin/Site
BP 50 to 99 = Mfg/Cup/FS/Spine
BP 100 to 150 = Global
BP 151 to 199 = Purchase only bid packages

late can have a tremendous impact on construction execution in the field, where time is truly money.

Going back to the previous discussion of the relationship between engineering and procurement, the construction schedule must also integrate procurement information related to anticipated equipment deliveries. The procurement schedule must be developed to support construction to the greatest extent possible. However, there will always be instances where unavoidable situations, such as equipment delivery delays due to customs inspections, or unanticipated fabrication delays due to material shortages or labor problems impact the final procurement schedule.

The integration of all engineering and procurement data on a regular basis into the master schedule is an important function of project control. The project manager must carefully evaluate the impact of delays and find means to work around situations where the schedule shows potential delays.

Building vs. turnover

Construction schedules are developed with one primary goal: build the building. But at some point in time, there must be a fundamental shift away from building the building to the turnover of systems to support facility start-up and commissioning. This shift in scheduling and execution philosophy must follow the logic that is developed for system start-up, in that plant systems will follow a logical sequence of start-up operations in order to support the start-up and validation of process systems.

In Figure 5.6, the schedule of start-up operations is shown for a biotech manufacturing facility. In it, the sequence follows the logic that plant utility systems will support clean utility systems; these clean utility systems will then support the process systems and components necessary to produce final product.

The construction schedule must integrate this information into the sequence of turnover activities to ensure that systems are complete when they are needed to support start-up. For example, the clean steam system cannot be tested for operation until the plant steam system is operational, which requires the boiler system to be complete. The same logic is applied to water systems; in order for the CIP system to be operational, the DI (deionized water) and/or WFI systems must be operational.

Integrating a construction schedule to meet start-up and commissioning activities is always a challenge, especially when validation activities follow close on the heels of commissioning. Knowing when to make the switch from a building focus to a system focus is important; it will involve the refocusing of resources, the development of detailed turnover schedules that include walk-down activities for the various systems, and a commitment from the owner to support these activities through appropriate levels of staffing during the final stages of project completion.

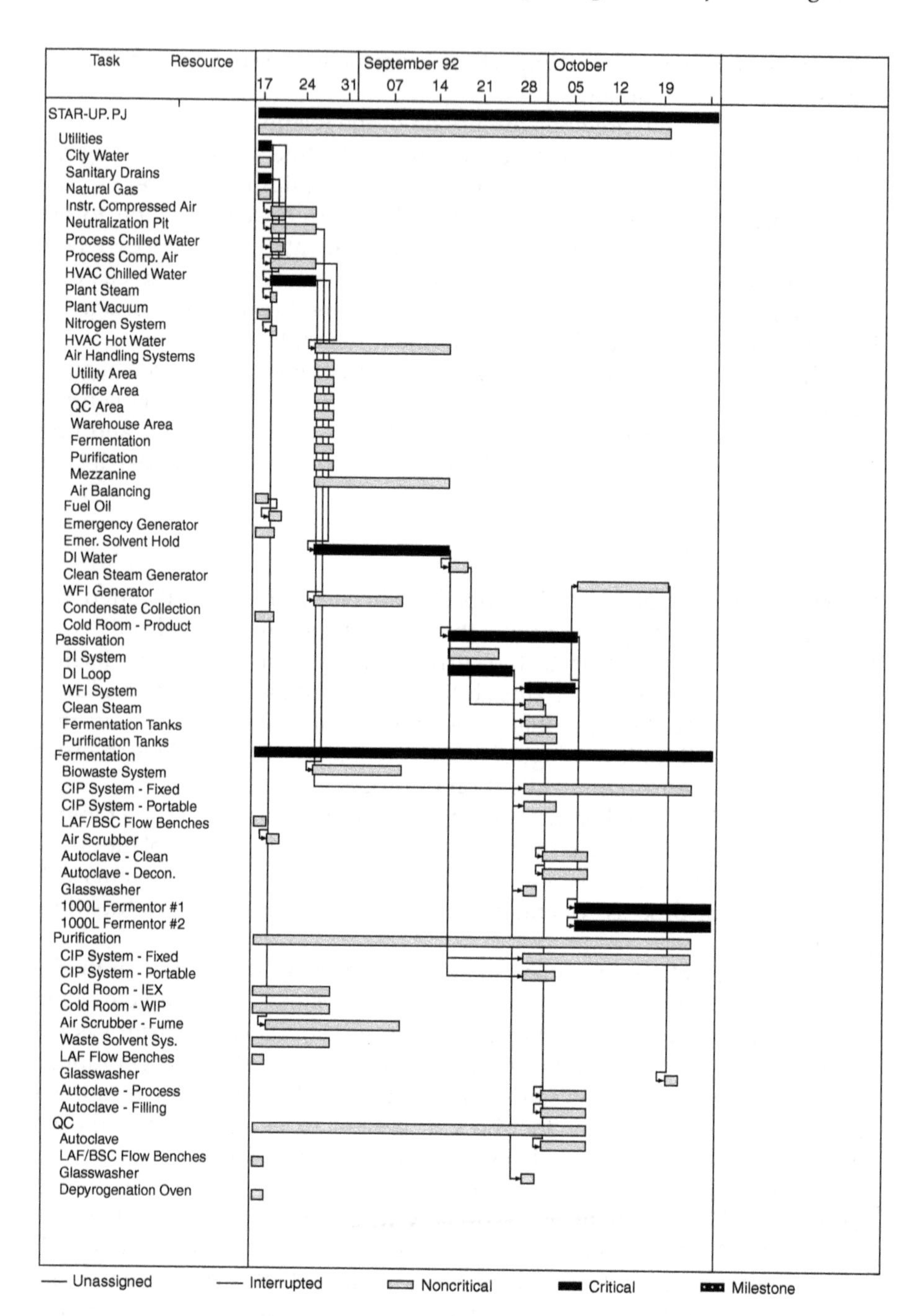

Figure 5.6 Manufacturing facility start-up schedule.

The project and the product

Facilities are built for one reason: to produce a product. The launch of a product to meet a market window of opportunity is the focus of the entire project effort. For this reason, it is important to realize that the actual design, construction, and validation of the facility are only single parts of the overall project puzzle. Project schedules for the facility must also be integrated into the overall schedule for the product.

The development of a biopharmaceutical product is a long, technically demanding process. It takes years to bring a product through the various phases of clinical trials to a point where final ELAs/PLAs (Product License Applications) are made to the FDA and approvals given to produce and sell a product on the open market. In many cases, this window of opportunity is very small due to a race with competitors to get to the market first with a potential blockbuster drug. The execution of design, construction, and validation activities must focus on this window.

Figure 2.1 (page 13) gives one example of how the proper integration of the product development schedule will lead to a verification of need dates for facility completion that will support consistency runs and, ultimately, product launch. Without going through this type of schedule exercise, corporate management will not have a sense of the need to start activities in a certain time frame or when funding must be made available. It also provides a good tool on staffing requirements and the anticipated submission time frame for documents to the FDA that will support final approval of the facility.

Change control

Change is an inevitable part of our everyday lives. It can upset even the best-laid plans, causing much frustration, aggravation, and stress. Many people will tell you the change is good; to the project manager trying to control cost and schedule on a project, however, change can be anything but good if it is not properly monitored and controlled.

Changes to a project will occur. The changes may come in the form of revised project requirements, changes to the design philosophy, technology changes, changes in the business climate, regulatory changes, or even changes in project team members. Knowing where these changes will likely come from and how they should be categorized is essential for managing the change control process.

Scope changes

Changes to the scope of work for this type of project can be generally described as any addition, deletion, or modification that is made to the design basis of the facility. The sources of such changes can be derived as follows:

- Revised project requirements that will commonly come from the client. An additional need may be uncovered or additional capabilities may be required to meet the operational requirements of the facility. One example could be the need to add additional hold tanks for buffers due to manufacturing cycle times; another could be the requirement to add cold WFI drops in certain suites to have necessary access for cleaning.
- Changes to the design philosophy identified by the A/E firm as a more effective way to meet project requirements. One example could be in the method that is used for producing process cooling water or the approach that is taken in designing clean rooms for manufacturing operations to reduce classified space requirements.
- Technology changes may involve new types of equipment to reduce the size of classified processing suites.
- The business climate may change, causing a rethinking of the project design basis. If your main competitor beats you to the market with his version of your product, you may very well decide to continue with the project but reduce the output of the process based on revised market projections.
- Regulations are continually changing. These can be changes that occur due to FDA findings, OSHA incidents, or state and local code revisions. One example could be the implementation of on-line TOC (total organic carbon) monitoring in water systems for biotech manufacturing that would comply with pending changes to USP requirements for WFI.
- In the life of any project, the people involved can change due to a variety of different circumstances. As the pharmaceutical industry continues to consolidate, many companies will find themselves the target of takeovers that can have a dramatic effect on plans for expansion and capital spending. Manufacturing capacity that is needed now may not be required if the new owner already has excess capacity. The project sponsor may be given a new position within the new organization, or worse, may be given early retirement! In either case, the change that is brought about in the scope of the project could be very significant.

Baseline changes

Every project is formed with certain assumptions that form the baseline of the project. These baselines deal with products to be manufactured within the facility, the capacity of the process, flexibility, the target cost for the project, or the schedule to meet market launch to name a few. These baseline criteria are first defined during facility programming and are carried throughout the design process by the A/E firm.

At any time during the life of a project, the client may find it necessary to change one or more of these baseline assumptions due to a variety of circumstances. In some cases, this may be due to information obtained during scale-up of the process; it may be due to results from clinical trials; it could be due to market conditions. Any change to the characteristics of the product or the method in which the product will be produced can be classified as a baseline change to the project.

For example, if the number of finished doses increases, the output of the process must increase based on optimization of the process design. Such an increase could cause an increase in the size of the production fermentor, the size and number of buffer tanks, the lyophilization capacity required for drying, and the size of utilities to support the process equipment. This increase in output could also be met by increasing the time the facility is in a manufacturing mode, such as a single shift per day to multiple shifts per day for an extended work week, provided there is sufficient time in the manufacturing schedule to accommodate proper cleaning, changeover, and QC hold durations for the process.

Another cause for a baseline change can occur when the estimated costs of the project exceed the anticipated funding level that gains board approval. In this case, items such as future capacity expansion, flexibility, and implementation of new process or control technologies may become a victim of the budget ax. Through a system of value engineering, these items can be analyzed in a manner that will allow for sound optimization of project alternatives that will yield the best "bang for the buck" in cases where cost becomes an issue.

There will also be cases where time is the main driver of the project. Many biotech and pharmaceutical companies are in a fierce race to the market, with the winner surviving and becoming profitable, and the loser facing possible corporate elimination. Most biotech and pharmaceutical projects are developed using a "fast track" method of implementation. As forecast dates extend beyond what is acceptable to corporate management for the completion of the project, alternatives must be addressed that will almost certainly add cost to the project. Again, it may become necessary to reduce the overall scope of the project by decreasing the size of the facility, by revising the method of execution of the project, or by looking at alternatives to manufacturing all or parts of the product in a single facility vs. some contracting arrangement with a contract manufacturer.

Managing change

Change management is an art; just ask any project manager who has been through a project that has become out of control due to constant change in the baseline and/or scope of the project. When this occurs, management usually goes on a witch hunt, looking for someone to blame. Chances are, he or she may no longer be a project manager!

Changes in scope must be identified, tracked, and evaluated through a systematic program of change control management in order to maintain focus on the project as a whole. If this type of "macro" view of change is not implemented, people will tend to focus on bits and pieces—small segments of the project scope that, individually, do not create severe impacts to project cost and schedule. But taken collectively, they can turn any project into a disaster by allowing costs to creep up and schedules to extend due to the extra work involved in implementing the total number of small, individual changes.

Objectives of change control

The development of change control procedures on any project is of paramount importance. The basis of the procedures should focus on some primary objectives:

- What is the process for identifying and submitting potential changes?
- How will changes be evaluated against the project baseline?
- What are the responsibilities of the project manager when a scope change occurs?
- How will changes be approved or disapproved?
- What is the process for developing the cost and schedule impact of the change to the project?

There are numerous methods to document and define changes to a project scope. Figure 5.7 is an example of how a change control procedure was implemented for a project, allowing for identification of the change, development of the change estimate, and approval by the client of the change.

Change control requires documentation. Formal change control involves the use of change notice forms that give all the necessary information related to the change that will be necessary for the client to evaluate the change and make a sound business decision. These forms indicate the scope of change, estimated time to implement the change (man-hours), cost impacts (additions or deductions), and any impacts to the overall project schedule. They may also provide information that may influence the final decision for acceptance or rejection, such as potential regulatory impacts or changes to the operation or maintenance philosophy that may be currently implemented.

All change notices should be maintained in a change control log, regardless of whether they are approved or disapproved. This log will become an important tool for project control as the design of the facility progresses. Knowing that issues have been addressed, who made the evaluation and approval, and what the estimates were for cost and schedule impact should become very important as management reviews the project scope against the project baseline, asking questions regarding why things have changed and whether those changes are in the best interest of the project.

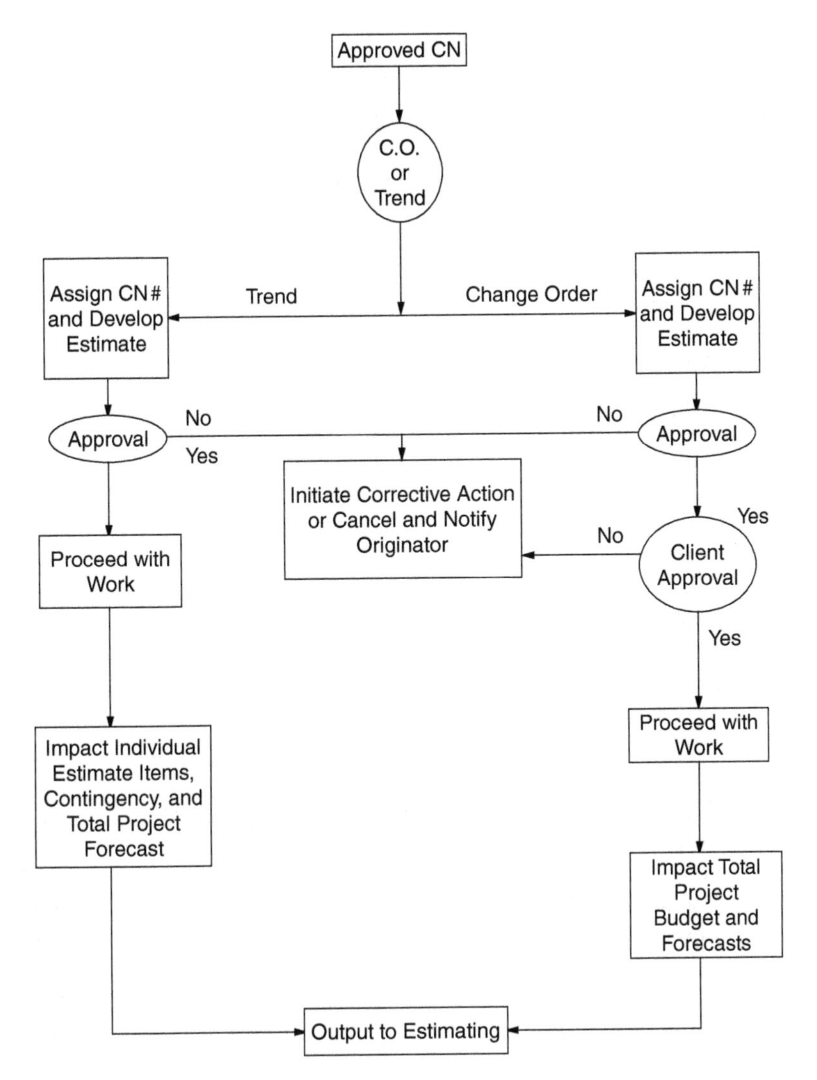

Figure 5.7 Engineering design change control.

Results

Proper change control management results in no surprises. The project manager will always know the current scope of the project and the status of the project in relation to cost and schedule. The client will know where the project stands in relation to the budget and the estimated completion date. Knowing where you are is half the battle in project management.

To give an example of how a well-developed change control produces verifiable results, let us look at a simple case: tracking budget trends against design changes.

Table 5.6 Change Notice Trend Log

CN No.	Description Value ($)	Construction Value ($)	Engineering Change ($)	Total
1	Site work revisions	888,000	0	888,000
2	Building size reductions	–8,770,000	0	–8,770,000
3	Value engineering revisions	–29,253,000	0	29,253,000
4	Automation—Fermentor valves	936,000	44,600	980,600
5	Add UF system	505,000	70,200	575,200
6	Tower water	-56,000	76,200	20,200
7	Caustic system modifications	-200,000	13,800	–186,200
8	WFI still size increase	200,000	0	200,000
9	Add filters for gas services	36,000	4,400	40,400
10	Add bulk CO2 storage	20,500	13,300	33,800
11	Extend pipe rack	15,000	11,600	26,600
12	Revise chiller loads	30,000	8,500	38,500
13	Structural revisions—Lab	62,400	0	62,400
14	Raise warehouse height	62,600	0	62,600
15	Add sand filter to city water	132,000	20,000	152,000
16	Add lift pumps	77,500	18,000	95,500
17	Use half-pipe jackets	59,600	0	59,600
18	Piping material escalation	358,000	0	358,000
19	Additional landscape	260,000	0	260,000
20	Additional structural girts	43,400	15,000	58,400
21	85 deg temp in utility chases	353,000	9,600	362,600
22	Delete tanks	–47,700	2,200	–45,500
23	Revise building exteriors	–404,000	39,600	–364,400
24	Add 3000 L tank	136,100	21,000	157,100
25	Escalation on structural steel	358,000	0	358,000
26	Increase in Engineering services	0	252,000	252,000
27	Increase acid/caustic sizing	55,000	10,000	65,000

— continued

Table 5.6 (continued) Change Notice Trend Log

CN No.	Description Value ($)	Construction Value ($)	Engineering Change ($)	Total
28	Add access to interstitial space	0	9,400	9,400
29	Additional permit drawings	0	4,800	4,800
30	Add recirculating pump	20,000	0	20,000
31	Add 3000 L tank—revised cost	31,100	0	31,100
32	Revise landscape	37,000	0	37,000
33	Revise layout drawings	0	8,500	8,500
34	Add DDC controls—utilities	2,000	3,000	5,000
35	Engineering Services	0	–120,000	–120,000

Case example — Preliminary engineering

Upon completion of conceptual engineering, Bio Company had a new manufacturing facility that was estimated at $205 million. At the completion of preliminary engineering, Bio Company wanted a revised cost estimate for the project in order to obtain final funding levels from the Board of Directors. At the start of preliminary engineering, A/E, Inc., implemented a change notice program to identify potential changes to the project scope, requiring client approval for any additional design effort to be expended. Through preliminary design, a total of 35 change notices were issued for review and approval, as shown in Table 5.6. These changes were primarily design and requirements changes brought about through design development.

For every change notice that was approved by the client, a trend was developed against the project budget to identify the changes to the overall cost. All of these costs were captured on a weekly basis in a trend log. Each month, the trends were summarized in a trend report that captured the total impact to the project. At the end of the preliminary engineering phase, the total impact of the incorporated change notices had trended the estimate up to $228 million.

Because project management was aware of this trend, they were confident that there would not be a great order of magnitude difference in the trended number and the final estimate number at the completion of preliminary engineering. In reality, the two numbers were within 6 percent of each other.

Change control during construction

Change control does not stop at the end of detail design. There will be numerous situations that will occur during the execution of work activities in the field that will lend themselves to a formal control of change that will add cost and/or time to the project. There will be instances where the detail design drawings do not reflect what the client had "visualized" in the review of the documents, creating a "that-is-not-what-I-thought-I-was-getting" situation that will, ultimately, lead to a change.

There will be cases where the design details have omitted information necessary to produce a finished product (e.g., trim details around built-in equipment or architectural finish details). There will be unforeseen conditions, such as interferences between HVAC duct and sloped sanitary process lines, that were not found during design development and that will require redesign. In each case, there will be a need to develop additional design information in order for the construction activities to be completed, resulting in a change in the scope of work, either through increased cost, schedule, or both.

For these changes, the question becomes, "Where will the money come from?" Contingency that is carried in a project's budget is generally used to cover these types of changes during construction. In many cases, the change is defined along a fine line between new scope of work and inadequate design information. Unfortunately, most projects cannot ask for more money; therefore, these changes are usually absorbed, with funds taken from contingency to cover the costs. Project managers must be aware of managing the contingency to cover costs. Project managers must be aware of managing the contingency fund against these types of changes so that the amount of funding in the contingency account does not fall to a level that is unacceptable or insufficient to support the project through completion.

Progress monitoring

As previously stated, before you can manage a project properly, you must know where you are, where you should be, and where you are going. Performance must conform to the project plan if the project is to be successful. The project manager must see to it that the goals and expectations of the project are met; to do so requires a system of monitoring and reporting progress.

Status

Project control requires systems to be in place that track the accomplishment of design and construction work activities and their associated costs. They must also provide an assessment of the overall productivity of the effort and the quality of the work being generated. The systems must provide for a method of collecting the required data, a method of analysis, and a means of presenting the data to project management.

Data for establishing project status come from a variety of sources. Scheduled status review meetings are primary tools used by project managers to obtain information regarding tasks started, progress toward milestone completion, activities behind schedule, and potential problems. Issues related to inadequate resources, delayed issue of information, or potential changes are open discussion items for the project team.

Personnel time sheets also provide important data related to the man-hours required to complete given tasks. Each person uses specified codes to record expended man-hours against activities that can then be consolidated into a master report to show expenditures.

Man-hour data can be collected by discipline and evaluated against the budget, based on a comparison of earned vs. expended man-hours for design. This information can then be graphically represented to show performance against budget and schedule, either for the entire project or on a by-discipline basis.

During construction it may also be beneficial to update status based on billings from contractors against the value of their contract. In this case, monthly billings are accumulated against the total contract value, including approved changes, to provide a percentage of completion for the contract. While simple to execute, this method requires diligence in verifying that billings correspond to actual work completed in the field. If a contractor takes credit for materials purchased, work installed, or testing completed, it must be verified that the materials are on-site, the installed work is complete to an acceptable level of quality, and that the test documentation has been signed-off and submitted. Once the physical percentage of completion for each contract has been established, the overall project status can then be plotted against a standard performance curve (Figure 5.8).

Many construction companies use quantity surveillance or verification systems for measuring physical progress in the field. In these systems, dedicated personnel are responsible for documenting a contractor's status related to material purchased, materials installed, and final testing and acceptance. On a system-by-system basis, benchmarks are established for levels of completion and a percentage assigned to each level. As work progresses, higher levels of completion will receive a cumulative percentage for completed work, resulting in a total of 100 percent when a given system is signed-off and accepted by the Owner.

Analysis

Progress monitoring requires comparison of planned to actual results so that any variances are identified. This comparison can then be applied to individual disciplines, systems, contracts, and/or the entire project to answer questions related to status: Are we ahead or behind schedule? Are we under or over budget? Are we as productive as we should be? Are our resource levels acceptable?

When variances are identified, it is important to determine the cause of variances before determining the corrective action. Variances have many possible causes:

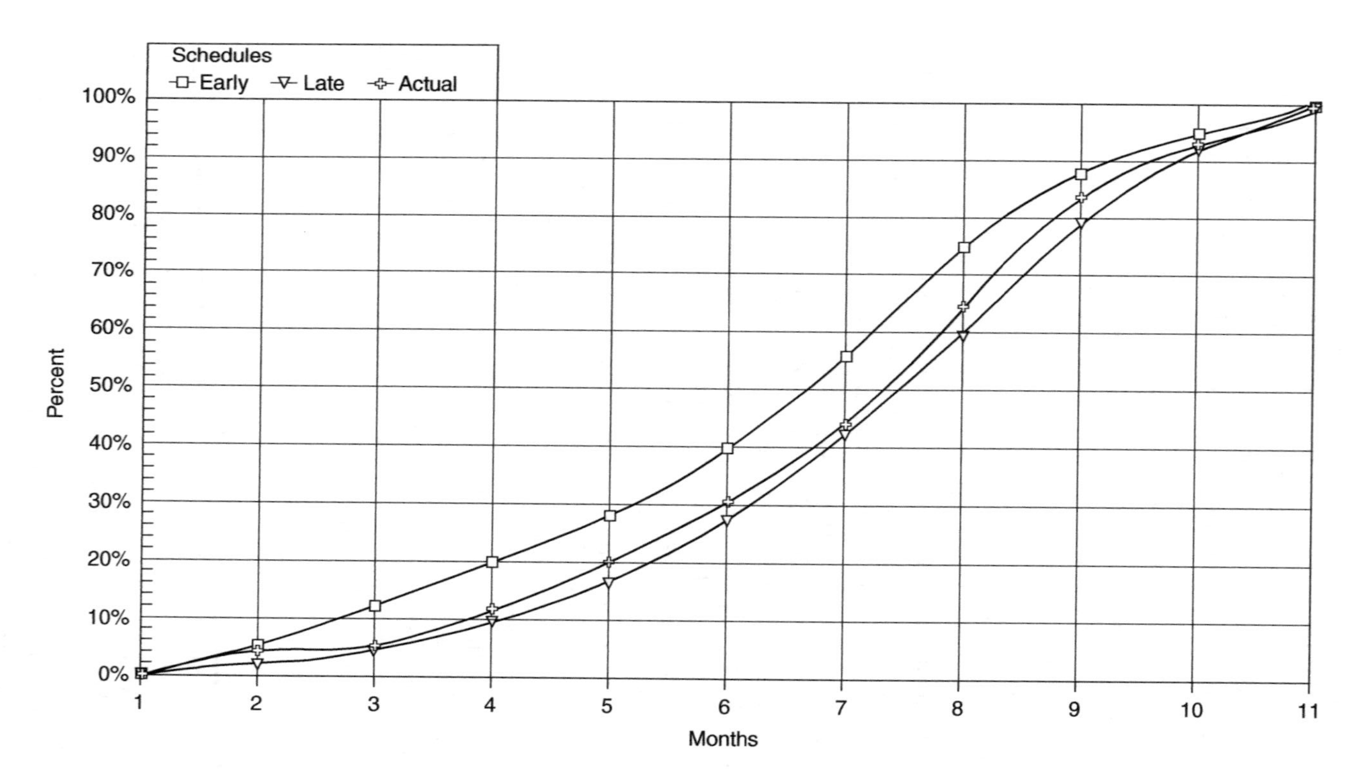

Figure 5.8 Physical progress curve.

- **Poorly defined scope** A contractor's scope of work did not clearly define all of the elements that were his or her responsibility to furnish and install. Drawings and specifications could have been omitted or supplemental details not forwarded through a document control system.
- **Inadequate general conditions** The items that are the responsibility of the contractor to furnish as a part of the work execution are critical. If material handling equipment, such as cranes or forklifts are required, it should be so stated in the general conditions. If there is a need for special measures to protect personnel or installed equipment during construction, these situations must be addressed.
- **Inadequate resources** Whether in design or construction, it takes people to execute a project. Improper staffing can cause serious schedule problems. Many times, insufficient budget will drive companies to try and skimp on resources in hopes of saving some money.
- **Poor estimating** If the budget has not been developed in a manner that clearly defines the scope of the project and allows sufficient contingency for the execution of the work, variances in cost will almost always occur. Omission of equipment items, inadequate allowances for escalation, misinterpretation of sales tax requirements, incorrect direct labor factors, or inadequate funding for inspection and testing are common areas of concern.
- **Delays** Delays in the delivery of equipment and materials can create monumental problems related to cost and schedule. Delaying engineering completion due to insufficient vendor information, delaying building closure due to late equipment deliveries, or delayed system completion due to delays in piping materials or process instruments create variances.

There are three courses of action that can occur when a variance is identified:

1. Do nothing.
2. Review the plan and make modifications to accommodate the variance.
3. Look for areas of trade-off that could be adjusted to alleviate the problem.

The analysis of any variance must include the cause of the problem, the course of action taken to put the project back on schedule or within budget, and a plan to ensure that the cause of the variance has been eliminated and will not reoccur. While most projects are very diligent in these first two steps, many fail to execute preventive measures once a problem has surfaced. There is no excuse for making the same mistake twice.

Value engineering

One of the primary goals of any project control effort is to gain the maximum value for every dollar spent on the project through every phase of the project life cycle. We, as consumers, follow the same principle when we shop for bargains in the vehicles we purchase, the food we eat, or the services we contract for. It is called looking for the "biggest bang for the buck."

Defining value

The value that is placed on a particular item or function is related to the perceived worth that comes from the item or function. When dealing with sterile manufacturing facility projects, it is important to understand that worth is based on the value of the necessary function or item, not on the cost of the specific design. For example, the production of WFI is a necessary function for product manufacture. It is dictated by both design and regulatory requirements. The method that is used to produce WFI can vary in the type and size of equipment, the design of the still, or the design of the distribution system, thus varying the project cost. The true worth is from the need to produce WFI quality water, not that one particular system design costs more than another.

Defining worth from a value engineering (VE) perspective means that the worth of an item or services equals the lowest cost for which the necessary or essential function(s) of that item or service can be produced or obtained.

Value engineering defined

Controlling project cost is critical. Every project manager must be diligent in looking for ways to maximize value while maintaining the goals of the project. The ability to influence cost over the life of a project is maximized best in the early stages of the project, when decisions are made that will have profound impact on the final cost of construction. A VE program conducted diligently during the early design stages will optimize cost and function by focusing on means and methods of design alternatives that provide better value while meeting project design requirements.

Value engineering can be defined as an organized effort at analyzing the function(s) of system and equipment, facilities, supplies, and procedures required to meeting the necessary functional requirements of performance, reliability, quality, and project safety at the lowest total cost of ownership. Value engineering is a formal system that focuses on identifying and removing unnecessary costs, while examining the total cost of owning and operating a facility.

The VE process

The conceptual design estimate or budget estimate will provide the first real look at anticipated project costs. At this early stage of design information, decisions are being made at a rapid pace, which often become included in the scope of the design without any real look at necessity and cost impact. In many instances, decisions to proceed down a particular design path are founded on past Owner practice, user group desires for the optimal solution, or functional "wants" that are above and beyond the "needs" of the project.

As long as the cost of a project stays below the limit imposed by corporate management, these items can, and usually will, stay in the project scope regardless of their true value. Unfortunately, that rarely happens in this time of increasing facility costs, higher production costs, and corporate budget tightening. What occurs is a need to reduce project cost while maintaining the original financial goals of the project.

The VE process begins by defining a team of personnel from the various organizations and disciplines with the project. The A/E firm, the constructor, and the Owner will participate. It is important that the various user groups from the Owner's side are all represented since their input and experience will become extremely important during later evaluation of alternatives. The team must have at its disposal the necessary information about the project. This will include design and cost information, operating philosphy, and any design constraints that may be identified.

Value engineering relies on ideas—creative ideas that will focus on alternative solutions. Brainstorming sessions are held to generate alternatives without worrying about the details or practicality of the idea. There should be no discussion or comment on any idea presented. The brainstorming process should be freewheeling, with a focus on creativity. Do not let regulations or people control the thought process.

Key questions that will help stimulate creative ideas for alternatives include the following:

- Who else will perform the basis function(s)?
- Can it be eliminated totally or partially?
- Is there any duplication of function or service?
- Is it too complex for our needs?
- Can a standard item be used?
- Does it cost more than it is worth?
- Is it worth the maintenance cost?
- Do uses have established procedures for its use?
- Do any requirements appear excessive?
- Is the safety factor unrealistic?

Alternatives that are presented to the group may be simplified or modified by others and possibly combined with other alternatives. They are then recorded into breakdowns by building and/or system. Example 5.6 provides some examples of VE items taken from actual sessions for biotech manufacturing facilities.

With each alternative, there must be an attempt made at estimating a rough order of magnitude (ROM) cost impact. This will be an "educated

Example 5.6 Selected Value Engineering Alternatives, ABC Biopharmaceutical Corporation

1. Reduce corridor widths in manufacturing building.
2. Delete space for future equipment in manufacturing building.
3. Maximize use of prefabricated piping assemblies.
4. Eliminate automation of WFI drops.
5. Eliminate penthouse.
6. Eliminate variable frequency drives on air handler motors.
7. Reduce air change rates in nonclassified areas.
8. Delete videotape records of boroscopic inspections.
9. Use tower water in lieu of process water for cooling.
10. Delete bulk gas storage.
11. Delete "gray space" concept.
12. Reduce CIP system automation.
13. Eliminate buffer tanks that hold only WFI.
14. Use eccentrically braced structural frame vs. a moment-resistant structural frame for the support buildings.
15. Use metal siding instead of precast concrete panels for all buildings.
16. Waive bonding requirements for proven contractors.
17. Eliminate any landscaping not required by the city.

guess" made collectively by the team based on the information available and presented in "round" numbers for comparison. For example, an alternative to eliminate redundant steam boilers could have a ROM savings equal to the equipment cost of the boiler, estimated at $300,000. Another example could be a decision to eliminate a service corridor, where the ROM cost would equal the building square footage to be eliminated at some estimated $/ft^2. If the estimated building cost is set at $500 per ft^2, a 2000 ft^2 reduction would receive a $1 million ROM estimate.

Along with an estimated cost must also come an assessment of the probability of acceptance; after all, not every proposed alternative will find approval from all members of the team. For example, an alternative might be to eliminate certain automation on the CIP system, producing a large cost savings in terms of equipment and instrumentation cost. However, the likelihood of acceptance by the user groups is extremely low due to the problems that would arise in performing manual cleaning operations and the potential problems with the validation of cleaning cycles. Another alternative might be to reduce the number of WFI drops within manufacturing suites. The probability of acceptance could again be low due to problems that could arise during the production schedule, causing increased time to procure WFI from other areas.

Do not be surprised at the number of alternatives that are generated. At the conclusion of the brainstorming sessions, the team may find hundreds of ideas, thus the greater the potential for good ideas that will be eventually implemented. One recent session of this type for a biotech manufacturing campus project generated over 400 VE items for discussion, with an estimated face value of over $150 million.

Table 5.7 Value Engineering Item Rankings

Savings Magnitude ($)	Rank	Probability of Acceptance
>$1 million	5	90%
>$700,000	4	70%
>$300,000	3	50%
>$100,000	2	30%
>$100,000	1	10%

With this volume of information to be reviewed, it is obvious that there will be a significant time constraint on the team to streamline the process. From such a large number of alternatives, there must be consensus on which items are worth a more detailed review. But which ones do you select?

Pareto's law

Simply stated, Pareto's law of distribution says that in any system, 80 percent of the total wealth will be controlled by 20 percent of the population. Applied to VE, the 80/20 rule states that: "80 percent of the essential function is satisfied by 20 percent of the total number of components." By identifying that 20 percent of alternatives that represent the most savings to be gained, the team can streamline the review process.

To do this, a simple ranking system, as shown in Table 5.7, is used. Every item will receive a savings value and acceptability rating as shown in Table 5.8. A rank from 1 to 5 is applied to every item based on potential savings and acceptability. This ranking is also a consensus of the team. Only those items that have a ranking of 3 or higher should then be evaluated for potential acceptance. In selecting these best alternatives, be sure to ask, "Will the alternative meet all the necessary requirements of the basic function?"

The best alternatives selected will then be developed. Detailed estimates will be generated, along with a written proposal that outlines the specifics of the alternative, any pertinent facts that would impact the decision, and any advantages or disadvantages associated with the alternative. The VE Proposal Form in Figure 5.9 shows how an alternative is documented.

Once all proposals are complete, the team will collectively choose those that will be presented to corporate management for a final decision. The results are issued in a formal VE Report that will include a VE Log (Figure 5.10) of proposed items, the detailed VE proposals, and a total cost summary of the potential savings from the session. Management then has all the necessary data to make final decisions related to cost reduction.

VE results

This type of VE exercise will take from three to five days to complete, but it is well worth the effort. A/E firms and construction companies that perform VE as a normal part of their operating philosophy can see significant savings passed on to their clients. Figure 5.11 provides an example of one

Table 5.8 Value Team Alternative Comparison

Project
Date
Item: Manufacturing Building
Function: Organize Process

	Selected Ideas	Savings	Accept	Advantage	Disadvantage
B1-55	Use Hot Water in Lieu of Steam for Sterilization	$0	0%	Delete steam lines	Hot water won't sterilize
B1-56	Lease the Equipment	>$1,000,000	10%	Defer capital expense	Simply alternative financing
B1-57	Put Locker Rooms on the First Floor	>$100,000	50%	Eliminate elevator	Redesign floor layouts
B1-58	Reduce Supply Corridors; Internalize Length	<$100,000	30%	Smaller building	Redesign floor layouts
B1-59	Terminate Supply Corridors at the Process Area	<$100,000	50%	Smaller building	Redesign floor layouts
B1-60	Reduce Ceiling Heights to Reduce Air Flow Quantity	>$300,000	10%	Smaller air handling & energy costs	Feasible with deleted mezzanine?
B1-61	Defer Locker Rooms; Provide Disposable Gowns	>$100,000	0%	First cost savings	Dress in car?
B1-62	Combine Buffer Prep with Buffer Holding	<$100,000	50%	Smaller building	Segregation
B1-63	Combine Media Prep & Fermentation	<$100,000	0%	Smaller building	Segregation
B1-64	Combine Recovery with Buffer Prep & Buffer Hold	>$100,000	0%	Smaller building	Combine classified & nonclassified
B1-65	Utilize Ice Storage to Provide Process Chilled Water	$0	90%	Use off-peak electricity for chilling	Costs more at start

	Selected Ideas	Savings	Accept	Advantage	Disadvantage
B1-66	Widen Chase Column Lines & Delete Supplemental Columns	>$1,000,000	90%	Smaller building	Redesign floor layout
B1-67	Use WFI Water Spray in Lieu of Cooling Jackets	$0	0%	Avoid cost of jackets	Need jackets anyhow
B1-68	Put Offices, Utilities, & Storage in a separate Building	>$1,000,000	50%	Lesser code requirements	Access
B1-69	Put Utility Equipment in a Basement	>$300,000	90%	Free up first floor for other uses	Major impact on excavation
B1-70	Reduce the Size of Service Chases	>$300,000	10%	Smaller building	Redesign floor layout
B1-71	Modularize Areas Surrounding Tanks	$0	90%	Less expensive piping installation	Can complicate architectural layout
B1-72	Reduce the Amount of Epoxy Flooring	>$300,000	10%	Savings	Need to select appropriate alternatives
B1-73	Reduce the Amount of High-Build Epoxy	<$100,000	50%	Savings	Need to verify cleanlines
B1-74	Reduce the Number of Service Chases	>$700,000	50%	Smaller building	Rearrange service access to piping
B1-75	Eliminate the Second Floor Service Chases	<$100,000	0%	Smaller building; serve from below	Need to verify piping arrangement
B1-76	Reduce Stainless Steel Wall Protection	<$100,000	50%	Savings	Limit to areas that don't require it

— *continued*

Table 5.8 (continued) Value Team Alternative Comparison

	Selected Ideas	Savings	Accept	Advantage	Disadvantage
B1-77	Eliminate One Stair	>$100,000	50%	Savings	Less convenient traffic
B1-78	Provide Floors at Service Chases; Eliminate 2-Hour Walls	>$100,000	90%	Savings	Being implemented
B1-79	Hang Buffer Hold Tanks in Recovery	>$100,000	30%	Savings	Maintain cleanliness?
B1-80	Reduce Height of the Second Floor	>$300,000	10%	Now the mezzanine is gone	Where to put equipment
B1-81	Defer Equipment Dedicated to HER-2	>$1,000,000	10%	Savings—has alternate production site	Impacts capacity
B1-82	Alter the Building Height/ Area to Eliminate Fireproofing	>$100,000	0%	Savings	Major building reconfiguration
B1-83	Put the Penthouse Air Handling Equipment in a Basement	>$700,000	70%	Savings	Basement impacts excavation schedule
B1-84	Vertically Compress the Fermenter Tanks	>$300,000	10%	Volume reduction	Cramps access for maintenance
B1-85	Move the Switchgear to a Mezzanine; Provide Transformers	>$300,000	90%	High voltage closer to load	Need to reinstate mezzanine
B1-86	Delete Space for Future Fermenters 9 thru 12	>$300,000	90%	Preference	Very expensive to add in future

VALUE ENGINEERING PROPOSAL

ITEM: Use Tower Water in Manufacturing Building Service Areas

PROPOSAL: S2-9
SHEET: 1 of 3

ORIGINAL DESIGN:

Refrigerated water is used to serve all process users in the Manufacturing Building, including the process refrigerated water chiller.

PROPOSED CHANGE:

Provide a tempered lop (cooled by Tower Water) in the Manufacturing Building to cool the pasteurizers, Primatone tanks and the process chiller. Increase the process chiller 200 tons. Delete an 1100 ton chiller in the Utility Plant.

COST SUMMARY	Total Labor & Material	Mark-Up	Total Cost	Life-Cycle Cost
Original Design	566,000		566,000	
Proposal	300,000		300,000	
SAVINGS			266,000	

Figure 5.9a Value engineering proposal.

national construction management firm's results of VE exercises on a variety of pharmaceutical and biotech projects in the United States over the past 10 years.

Specific benefits from VE sessions can be found as the trending of project costs continues through the design effort. In many instances, more than one VE session will be conducted. Table 5.9 provides an example of how VE helped bring down the estimated costs of a proposed biotech manufacturing complex during the design phase. Again, by taking a proactive approach to cost control early in the project life cycle, a client was able to keep the project under predetermined capital spending limits.

Reporting

Communicating project status is an important part of project control and project management. Not only do the project team members and contractors need to know where the project is in terms of compliance to budget and schedule goals, but management will also have a keen interest in the status of the project. The information that contains the data necessary to evaluate a project's status must, therefore, be issued in a timely and informative manner to all project participants.

COST WORKSHEET

ITEM: Use Tower Water in Manufacturing Building Service Areas

PROPOSAL: S2-9
SHEET: 2 of 3

CONSTRUCTION ITEM		ORIGINAL DESIGN			PROPOSED CHANGE		
Line Item	U/M	Quantity	Unit Cost	Total	Quantity	Unit Cost	Total
				—			—
Chiller	Each	1	######	491,000			—
Electrical Service	LS		######	15,000			—
Process Chiller (200 Ton)	Each	1	######	60,000			—
				—			—
Heat Exchanger	Each			—	1	######	150,000
Process Chiller (400 Ton)	Each			—	1	######	100,000
Piping & Pumps	LS			—		######	50,000
				—			—
				—			—
				—			—
				—			—
				—			—
				—			—
				—			—
				—			—
				—			—
				—			—
				—			—
				—			—
				—			—
				—			—
				—			—
				—			—
				—			—
				—			—
				—			—
				—			—
				—			—
				—			—
				—			—
				—			—
				—			—
				—			—
				—			—
				—			—
TOTAL				$566,000			$300,000

Figure 5.9b Value engineering proposal.

Project status

A project status report must include answers to some fundamental questions:

1. Where is the project today?
2. What is the position of the project related to the overall budget?
3. Where will the project be at the time of the next report?

VALUE ENGINEERING SKETCH DETAIL

ITEM: Use Tower Water in Manufacturing Building Service Areas

PROPOSAL: S2-9
SHEET: 3 of 3

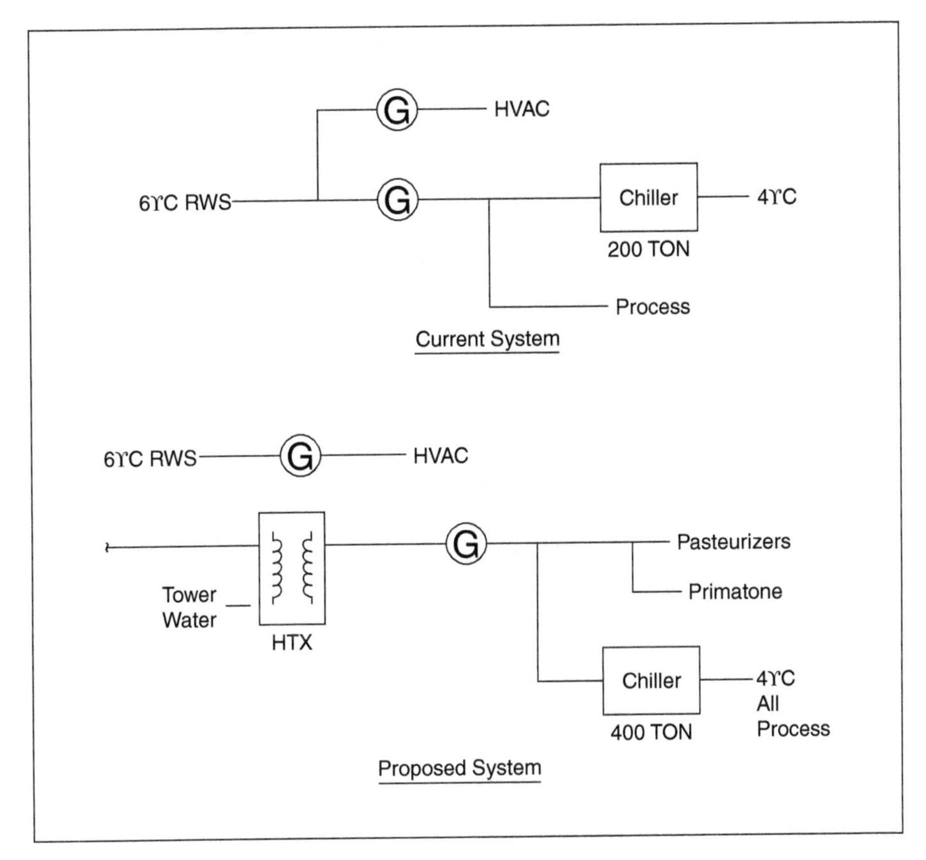

Figure 5.9c Value engineering proposal.

4. What are the areas of concern that need to be addressed before the next report?

Summary

Status reports should begin with a general summary of the progress made during the reporting period. The focus of effort, milestones met, and overall project accomplishments should be highlighted. This summary should remain very broad in its description of the project status.

VALUE ENGINEERING LOG

STATUS CODES
A = Accepted OF = Other Funding Sources
F = Further Evaluation E = Estimate Correction
RR = Rejected (Risk) RS = Rejected (Schedule)
D = Defer (Design to be installed in the future)

DOCUMENTS CODES
D = Drawing Required
S = Specification Required
N = None Required

ITEM #	DESCRIPTION	STAT CODE	INITIAL COMMENTS	DOC CODE	INITIAL VALUE	INCORPORATED
P2-6b	Eliminate Four 20 Litre Fermenters and One Cell Culture Suite	RR	May need to add one 80 litre fermenter for maximum efficiency		1,616,520	
	INOCULA FERMENTATION					
P3-1a	Eliminate Extra Space for 5 Inocula Trains	RR	Future expansion to 12 fermenters Delete "bump" on building		178,500	
P3-1b	Delete "Bumps" That Mimic Inocula Train Expansion Space	A	Exterior aesthetics	D	535,500	$535,500
P3-2	Provide Only Three Inocula Fermenter Trains (Now Four)	D	Defer first year while learning actual vs. calculated production capacity Will defer 20 l, 80 l, 400 l, Will install 2000 l but not the piping for it	D,S	1,491,100	$1,491,100
P3-6a	Eliminate Agitator Variable Frequency Drives at Prep Tanks	A	Cannot turn speed down at low tank levels	D,S	94,500	$94,500
P3-6b	Delete or Defer Agitators (Assume Tri-Blenders Remain)	RR	Not certain will mix at all volumes		413,895	
	PRODUCTION FERMENTATION					
P4-4	Delete or Defer Tri-Blenders	RR	Addition of solids more difficult. Add later if found needed		1,050,000	
P4-9	Defer "Gray Space" Piping for 2 Inocula Trains & 4 Fermenters	RR	Plant piped to wall but instruments, heat exchangers and portable tanks deleted. Cuts capacity in half.		2,952,000	
P4-10	Use Filters but Delete Pasteurizers	RR	Need to verify viral filtration is effective		863,000	
P4-12	Remove RTD's from Steam Traps in Fermentation	RR	Temperature monitoring performed manually		996,000	
P4-13a	Delete One Carbonate Tank and Two Headers	RR	Fill portable stainless steel tanks		(77,469)	
P4-13b	Delete One Carbonate Tank and Two Headers	RR	Fill portable Nalgene tanks		156,520	
P4-14	Eliminate or Defer Harvest Diafiltration Equipment	F	Lose 3% – 5% of product. Add later on demand to improve		200,700	

Figure 5.10 Value engineering log.

Period progress

Details related to specific progress made during the period are a vital part of the report. Detailed descriptions of work activities and the focus of manpower resources will paint a clear picture of where the emphasis has been placed.

Graphic representations of work progress are useful to help individuals retain the information presented in reports and in progress review meetings. Engineering performance curves and construction physical progress curves by discipline/contract are examples (Figure 5.12a and b).

The report should also include a status schedule, defining the current timeline and how each activity has progressed against the planned duration established in the baseline schedule of the project.

Simple line graphs or bar charts are the most useful method of defining budget status. In some cases, a simple spreadsheet identifying expenditures versus projections may also be included. Tables with endless columns of

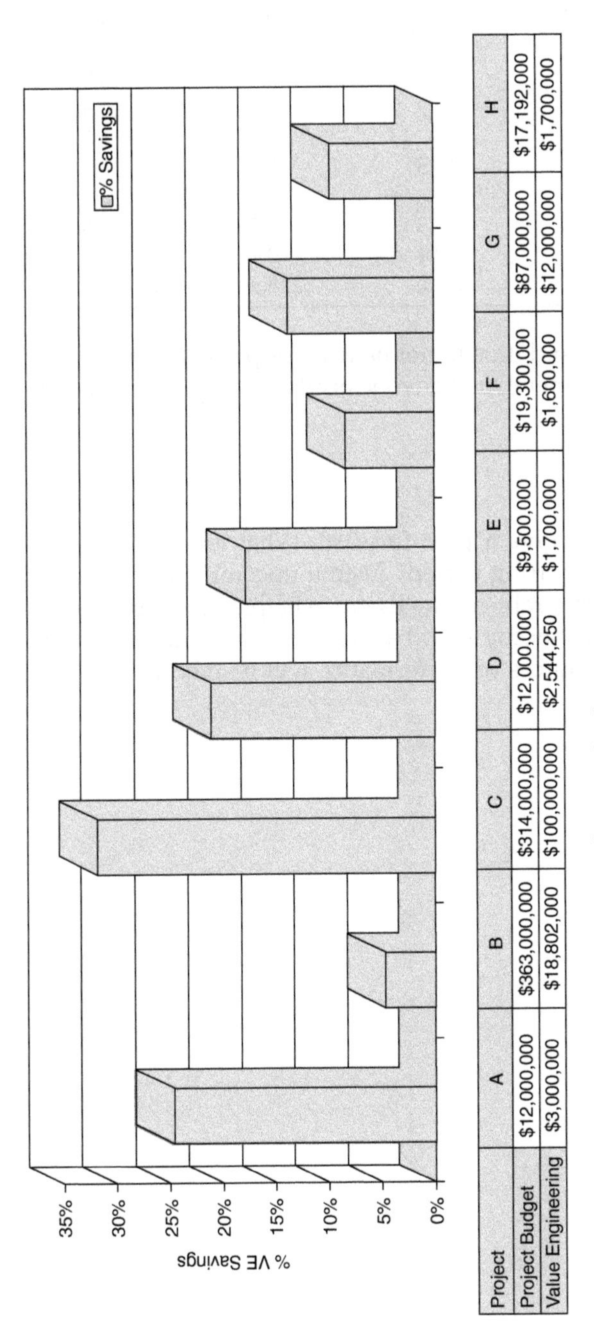

Project	A	B	C	D	E	F	G	H
Project Budget	$12,000,000	$363,000,000	$314,000,000	$12,000,000	$9,500,000	$19,300,000	$87,000,000	$17,192,000
Value Engineering	$3,000,000	$18,802,000	$100,000,000	$2,544,250	$1,700,000	$1,600,000	$12,000,000	$1,700,000

Figure 5.11 Value engineering savings as a percent of project budget.

Table 5.9 ABC Biotech Project Cost Trends ($MM)

Project Area	Conceptual Estimate	Budget Estimate	1st Value Engr	2nd Value Engr.
General	49.7	63.6	63.6	60.7
Site	19.3	23	20.7	19.7
Manufacturing	101.8	108.9	105.9	80.5
Central Utilities	8.6	8	7.3	6.3
Lab/ Administration	11.7	12.5	13.1	12.3
Warehouse	7.2	14.3	12.3	11.6
Miscellaneous	0.5	0.4	0.4	0.4
Total	205.1	237.2	228.9	197.4

numbers and volumes of cost data will not be read, or possibly understood, by most of the reviewers. Remember, if anyone needs more specific cost data, they will ask for it.

Where to next?

The progress report should give a path forward. What are the goals and expectations for the next reporting period? Listing upcoming activities and identifying areas of potential problems are useful in helping everyone understand the planning that must occur. It will also raise a flag concerning areas where special focus on resources or performance will be required in order to maintain the schedule.

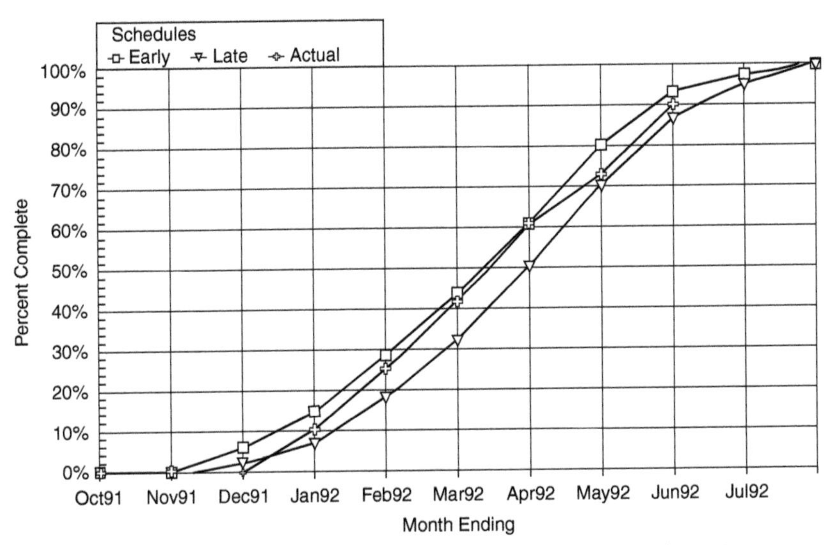

Figure 5.12a Main manufacturing: physical progress curves for stainless steel piping

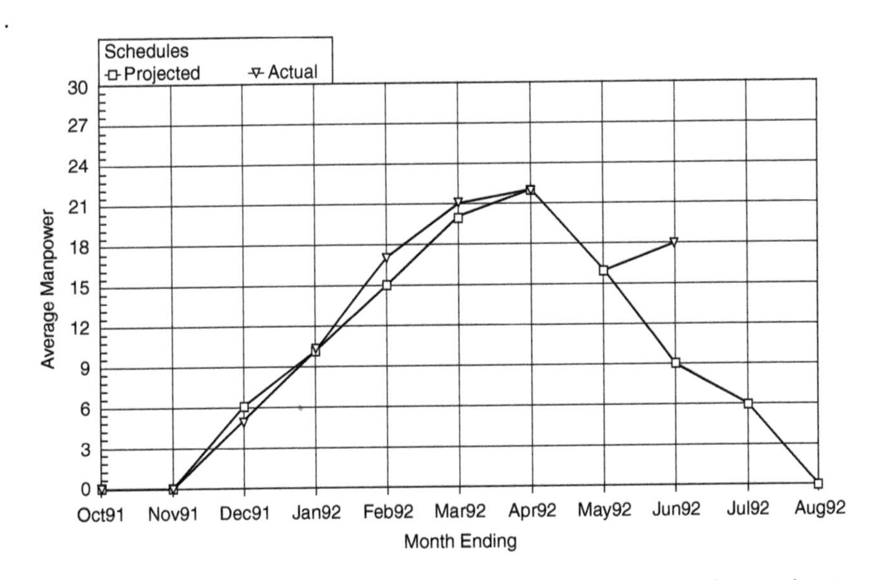

Figure 5.12b Main manufacturing: Project manpower loading for stainless steel piping.

chapter 6

Current good manufacturing practice: Project impacts

In the United States, the Food, Drug, and Cosmetic Act requires that pharmaceutical and biopharmaceutical products be manufactured in accordance with the current Good Manufacturing Practice (GMP) Regulations for Finished Pharmaceuticals, under 21 CFR Parts 210 and 211. The guidelines that appear in the various subparts of these regulations define the requirements for process validation and the concepts that make the design and construction of biotech facilities such a challenge to execute. Any company involved in pharmaceutical production, including firms that produce clinical trial materials, is subject to GMP regulations.

These GMP regulations are not limited to production within the United States. Canada, Japan, the European Union, and Great Britain all have regulations that govern compliance to a set of guidelines for facility design and construction.

Interpretation of GMP regulations comes from a wide variety of areas. The different centers within the FDA issue documents representing "Guidance for Industry" and internal guidance for FDA inspectors. Along with the FDA, organizations such as the International Society of Pharmaceutical Engineering (ISPE), the Parenteral Drug Association (PDA), the International Conference on Harmonization (ICH), and the International Standards Organization (ISO) also issue supplemental guidance documents that provide interpretation of regulatory expectations. Many of these documents, such as ICH Q7A, are becoming important as projects are defined and developed.

To understand the impact of GMP on a project requires an understanding of the definition of validation and the basis for such a program. The basic principles of quality assurance have as their fundamental goal the production of products that are fit for their intended use. Simply stated, these principles require the following:

- Quality, safety, and effectiveness must be designed into the product.
- Quality cannot be inspected or tested into the finished product.
- Each step of the manufacturing process must be controlled to maximize the probability that the finished product meets all quality and design specifications [1].

Based on these principles, an accepted definition of process validation would be documented evidence that provides a high degree of assurance that a specific process consistently produces a product of a certain predetermined quality. This definition requires a systems approach, including emphasis on design as well as operation, in order to meet the requirements of GMP. Through careful design and validation of the process, a manufacturer can establish the required level of assurance that the process will consistently produce an end product from successive batches that has the same quality and effectiveness.

It is also important to note that the GMP regulations are interpretive, not substantive; they are designed more to tell a manufacturer what it must do, not necessarily how to do it. Because of the great variety in products, processes, and manufacturing facilities, it is not possible to identify specifics for compliance. There are no specific facility standards that must be met to design, construct, commission, start up, and operate a facility; there is no universal GMP answer. However, the pharmaceutical industry and its member manufacturers have established general criteria that are recognized as acceptable to the FDA and global regulatory agencies for meeting process validation requirements. Our discussion will focus on these criteria.

What does the FDA look for?

One way that the FDA carries out its responsibilities is by inspecting companies and their facilities to see whether they are in compliance with GMP. By looking at the areas on which the inspectors focus their efforts, the project impacts of GMP will be better understood.

Buildings and facilities

- Is the facility suitable for the operations being carried out?
- Is the facility readily cleanable?
- Are there proper controls against cross-contamination?
- Is there adequate ventilation while still keeping out sources of contamination?
- Are there adequate sanitary facilities?
- Are the operational areas separate to prevent mix-ups and cross-contamination?
- What is the source of the water supply?
- Are there adequate systems for the handling and disposal of waste?

Materials handling and storage

- Can all physical factors, such as temperature and humidity, be monitored and controlled properly?
- Is there proper segregation for incoming and released components?
- Is there adequate storage space under the required environmental conditions?
- Are in-process materials properly stored?
- Are containers suitable for raw materials and intermediate product?

Equipment

- Is the facility equipment suitable for the intended use?
- Is equipment designed to facilitate cleaning?
- Are there proper filtration systems adequately designed and properly functioning?
- Does equipment design prevent contamination from external sources?
- Is equipment clearly and uniquely identified?

Engineering design

Because sterile drug product manufacturers must seek and gain regulatory approval for their facilities to produce materials for sale, designing a facility with GMP in mind requires more than just an approach. It requires the careful planning, coordination, monitoring, and control of many different design activities over the complete scope of engineering disciplines associated with the project. There is also an expectation that individuals involved in the engineering design process should have sufficient training, education, and experience to advise in the area to which they are assigned.

The first step is to define those systems and attributes that will require GMP review on a discipline basis. This should be a structured review of the facilities, utilities, and equipment within the scope of the project. A listing of document types can then be developed, showing where the emphasis on GMP will be focused. Table 6.1 provides a sample compliance review matrix for common biotech document types and the associated disciplines that will have review and coordination responsibilities.

Along with drawings, the personnel involved in the compliance review process must also focus on the equipment that will require special design considerations in order to meet the GMP guidelines identified in subpart D of 21 CFR 211. A listing of typical GMP equipment items follows:

- Buffer preparation material handling equipment
- Media preparation material handling equipment
- Fermentation preparation material handling equipment
- Hygienic vessels and tanks
- Hygienic pumps

Table 6.1 Overall Compliance Matrix

Item	Engineer	Compliance Review By
PFD	Process	Process
P&ID	Process	Process, Piping, Equipment, Control Systems
Overall Layout	Architect	Process, Process Architectural, Piping
Equipment Arrangements	Piping	Piping, Process Architectural, HVAC, Equipment, Process
Piping Layout	Piping	Process, Piping, Process Architectural, Equipment as necessary
Clean Rooms (Architectural)	Various	Process, Piping, Process Architectural, HVAC, Control Systems, Electrical, Equipment
Equipment Data Sheet/ Specification	Equipment	Equipment, Process, Piping, Control Systems
Control Systems	Control Systems	Control Systems, Electrical
Electrical	Electrical	Electrical, Control Systems
HVAC	HVAC	HVAC, Piping, Process Architectural, Structural (as necessary), Control Systems
Biocontainment	Various	Process, Piping, Process Architectural, HVAC
TOP Matrix	Validation	
Validation Master Plan	Validation	

- Hygienic filters and strainers
- Hygienic heat exchangers
- Autoclaves and sterilizers
- Depyrogenation ovens
- Glass washing machinery
- Centrifuges
- Homogenizers
- Clean steam generators
- Water for injection stills
- Reverse osmosis units
- Deionization units
- Ultrafiltration and homogenization units
- Clean-in-place units
- Lyophilizers
- Biocontainment/inactivation
- Chromatography systems
- Aseptic filling/processing equipment
- Electrical and instrument boxes and panels
- Conveyor systems

Once these items have been established as coming under the GMP review process, the quality management program of the A/E firm will require the development of specific design checklists to ensure that specific details of the design effort address all the necessary GMP issues related to the system and equipment items. Checklists are reviewed in Chapter 10.

Cleanliness

End products of any parenteral manufacturing effort are produced in a stringently controlled, clean environment that must be monitored and maintained in accordance with very strict standards. Therefore, the issue of cleanliness must be addressed from two viewpoints: (1) the design of the facility in order to maintain an acceptable GMP manufacturing environment (clean/sterile) and (2) the physical construction work during the execution of field activities.

The design issues related to cleanliness are well defined in terms of equipment and facility design, gowning and access control, HVAC system design, and cleaning protocols. These are discussed in other chapters within this book. Let us take a look at the not-so-obvious aspects of GMP related to facility construction.

General construction cleanliness

Construction is not, by nature, a clean activity. The majority of construction activities occur in an environment that actually contributes to the deterioration of clean conditions. In the construction of a sterile product manufacturing facility, it is important that all contractors and site employees understand the implications of GMP on construction activities and take the necessary actions to minimize the impacts of this fact.

There also must be a focus on general construction cleanliness for work areas, especially in cleanroom environments for manufacturing. Site work rules should include the elimination of any food, drink, or tobacco products from the immediate construction area. Concerns over rodents, insects, or food scraps becoming hidden from view in inaccessible areas require this special measure. An old chicken bone or piece of leftover sandwich inadvertently sealed within a wall can create a tremendous problem during the certification of cleanrooms. For this same reason, the use of wood products, especially small wooden shims or wood packing materials, should also be carefully controlled.

In warm climates, where workers still use salt tablets, it is also necessary to control the exposure of stainless steel components to salt water or salt compounds. It is also good practice to eliminate the use of hydrocarbon-based fuels and products, such as cutting fluids or gasoline/kerosene. Concrete, during curing, has a tendency to absorb these compounds, as do drywall and other architectural finish materials.

Preventive measures for these situations would include a program of cleaning, inspection, and sign-off of wall sections prior to closure with drywall, the use of electric or forced-air heaters instead of kerosene burning units, and the removal of any cutting operations from the building pad.

Many construction projects involve the remodeling or renovation of existing facilities where production operations are ongoing. In this case, prevention of product contamination and maintaining material and product flow are of paramount importance. Careful planning in moving workers and materials, controlling dust, debris, and noise, and ensuring plant utilities and services remain in operation is critical. Extensive preplanning measures must be undertaken by the operations staff and the construction organization to minimize the potential for plant shutdown or product loss.

Piping, HVAC, and equipment

Piping and HVAC systems require special attention during fabrication and installation in order to maintain system cleanliness. Compliance with the criteria identified in the design documents for material cleanliness, fabrication, and welding is an extremely important aspect of the construction operation. Contractors must address this during the early bid stages, as well as during field execution, to ensure that validation requirements are addressed without potential cost and schedule deviations.

Piping system fabrication specifications should be developed as "womb to tomb" documents regarding cleanliness. The specification must address material control, material segregation, material handling, cleaning, weld fit-up, performance criteria, and inspection requirements.

Stainless steel tubes, fittings, and components should be maintained in a clean, segregated area where dust and debris are minimized. End caps on all items should be maintained and checked for damage. Many projects require a "bag and tag" program where specific components for fabrication spool pieces are sealed in plastic bags and tagged with the appropriate fabrication drawing number or spool piece number, minimizing the potential for contamination and material mix-ups.

In handling stainless materials, it is customary to require personnel to wear clean, white cotton gloves to avoid transmitting potential contaminants from hands to the material surface. Handling devices such as clamps, jigs, or carts should have compatible material surfaces to come in contact with the piping components. Using bare carbon steel is unacceptable. Any tools that come in contact with the material surface, such as wire brushes, should also be of compatible material.

All temporary marking on stainless materials should be done with an etching tool, not with markers or grease pencils. Any cloth used to wipe down surfaces should be clean and not treated with any cleaning chemicals.

Fabrication activities should be performed to the greatest extent possible in an environment that is conducive to the end product — clean, warm, dry, and segregated from external contamination sources.

Ductwork for HVAC systems is usually fabricated at some remote location from its final installation point, due to the size and type of equipment required for the fabrication process. Transporting duct sections involves the use of end caps or seals to prevent internal contamination of the duct surface. Nesting of duct sections (pieces within pieces) is not an acceptable practice.

As duct sections are installed, an inspection program to verify debris removal is a good practice. Tools, materials, or debris left inside a duct section can not only damage the system, but can also cause severe problems during final balance and certification of the system.

Final system testing of air handling systems begins with a blowdown of the system. Over the duration of construction, regardless of the diligence of the installing contractor, duct systems will become dirty. To avoid possible contamination of rooms and equipment, be sure that temporary filters are installed at all supply and return grills. This will prevent contaminants from being sucked into the system, as well as any contaminants from the system being dispersed into the facility.

Schedules may also dictate the need to install some large equipment items early during construction, exposing them to daily dirt and construction activity. When this occurs, a program of ongoing equipment protection and inspection should be established that focuses on the physical protection of the equipment, but also addresses the integrity and maintenance of clean interior conditions. Ensuring that flanges remain sealed and protected, pass-through equipment remains sealed, and nozzle connections are not exposed will be time and money well spent.

Documentation

The concept of validation is foreign to most construction projects —while some form of documentation may be required for quality inspections and as-built conditions, no requirement for proof of compliance is dictated by law. Only the food industry and the nuclear power industry have compliance programs similar to the pharmaceutical industry's.

Earlier, we defined the basis of validation as "documented evidence." Throughout the design and construction process, the concept of documented evidence is prevalent in the development of specifications, contractual documents, vendor design submittals, testing and inspection, and commissioning. The documentation requirements for the facility will follow a well planned road map, called the Validation Master Plan. It is here that the documentation requirements are first defined and the process of GMP compliance begins.

Qualification and validation

Equipment and system qualification and validation are expressly required by the FDA. Qualification establishes the confidence that process equipment and ancillary systems are capable of consistently operating within established limits and tolerances, and that finished products produced through a specified process meet all release requirements. Validation, as we have already defined, provides the "documented evidence" of that fact.

In 2002, both CBER and CDER adopted the ICH Q7A Good Manufacturing Practice for Active Pharmaceutical Ingredients. This document defines an additional qualification area, that of Design Qualification:

"Design Qualification (DQ) is documented verification that the proposed design of the facilities, equipment, or systems is suitable for the intended purpose" [2].

To provide this evidence, every project will require a comprehensive documentation package that should consist of the following documents:

- Validation Master Plan
- Design Qualification (DQ)
- Installation Qualification (IQ)
- Operational Qualification (OQ)
- Performance Qualification (PQ)
- Process Validation

During the validation process, the responsibility for the development and issuance of OQ and PQ documents depends on the execution philosophy developed by the owner — whether to execute this work using in-house resources or contract the development of qualification documents to a third party. This is usually dependent on availability of resources and the experience level of these resources.

In regard to the design and construction of the facility, the focus of the validation effort will be on the Master Plan, the DQ, and the IQ. In these documents, the impact of GMP on the design and construction of the project will be most evident.

Master plan

The Master Plan is written as an outline of the overall validation program. This document, developed from a vast array of information sources (Figure 6.1), provides specific validation tasks and documents that are necessary for the facility, systems, and processes to meet the guidelines established for GMP.

The sections of a typical Master Plan (Example 6.1) define the scope of the project and the validation effort, the approach to the validation effort, the validation documents for each system, and the systems and equipment to be validated. Each section provides vital information related to the documentation requirements that are of interest to the FDA during the inspection of facilities prior to licensing and during ongoing operational inspections.

Appendix 6.1 provides a generic example of a Master Plan, formatted for a fermentation facility.

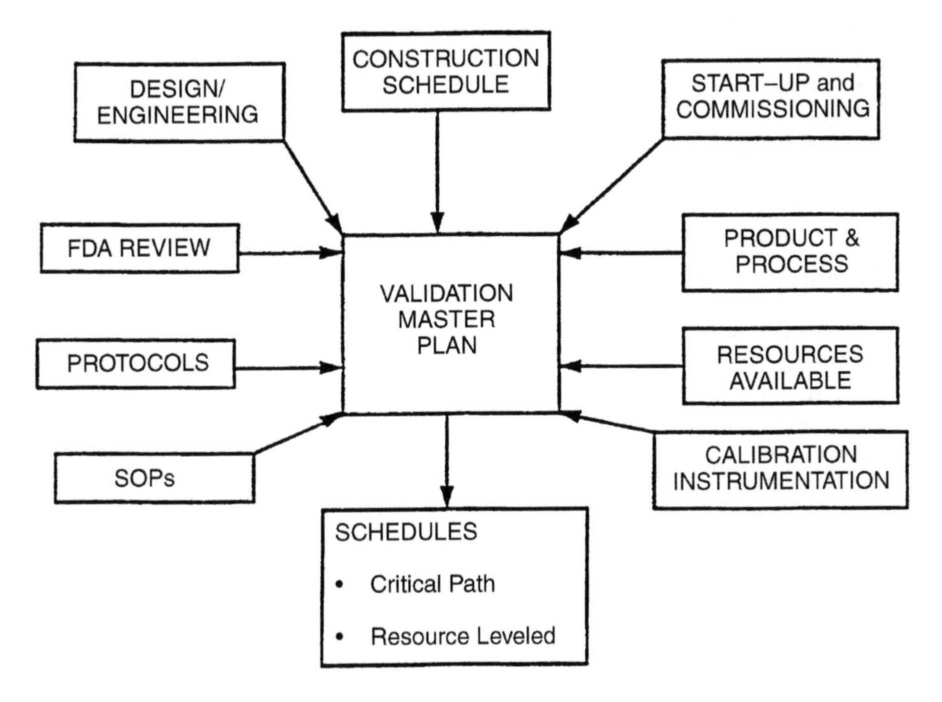

Figure 6.1 Development of a Validation Master Plan.

Design qualification

The concept of Design Qualification is based on ensuring that the design meets intended requirements. The ISPE Commissioning and Qualification Baseline Guide, ICH Q7A, and EC Annex 15, all embrace this process and focus on enhanced design reviews as a means of implementing DQ. For DQ to add value, the focus must be on documented verification of the design's meeting well defined user requirements, as defined in a formal Basis of Design (BOD) document. The requirements must be comprehensive to include each and every requirement that could impact the manufacture of the product. These would include topics from

- Process definition
- Maintenance
- Facility
- Quality
- Capacity definition
- Safety
- Operability
- Environmental issues
- Regulations

Example 6.1 Sample Table of Contents for a Validation Master Plan

I. Introduction
- A. Scope
- B. Master Plan
  1. Purpose
  2. Validation Approach
  3. Acceptance Criteria
  4. Validation Responsibilities

II. Process Description
- A. Facility
  1. People Flow
  2. Product Flow
  3. Architectural
  4. Environmental
- B. Utilities
- C. Process
  1. Process Overview
  2. Unit Operations

III. Validation Requirements
- A. Document Identification
- B. Facilities
- C. Utilities
- D. Process

IV. Standard Operating Procedures

V. Turnover Packages
- A. TOP Responsibilities
- B. TOP Documentation Requirements
- C. TOP Matrices

VI. Validation Schedule
- A. Validation Sequence
- B. Summary Schedules
  1. Final Report Schedule
  2. TOP Schedule
  3. IQ/OQ Schedule
  4. PQ/VQ Schedule

VII. Appendices
- A. Terms & Definitions

Installation qualification

Installation Qualification provides field verification that the proper equipment has been installed, that the equipment has been installed in accordance with the manufacturer's specifications, and that the required installation and testing procedures were followed during installation activities. Each system is evaluated to determine proper installation and connection of the supporting services (air, water, steam, power) and components (valves, filters, instruments, piping, controls). All data collected during the execution of IQ is packaged and summarized for review, approval, and archive.

While the Master Plan provides general system descriptions, IQ provides a more detailed system description that includes specific equipment information. Manufacturer's name, model number, capacities, and materials of construction are included.

Piping system components are also checked as a part of IQ, including materials, welding records, slope verification, final pressure testing, and passivation. Hangers, supports, and insulation type and integrity are also verified. Critical and noncritical instrumentation is identified, and calibration is verified as required.

Appendix 6.2 includes a sample IQ for a WFI system addition.

Turnover package

If a facility is to be validated, it is required that all installation documents — drawings, specifications, certifications, and vendor data — be collected. All of this information should be collected in a single place. The turnover package (TOP) is the tool used to capture all of the design and construction documents.

Each validated system will have a designated TOP. The TOP will generally break down the system documentation into different categories or sections — general information, equipment, piping, HVAC, electrical, instrumentation, controls (programmable logic), and architecture. Sections that are required for each system are identified by a matrix, similar to that shown in Example 6.2.

For each section, another matrix is then developed that defines each specific document, by the type of form, which must be submitted for inclusion in the IQ. Example 6.3 provides an example of a WFI system section for piping documents. Many of the forms to be submitted will come from the equipment manufacturer or the installing contractor of the system. Appendix 6.3 includes examples of sample documentation forms.

By collecting documents in this manner, engineers and validation personnel have an easy system for verifying that all submittal requirements have been met. It also provides easy access to documents for FDA inspectors during site inspections.

Operational qualification

Operational Qualification verifies that a specific piece of equipment or system operates as specified and meets the design requirements for control of operating parameters. OQ describes the operational tests, measurements, and control tolerances of key parameters that are critical to the proper operation of the system.

OQ testing is carried out before the actual product is committed to the system. In the case of process steps, OQ testing may utilize a suitable placebo

Example 6.2 Turnover Package Documentation Requirements by System

Validation System/ System Number	Architecture /01	Ion Exchange Cold Room /02	In Process Cold Room /03	Cold Vail Strg. Cold Room /04	Drains /05	HVAC /06	Emergency Solv./ Firewater Coll /07	Fermentation CIP /10	Purification CIP /11	Portable Tank CIP /12	Portable Tank SIP /13	DI Water /14	WFI /15	Clean Steam /16	Process Air /17	Process Chilled Water /18	Nitrogen Gas /19	Waste Solvent System /20	Instrument Air /21
Section 1: General Information		X	X	X	X	X	X	X	X	X	X	X	X	X	X	X	X	X	X
Section 2: Equipment		X	X	X		X	X	X	X	X	X	X	X	X	X	X	X	X	X
Section 3: Piping					X		X	X	X	X	X	X	X	X	X	X	X	X	X
Section 4: Electrical		X	X	X		X		X	X	X		X	X	X	X	X		X	X
Section 5: Instrumentation		X	X	X		X	X	X	X	X	X	X	X	X	X	X	X	X	X
Section 6: PLC								X	X	X	X	X	X	X		X			
Section 7: HVAC						X													
Section 8: Architecture	X	X	X	X															

Example 6.3 Turnover Package Documents

	System	S 721
TOP NO. 15 TOP SYSTEM-WFI		
SECTION 3-PIPING		
3A. PIPE SPECIFICATION	E	-
3B. PIPE LINE INDEX	E	-
3C. CLEANING CERTIFICATION	C	V
3D. PIPE MATERIAL CERTIFICATION	C	V
3E. PRESSURE TEST CERTIFICATION	C	V
3F. WELDING CERTIFICATION	C	V
3G. PASSIVATION CERTIFICATION	C	V
3H. LINE/DEVICE VERIFICATION	C	-
3I. LINE/DEVICE CHECKLIST	C	-
3J. LINE LABEL VERIFICATION	C	-
3K. LINE SLOPE VERIFICATION	C	-
3L. LINE SUPPORT VERIFICATION	C	-
3M. ISOMETRIC DRAWING LIST	C	-
3N. ORTHOGRAPHIC DRAWING LIST	-	-
3O. WELD INSPECTION RECORDS	C	V
3P. VALVE & SPECIALTY DATA SHEET	E	-
3Q. SAFETY VALVE CERTIFICATION	C	-
3R. DYE LEAK TEST RESULTS	-	-

LEGEND:
C: CONSTRUCTION DOCUMENT
E: ENGINEERING DOCUMENT
O: OWNER DOCUMENT
V: VENDOR DOCUMENT

APPROVAL STAMP:______________
REVISION NO.: 0
REVISION DATE: 3/8/91

or water. OQ testing may not necessarily follow the process. It is designed to establish a performance baseline to provide assurance that the system can operate as intended and for future troubleshooting.

For utility systems, OQ testing must also verify that the system can adequately service all users at the normal peak design load. To do so, worst-case challenges will be defined and incorporated into the testing strategy for the system. A worst case is a situation that may occur in normal operation. For example, the stress on a WFI system when several points of use are used at once would be a common worst-case occurrence.

Any exceptional conditions encountered during the execution of OQ are identified for evaluation and review. The exceptional conditions will then be investigated and corrected, or requalification studies for the system will be initiated by the validation group. Again, all data collected during the execution of OQ is summarized and packaged for review, approval, and archive.

Performance qualification

Performance Qualification is required for critical systems, such as WFI, DI water, or clean steam, where performance or process parameters are known and would affect the product if they were not met. It is performed after the completion of IQ and OQ activities.

In PQ, satisfactory performance of the system throughout the proposed operating conditions is checked over an extended period of time. Equipment cleaning procedures, aseptic processing, clean steam quality, sterility, and filter integrity are areas where PQ studies are performed.

Execution

Compliance with GMP is not something that starts well into the project; it must begin at the start of conceptual design and continue to be a focal point of execution throughout design, construction, and commissioning. Many projects fall into the trap of assuming that GMP compliance is a design issue; if the design meets GMP, then all validation requirements will be met. All that remains to be done is to build per the design and document what you have done. But the impacts go well beyond that simple belief.

During design, there will be a need for continued enhanced design review and coordination between disciplines to maintain overall project GMP compliance. Each discipline must coordinate design development to ensure that compliance is not only by system or individual project element, but that a synergy between all disciplines produces a complete facility that meets GMP.

To accomplish this, frequent GMP review meetings should occur, with a focus on reviewing GMP issues by area or system. Identifying a project GMP coordinator is a good idea. This will provide a focal point for communicating GMP questions not only within the design organization, but also with the owner and the FDA.

Validation and QA personnel should also be involved early in developing the overall validation philosophy with the owner, including the outline for the Master Plan and the execution philosophy for validation and qualification.

During design, it is important to ensure that the BOD is placed under the project change management system to allow for tracking changes to the BOD as user requirements are identified and defined. Before the design is implemented in the field, the DQ should give all reviewers a high degree of confidence that every requirement has been factored into the design.

During construction, execution efforts must focus on maintaining project cleanliness, development and execution of a quality assurance and control program that addresses GMP, and development and control of required documentation. Document control will include not only normal construction data, but also all TOP documents identified by the Master Plan.

It is during project commissioning that the impacts to execution will be felt most in terms of project resources. The execution of system walk-downs, the review of TOP documentation, the development of SOPs and protocols, and the execution of IQs and OQs will require significant personnel resources and time. Project budgets and schedules must accurately reflect this fact; cutting corners risks overruns and delays.

Project budgets

GMP impacts on design budgets will be reflected in the person hours required to execute the design. Most A/E firms calculate discipline person hours that include an allowance, whether stated or factored, for design development to meet GMP through review meetings, checklist development, or special documents, such as pre-BLA packages and Master Plans.

The effect of GMP on construction budgets occurs in a variety of different ways. First, contractors with little or no previous sterile product facility experience will tend to use past-experience cost information in developing budget estimates. These estimates are generally low and carry with them a very high level of risk, not reflecting the impact of cleanliness issues, increased inspection, or GMP documentation. These budgets also do not reflect the decrease in productivity due to added inspection requirements, congested working conditions traditionally found in biotech facilities due to the massive mechanical and HVAC systems, and the complexity of the automation/control interfaces between systems.

Another area of budget impact is defining the allowance for validation. Validation costs are difficult to estimate, due to the nature of the expenditures and the long period of time involved in their completion. Typical allowances in budgets will be in the range of 5 to 7% of TIC. This can also be as high as 12 percent on some projects, a significant number for many companies.

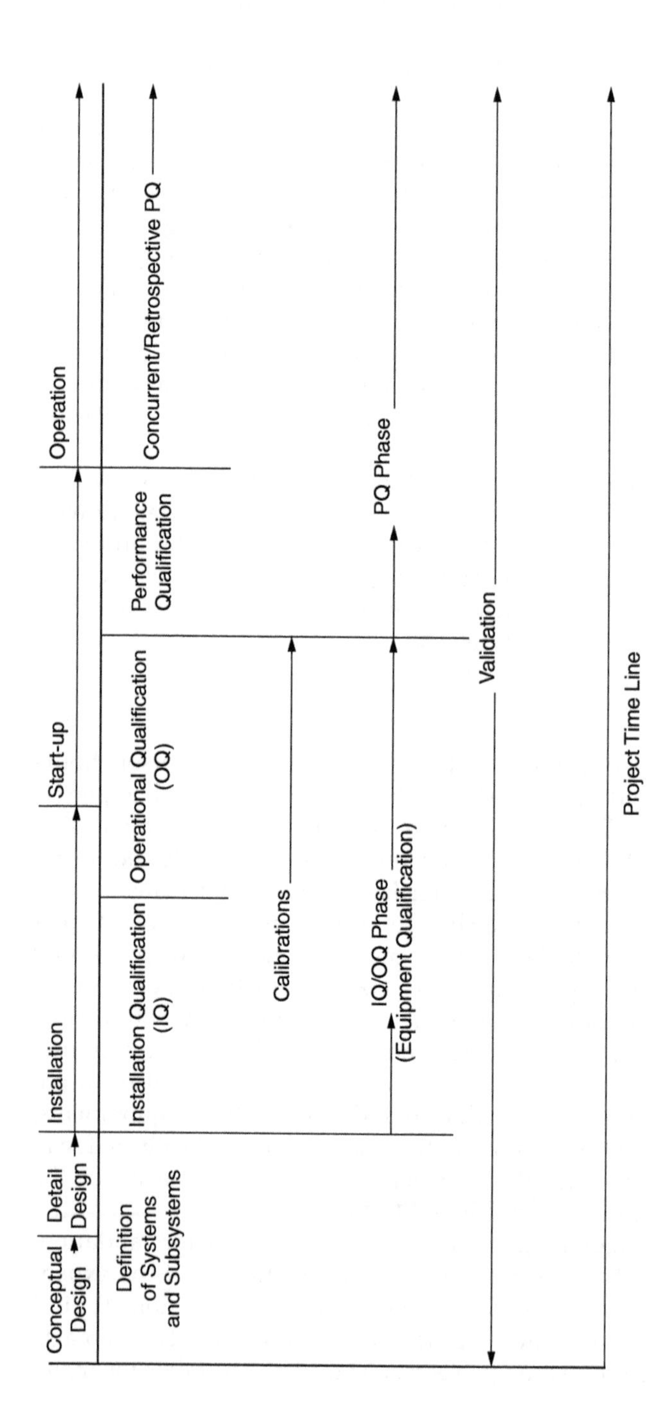

Figure 6.2 Project timeline.

Schedules

The major GMP impact on project schedules comes from the effort associated with the execution of project validation. Shown graphically in Figure 6.2, validation activities should begin early in the life of the project and continue throughout the execution of design and construction and throughout the life of the facility. Failure to realize the impact of meeting GMP qualification requirements when developing a project schedule can lead to unrealistic expectations regarding turnover, cash flow, and resource needs.

References

1. FDA. 1987. *Guideline on General Principles of Process Validation.* Washington, DC: Department of Health and Human Services, Food and Drug Administration.
2. FDA. 2001. Q7A Good manufacturing practice guide for active pharmaceutical ingredients. *Guidance for Industry* (August).

Appendix 6.1

Validation master plan

The purpose of this example is to provide a generic format for a Validation Master Plan, using a fermentation facility as the case facility. The intent is to show basic format and samples of content for selected applicable sections of the Plan.

I. Purpose: This section provides a basic intent of the Plan scope.

The purpose of this document is to present the Validation Master Plan for the new biologics manufacturing facility for ABC Biotech, located in Small Town, USA. The intent of the validation activities is to assure that the facility meets corporate and regulatory requirements for validation prior to facility operation. The facility is scheduled for completion in March 1997, with validation completion scheduled for November 1997.

II. Background: This section provides an overview of the facility and its operation basis.

This fermentation facility is designed to manufacture bulk pharmaceutical products by recombinant DNA technology. The manufacturing process consists of a single production train, composed of a fermentation and recovery system. The fermentation system consists of a seed fermentor, a production fermentor, a nutrient delivery system, and the associated piping and control systems to support the process.

The recovery system consists of a centrifuge, an ultrafilter, and a product isolation chromatography column. The final product is filtered through 0.2 mm filter into transfer containers.

III. Description: This section provides a more detailed description of the facility and the main components of its design, namely the overall facility, the architectural finish philosophy, the HVAC system, the process operations, the utility systems, and the process control system. Reference drawings of the facility layout, equipment locations, and HVAC zoning would be included.

A. Facility Description: The new fermentation facility will occupy approximately 60,000 ft^2. The facility is designed to current GMP guidelines and will include separate areas for production, utility support, laboratories, warehousing, and administrative offices.

B. Architectural Finishes

C. HVAC System

D. Process Description

D.1.Media Preparation

D.2.Inoculum Preparation and Seed Fermentation

D.3. Production Fermentation
D.4. Harvesting: When the fermentation operation is completed, the fermentation broth is transferred to a continuous desludging centrifuge. The yeast cells are discharged to the solids receiver. The supernatant, which contains product, is transferred to the harvest tank. The supernatant in the harvest tank is cooled and temperature controlled by circulation through an external loop using a glycol-chilled heat exchanger. During cooling, the supernatant pH is adjusted by a phosphoric acid solution.
D.5. Ultrafiltration
D.6. Buffer Preparation
D.7. Product Isolation Chromatography
D.8. Bulk Filling
D.9. Biowaste Inactivation
D.10. Autoclaves

E. Utility Description
E.1. Purified Water System
E.2. Clean Steam
E.3. Clean-In-Place System: A CIP skid is provided to clean and sanitize permanently installed fixed equipment in the Media Preparation, Fermentation, and Harvesting areas. The CIP system consists of a prerinse tank, acid wash tank, caustic wash tank, motive tank, CIP supply pump, CIP motive pump, various CIP spray balls, and CIP return pumps. CIP of the ultrafiltration system is accomplished by a separate system.
E.4. Compressed Air
E.5. Plant Steam
E.6. Chilled Glycol
E.7. Cooling Water
E.8. Waste Water Neutralization

F. Process Control

IV. Validation: This section provides a description of the sequence of validation activities for a typical validated system. Each system or equipment item will be treated uniquely during validation. The systems and/or equipment items are validated using approved protocols developed and approved by ABC Biotech.

A. Turnover Packages (TOPs): This section will provide all of the engineering and construction documentation required for validation in single-source packages defined by system. Each TOP is divided into the following eight sections:
A.1. General Information
A.2. Equipment
A.3. Piping
A.4. Electrical

A.5.Instrumentation/Controls
A.6.Automation
A.7.HVAC
A.8.Architectural

B. Installation Qualification/Operational Qualification
C. Equipment/Utilities Performance Qualification
D. Process Validation
E. System Testing and Monitoring
F. Acceptance Criteria

V. Documentation: This section defines the protocols and validation requirements for the facility.

VI. Responsibilities: This section defines the organization functions and designated personnel responsible for the preparation, control, and approval of all protocols, procedures, test data, and control and issue of the final report.

Appendix 6.2

Sample installation qualification (IQ) for WFI system

1.0 Document Approval
2.0 Purpose
3.0 System Description
4.0 Installation Checklist
5.0 Reference Documentation
6.0 System Summary

1.0. Document approval

A. This Validation Protocol has been approved for use by the following responsible parties:

	Name	Signature	Date
Project Engineer			
Plant Engineer			
Technical Services Mgr.			
Area Manager			
Plant Manager			
Quality Manager			

B. The results within this Validation Protocol have been reviewed and approved by the following responsible parties:

	Name	Signature	Date
Project Engineer			
Plant Engineer			
Technical Services Mgr.			
Area Manager			
Plant Manager			
Quality Manager			

2.0. Purpose

The purpose of this Protocol is to ensure that the WFI Loop Addition and Modifications are installed per the requirements set forth in the design specifications and drawings. Also, this document verifies that all appropriate GMPs were followed during construction of the circulation loop.

3.0. System description

Thus description covers the additions and modifications that were installed on the previously validated WFI System. The additions to the WFI System included installation of a heated WFI storage tank, two distribution pumps (one for the existing Clean Area B Loop and one for the new Clean Area C Loop), addition of a use point to Clean Area B Loop, distribution piping for the Clean Area C Loop, and DCS controls.

Loop A circulates WFI from the still to Hold Tank B (formulation) and Hold Tank C (production).

Clean Area B Loop circulates WFI from Hold Tank C (T-100) to a number of existing points of use through a validated distribution loop. A point of use, supplying WFI to Service Autoclave III, has also been added to Clean Area B Loop.

Clean Area C Loop circulates WFI quality water on a continuous basis from Hold Tank 3C (T-100) to the following areas:

1. SVP Class Wash (Rm. 101)
2. Marshalling Sink (Rm. 121)
3. TL Syringe Wash (Rm. 123)
4. Hold Tank #6 (Rm. 130)
5. Hold Tank #7 (Rm. 130)
6. WFI Heat Exchanger (Rm. 150)
7. Stopper Washer (Rm. 150)
8. Stopper Rinser (Rm. 148)
9. Cleaning Drop (Rm. 201)

The water that feeds the loop described above is supplied at 80°C, ±10°C, and is maintained in that temperature range by a steam jacket on the WFI Storage Tank and an insulated distribution loop. Water is circulated in the loop at a minimum supply velocity of 5 ft/sec, and a minimum return velocity of 3 ft/sec. System back pressure is maintained at a minimum pressure of 40 psi.

4.0. INSTALLATION CHECKLIST

A. WFI Storage Tank

Equipment Number: ____________ PO Number: ____________ Room: __________

	Specified	Installed
Manufacturer	Precision Stainless ____________	____________
Serial Number	____________	____________
Capacity (gal)	____________	____________
Material of Construction	316L ____________	____________
Internal Finish (Ra)	____________	____________
Operating Pressure (psig)	____________	____________
Operating Temperature (°C)	80 ____________	____________
Gasket Material	Silicone ____________	____________

General Observation:	**Yes or No**
1. Is the tank labeled clearly?	____________
2. Does the tank have an ASME code stamp?	____________
3. Is the tank fully drainable?	____________

All design specifications have been reviewed and a documentation of any deviations is attached.

Verified by: ______________________________ Date: ____________

B. Tank Rupture Disk

Manufacturer: ______________ PO Number: ______________ Room: ______________

	Specified	**Installed**
Model	______________	______________
Serial Number	______________	______________
Material of Construction	______________	______________
Internal Finish (Ra)	______________	______________
Pressure Rating (psig)	______________	______________

General Observation:	**Yes or No**
1. Is the rupture disk labeled clearly?	______________
2. Is the rupture disk installed and verified per manufacturer's instructions?	______________

All design specifications have been reviewed and a documentation of any deviations is attached.

Verified by: ________________________________ Date: ______________

C. Clean Area B Distribution Pump

Equipment Number: P-105 PO Number: ________ Room: ________

	Specified	Installed
Manufacturer	Tri-Clover	
Model		
Serial Number		
Flow Rate (gpm)		
Impeller Size (in.)	6 3/4	
Impeller Material	316L	
Casing Material	316L	
Wetted Parts Finish (Ra)	30	
RPM	3500	
Seal Manufacturer	Tri-Clover	
Seal Type	Mechanical	
Lubricant	Grease	
O-Ring Material	Viton	

General Observation:	**Yes or No**
1. Is the pump labeled clearly?	________
2. Does the pump have a casing drainage valve?	________
3. Is the pump installed level?	________
4. Is the pump easily disassembled for cleaning?	________

All design specifications have been reviewed and a documentation of any deviations is attached.

Verified by: ______________________ Date: ________

D. Clean Area C Distribution Pump

Equipment Number: P-181 PO Number: ______ Room: ______

	Specified	Installed
Manufacturer	Tri-Clover	
Model		
Serial Number		
Flow Rate (gpm)		
Impeller Size (in.)	6 3/4	
Impeller Material	316L	
Casing Material	316L	
Wetted Parts Finish (Ra)	30	
RPM	3500	
Seal Manufacturer	Tri-Clover	
Seal Type	Mechanical	
Lubricant	Grease	
O-Ring Material	Viton	

General Observation:	**Yes or No**
1. Is the pump labeled clearly?	______________
2. Does the pump have a casing drainage valve?	______________
3. Is the pump installed level?	______________
4. Is the pump easily disassembled for cleaning?	______________

All design specifications have been reviewed and a documentation of any deviations is attached.

Verified by: ________________________________ Date: ______________

E. Vent Filter

Equipment Number: ______________ PO Number: ______________ Room: ____________

	Specified	Installed
Housing		
Manufacturer	Millipore	
Model		
Serial Number		
Material of Construction	316L	
Interior Finish (Ra)	25	
Jacket Pressure Rating (psig)	50	
Gasket Material		
Filter Cartridge		
Manufacturer	Millipore	
Material of Construction	PTFE	
Effective Filter Area	7.0 ft^2	
Maximum Flow Rate (cfm)	50	
Maximum Clean DP (psi)	1.0	
Seal Material	Silicone	

General Observation:	**Yes or No**
1. Is the filter labeled clearly?	________
2. Is the housing installed so it is drainable?	________
3. Are steam and condensation lines installed properly?	________

All design specifications have been reviewed and a documentation of any deviations is attached.

Verified by: ______________________ Date: ________

F. Distribution Piping

Note: In the case of multiple material vendors, this page should be copied, completed with the appropriate information, and inserted into the document.

Installing Contractor: ______________ PO Number: ______________

	Specified	Installed
Tubing		
Manufacturer	______________	______________
Type	______________	______________
Material	316L	______________
Interior Finish (Ra)	20	______________
Material Lot Numbers	______________	______________
Fittings		
Manufacturer(s)	Tri-Clover	______________
Type	Auto-Weld	______________
Material	316L	______________
Interior Finish (Ra)	20	______________
Ferrules		
Manufacturer(s)	Tri-Clover	______________
Type	Auto-Weld	______________
Material	316L	______________

Interior Finish (Ra)	20	______

General Observation:	**Yes or No**
1. Is piping installed so that all dead legs are less than 6 pipe diameters?	______
2. Does all pipe slope a minimum of 1/8 per foot?	______
3. Is the system completely drainable?	______
4. Has all piping been installed per drawings?	______
5. If field changes were necessary, have they been noted on as-built drawings?	______
6. Have all welds been identified on the weld isometric and the tubing?	______
7. Has the daily welding log been reviewed for compliance?	______
8. Has the passivation log been reviewed for compliance?	______
9. Are manufacturer's certificates of material compliance on file for piping, ferrules, and fittings?	______
10. Were the piping and accessories received, stored, and distributed in accordance with approved procedure?	______
11. Has the piping system been boroscoped?	______
12. Has the system been hydrostatically tested by an approved method?	______

Identify the location of the following information:

As-Built Drawings	______
Isometric	______
Daily Welding Log	______
Passivation Log	______

Material Certifications ____________________________

Welder Certification ____________________________

Sample Welds ____________________________

Welding Procedures ____________________________

Weld Inspection Reports ____________________________

Hydro Test Results ____________________________

All design specifications have been reviewed and a documentation of any deviations is attached.

Verified by: __________________________ Date: ____________

G. Diaphragm Valve Worksheet No. ________

Repeat information set for each valve manufacturer. Copy this worksheet as necessary and insert into the Qualification Document.

Manufacturer: ____________ Model Number: ____________

	Specified	**Installed**
Body Material	316L	
Bonnet Material	Nylon Coated	
Diaphragm Material	EPDM	
Gasket Material		
Interior Finish (Ra)	25	

General Observation:	**Yes or No**
1. Are all valves labeled clearly?	
2. Are all valves installed per manufacturer's recommendations for drainability?	
3. Are valved installed per the design drawing?	
4. Are the manufacturer's certificates for material compliance on file?	

5. Where is the manufacturer's data located? ______________________________

All applicable specifications have been reviewed and a documentation of any deviations is attached.

Verified by: ______________________________ Date: ____________

H. Pressure Gauge Worksheet No. ________

Repeat the following information set for each gauge manufacturer. Copy this worksheet as necessary and insert into the Qualification Document.

Manufacturer: ____________ PO Number: ____________ Model Numbers: ________

	Specified	Installed
Gauge/Diaphragm		
Product Contact Material	316L ____________	____________
Gasket Material	____________	____________
Pressure Transfer Fluid	Silicone ____________	____________

General Observation: **Yes or No**

1. Are pressure gauges clearly labeled? ____________
2. Are pressure gauges installed as indicated on the instrument location plans? ____________
3. Are pressure gauges calibrated and tagged with calibration stickers? ____________
4. Are calibrations organized and in place per SOP C-1? ____________

All applicable specifications have been reviewed and a documentation of any deviations is attached.

Verified by: ______________________________ Date: ____________

I. Pressure Indicating Transmitter Worksheet No. ________

Repeat the following information set for each transmitter manufacturer. Copy this worksheet as necessary and insert into the Qualification Document.

Manufacturer: ______________ PO Number: ______________ Model Numbers: ________

	Specified	Installed
Transmitter		
Product Contact Material	316L	________________
Gasket Material	________________	________________
Pressure Transfer Fluid	Mineral Oil	________________

General Observation: **Yes or No**

1. Are pressure transmitters clearly labeled? ________________
2. Are pressure transmitters installed as indicated on the instrument location plans? ________________
3. Are pressure transmitters calibrated and tagged with calibration stickers? ________________
4. Are calibrations organized and in place per SOP C-1? ________________

All applicable specifications have been reviewed and a documentation of any deviations is attached.

Verified by: ________________________________ Date: ______________

J. Temperature Gauge Worksheet No. __________

Repeat information set for each gauge manufacturer. Copy this worksheet as necessary and insert into the Qualification Document.

Manufacturer: ______________ PO Number: ______________ Model Numbers: __________

	Specified	Installed
Gauge/Well		
Product Contact Material	316L	
Thermal Conducting Medium	Flor-a-lube or Equivalent Substance	

General Observation: **Yes or No**

1. Are temperature gauges clearly labeled? ______________
2. Are temperature gauges installed as indicated on the instrument location plans? ______________
3. Are temperature gauges calibrated and tagged with calibration stickers? ______________
4. Are calibrations organized and in place per SOP C-1? ______________

All applicable specifications have been reviewed and a documentation of any deviations is attached.

Verified by: ______________________________ Date: ______________

K. Temperature Element Worksheet No. ________

Repeat information set for each element manufacturer. Copy this worksheet as necessary and insert into the Qualification Document.

Manufacturer: ____________ PO Number: ____________ Model Numbers: ________

	Specified	Installed
Element		
Product Contact Material	316L ____________	____________
Gasket Material	____________	____________
Thermal Conducting Medium	Flor-a-lube or Equivalent Substance ____________	____________

General Observation: **Yes or No**

1. Are temperature elements clearly labeled? ____________
2. Are temperature elements installed as indicated on the instrument location plans? ____________
3. Are temperature elements calibrated and tagged with calibration stickers? ____________
4. Are calibrations organized and in place per SOP C-1? ____________

All applicable specifications have been reviewed and a documentation of any deviations is attached.

Verified by: ______________________________ Date: ____________

L. Level Element Worksheet No. __________

Repeat information set for each element manufacturer. Copy this worksheet as necessary and insert into the Qualification Document.

Manufacturer: ______________ PO Number: ______________ Model Numbers: ________

	Specified	Installed
Element		
Product Contact Material	316L ________________	________________
Gasket Material	________________	________________
Thermal Conducting Medium	Mineral Oil ________________	________________

General Observation:	**Yes or No**
1. Are level elements clearly labeled?	________________
2. Are level elements installed as indicated on the instrument location plans?	________________
3. Are level elements calibrated and tagged with calibration stickers?	________________
4. Are calibrations organized and in place per SOP C-1?	________________

All applicable specifications have been reviewed and a documentation of any deviations is attached.

Verified by: ____________________________________ Date: _______________

M. Conductivity Element Worksheet No. _________

Repeat information set for each element manufacturer. Copy this worksheet as necessary and insert into the Qualification Document.

Manufacturer: _____________ PO Number: _____________ Model Numbers: ________

	Specified	**Installed**
Element		
Product Contact Material	Titanium & Ryton ______________	______________
Gasket Material	______________	______________

General Observation:	**Yes or No**
1. Are conductivity elements clearly labeled?	______________
2. Are conductivity elements installed as indicated on the instrument location plans?	______________
3. Are conductivity elements calibrated and tagged with calibration stickers?	______________
4. Are calibrations organized and in place per SOP C-1?	______________

All applicable specifications have been reviewed and a documentation of any deviations is attached.

Verified by: ________________________________ Date: ______________

N. Control Valve Worksheet No. ________

Repeat information set for each valve manufacturer. Copy this worksheet as necessary and insert into the Qualification Document.

Manufacturer: ____________ PO Number: ____________ Model Numbers: ________

	Specified	**Installed**
Element		
Type	Sanitary Globe	____________
Product Contact Material	316L	____________
Gasket Material	____________	____________

General Observation: **Yes or No**

1. Are valves clearly labeled? ____________
2. Are valves installed according to P&IDs and piping drawings? ____________
3. Are valves properly oriented to prevent pockets? ____________
4. Are calibrations organized and in place per SOP C-1? ____________

All applicable specifications have been reviewed and a documentation of any deviations is attached.

Verified by: ______________________________ Date: ____________

0. Utilities and Services

1. Electrical

Service to Item	Service Voltage	Rated Amps	Breaker		
			Rating	Location	Labeled
P-181	460V				
P-105	460V				

All applicable specifications have been reviewed and a documentation of any deviations is attached.

Verified by: ______________________ Date: __________

2. Plant Steam

Service to Item	Pressure (psig)	Flow (lb/hr)	Line Size (in.)	Line Material	Labeled
T-100	50			Carbon Steel	
Vent Filter	50			Carbon Steel	

All applicable specifications have been reviewed and a documentation of any deviations is attached.

Verified by: ______________________________ Date: ____________

3. Clean Steam

Clean Steam Generator Number: ______________________

Operating Pressure: ______________________

Maximum Flow Rate: ______________________

Service to Item	Pressure (psig)	Flow (lb/hr)	Line Size (in.)	Line Material	Labeled
T-100	50			Carbon Steel	

All applicable specifications have been reviewed and a documentation of any deviations is attached.

Verified by: ______________________ Date: ____________

5.0. REFERENCE DOCUMENTATION

A. Design and Operating Specifications

Title	Rev.	Date	Location
Manufacturer's Manuals			
________	________	________	________
________	________	________	________
________	________	________	________
________	________	________	________
________	________	________	________
________	________	________	________
________	________	________	________
________	________	________	________
________	________	________	________
Design Specifications			
________	________	________	________
________	________	________	________
________	________	________	________
________	________	________	________
________	________	________	________
________	________	________	________
________	________	________	________
________	________	________	________
________	________	________	________

Verified by: ____________________ Date: __________

B. Construction Drawings

The following drawings have been inspected and accurately represent the system as-built.

Title	Rev.	Date	Location
P&ID			
Isometrics			
Orthographic			

Verified by: ______________________ Date: __________

C. Operating Procedures

The following drawings have been inspected and accurately represent the system as-built.

SOP Number	Rev.	Date	Title
Operation			
Maintenance			
Calibration			

Verified by: ______________________________ Date: ____________

6.0. SUMMARY

Validation Specialist

Validation Project Manager

Appendix 6.3

Turnover Packages

Line Support Verification

CLIENT: ____________________ LOCATION: ____________________

PROJECT NAME: ____________________ PROJECT NO.: ____________________

TOP SYSTEM: ____________________ TOP NO.: ____________________

This is to certify that the ____
system, TOP # __________, has been inspected and the support of the utility and process piping is adequate and conforms to Piping Specification Section No. ______________:
Titled: ____.

(NOTE: This inspection can be completed together with the "Line Slope Verification.")

REMARKS: ____

____

____

____

____

____

____

____

____

Performed By: ____________________ Date: ____________________

Checked By: ____________________ Date: ____________________
General Contractor

Client: ____________________ Date: ____________________

Line Slope Verification

CLIENT: ____________________ LOCATION: ____________________

PROJECT NAME: ____________________ PROJECT NO.: ____________________

TOP SYSTEM: ____________________ TOP NO.: ____________________

This is to certify that the ____
system, TOP # __________, has been jointly inspected and line slopes are proper. This includes a check for (1) high point vents; (2) low point drains; (3) traps in critical system lines; (4) proper diaphram valve angle; and (5) eccentric reducers as required.

(NOTE: This inspection can be completed together with the "Line Support Verification.")

The above was inspected and the following needs to be completed before final certification:

____ Date: ____________________

____

____

____

____

____

____

____

____

____

____

____

Above Work
Authorized By: ____________________ Date: ____________________

Performed By: ____________________ Date: ____________________

Checked By: ____________________ Date: ____________________
General Contractor

Client: ____________________ Date: ____________________

Passivation Certification

CLIENT: ____________________ LOCATION: ____________________

PROJECT NAME: ____________________ PROJECT NO.: ____________________

TOP SYSTEM: ____________________ TOP NO.: ____________________

Date Prepared: ____________________

This is to certify that the ____,
TOP # __________, has been cleaned per the Piping Specification No. ______________,
Titled: ____.

REMARKS: ____

____

____

____

____

____

____

____

____

____

____

____

Performed By: ____________________ Date: ____________________

Checked By: ____________________ Date: ____________________
General Contractor

Client: ____________________ Date: ____________________

Equipment Checklist: HEPA Filter

OWNER:	____________	LOCATION:	____________
PROJECT NAME:	____________	PROJECT NO.:	____________
TOP SYSTEM:	____________	TOP NO.:	____________
Equipment No.:	____________	Manufacturer/Model:	____________
Description:	____________	Serial Number:	____________

		YES/NO	INITIALS/DATE
I.	**PRE-INSTALLATION INSPECTION**		**INITIALS/DATE**
1.	Nameplate: Check against certified vendor drawing(s)		________
2.	Check for visible evidence of damage, particularly on seal surfaces		________
3.	Check that all parts are included with shipping		________
II.	**INSTALLATION INSPECTION**	**YES/NO**	**INITIALS/DATE**
1.	Securely and properly installed per vendor specifications	______	________
2.	Verify all duct connections against HVAC drawing	______	________
III.	**OPERATIONAL INSPECTION**	**YES/NO**	**INITIALS/DATE**
1.	Verify integrity of HEPA filters by DOP testing and attach results	______	________
2.	Record pressure drop across filter bank ________	______	________

REMARKS: ____

____

____

APPROVAL SIGNATURES:

Construction: ____________________ Date: ____________

Owner: ____________________ Date: ____________

chapter 7

Mechanical systems

Sterile product manufacturing facility projects are risky undertakings. In order to produce a finished product that meets the requirements defined in the GMP guidelines issued by the FDA, the facility and the processes that are used to produce drug components must rely heavily on the ability of the mechanical systems and equipment to produce and maintain sanitary conditions. To say that these facilities are mechanically intensive is an understatement. The massive amounts of piping and ductwork required to support sterile manufacturing cause many kinds of complications for designers who design facilities, craftsmen who build them, and operators who have the responsibility to maintain product that meets GMP.

It is not uncommon for the average costs of a sterile product manufacturing facility to show mechanical systems and equipment representing 35 to 50 percent of total installed cost (TIC) (Figure 7.1). This also represents the largest component of risk associated with the project, since project labor distribution, complexity, and schedule milestones will almost always focus on mechanical systems.

The various mechanical systems in a parenteral facility do have some common elements in relation to other types of process-driven facilities. Plant utility systems for steam, condensate, chilled water, and compressed air are similar in nature. The same can be said for normal sanitary plumbing and drainage systems. HVAC systems for office support and nonmanufacturing support will also see many similarities. But that is usually where the comparisons end.

In looking at mechanical systems in a sterile product facility, the focus should be on two primary areas: sanitary process piping and classified-area HVAC systems. These are the prime concerns of designers and builders, validation and quality teams, and operations and maintenance organizations. Within these system classes are the challenges and concerns related to GMP compliance and the real test of a facility design in minimizing contamination risks, maintaining operational philosophy, and meeting quality standards.

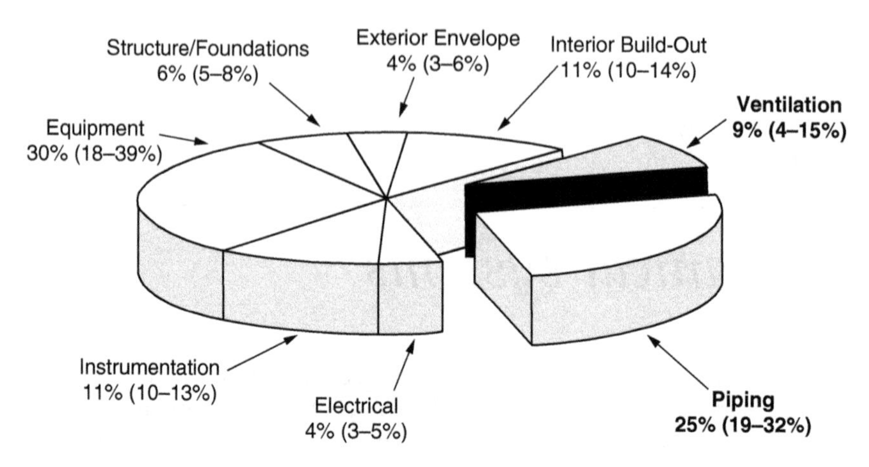

Figure 7.1 Mechanical systems costs.

Process piping

Some call it hygienic or high purity; others simply call it sanitary process or process. But the systems are the same: WFI water distribution, clean steam and condensate, biowaste, cleaning, media and process distribution, or product. These are the systems that must meet strict criteria for materials of construction, cleanliness, drainability, integrity, and cleanability established by the FDA for the systems whose components come into contact with the final drug product or a component thereof.

The development of sanitary process piping systems that formed the foundation for the pharmaceutical industry has its roots in the 1970s when, due to product contamination concerns and product recalls, large volume parenteral (LVP) manufacturers saw the FDA define the most specific criteria ever imposed on process systems. The "six-diameter" rule, Figure 7.2, came into existence, and automatic welding of tubing systems became the norm. As validation became more important in the early 1980s, the industry moved to more extensive documentation requirements and the exclusive use of materials and fabrication methods that had shown success in producing higher-quality finished systems that met GMP.

Today, these systems represent some of the most complex and difficult systems to design and fabricate. Recognition of this fact is the first step in producing a design that is both functional and buildable.

Design issues

Sanitary process system design focuses on the FDA guideline that a process must be repeatable [1]. To be so, process systems must be designed to prevent cell growth that could contaminate product or product components. This means that the systems must be cleanable, fully drainable, easily sanitized and sterilized, and designed to maintain integrity during the full extent of

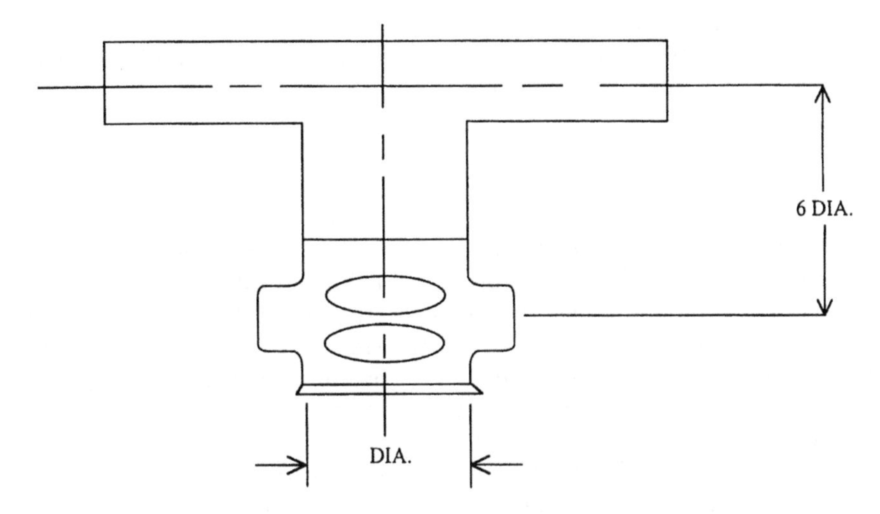

Figure 7.2 Maximum length for dead legs is specificed as six times the branch-run diameter by the "6-D" rule.

the manufacturing operation. To prove this, the system design must also incorporate the necessary measures for documenting compliance to GMP regarding the fabrication and installation of the system.

Many of the design attributes of sanitary process systems are well known:

- Materials of construction, usually 316L stainless steel, are corrosion resistant and easily fabricated.
- Surface finishes lend themselves to easy cleaning, normally a minimum 180 grit finish (20–25 Ra).
- Installations are fully drainable so that microbial growth cannot occur in residual pockets of process fluid. This is accomplished by eliminating dead legs to ensure continuous access of the cleaning/sterilizing medium and defining minimum slope requirements of 1/8 inch per foot (1 percent) for the installed system.
- Turbulent flow is used to inhibit the colonization of microbes on surfaces.
- Components, such as valves and seals, are designed for drainability and cleanability.
- Support systems allow for material expansion due to the cleaning regime while maintaining the slope for drainability.
- "Closed" systems prevent cross-contamination.
- Documentation requirements address each and every aspect of the installation to prove system compliance with mandated regulatory guidelines.

Beyond these "golden rules" are other issues that have a tremendous impact on system cost and the time required to fabricate and install the various components of the system.

Fabrication and installation

System cost

It is not uncommon for the installed cost of sanitary process systems to be in a range of $150 to $225 per foot. As Table 7.1 indicates, systems can fall into many cost ranges, depending on their size, material requirements, configuration, and inspection philosophy. As previously mentioned, because of this high percentage of cost in relation to overall construction cost (Figure 7.3), the task of minimizing risk becomes an important goal for the project manager.

The project team should take numerous steps to identify and minimize the risk associated with sanitary process piping installation. Many of the measures discussed here will also help to ensure compliance with GMP and produce an end product with minimal delays and cost variations.

Constructability reviews

Before the start of any installation work, the project team should review the piping design for compliance with GMP and for established sanitary design practices common to the industry and to the internal standards of the client. This review should focus on proper line slopes, identification of any dead-leg conditions, sample connection locations, steam inlet locations for sterilization, support locations and configurations, hanger spacing, proper discharge angles for equipment, proper valve port orientations for drainage, and the issues identified by operations personnel for routine maintenance access and product operations.

One of the most difficult aspects of sanitary piping system design is in defining diaphragm valve locations, configurations, and orientations to meet

Table 7.1 Installed Cost for Sanitary Process Piping Systems

Project Type	Total Cost (TIC) ($MM)	$/Installed ft
Biotech Manufacturing	3.2	165
Biotech Manufacturing	9.1	178
Biotech Manufacturing	17.0	220
Cell Culture	7.2	143
Utility Renovation	1.1	155
Sterile Fill/Finish	3.0	189

Note: TIC includes all material, welding, inspection, documentation, passivation, and overhead/profit.

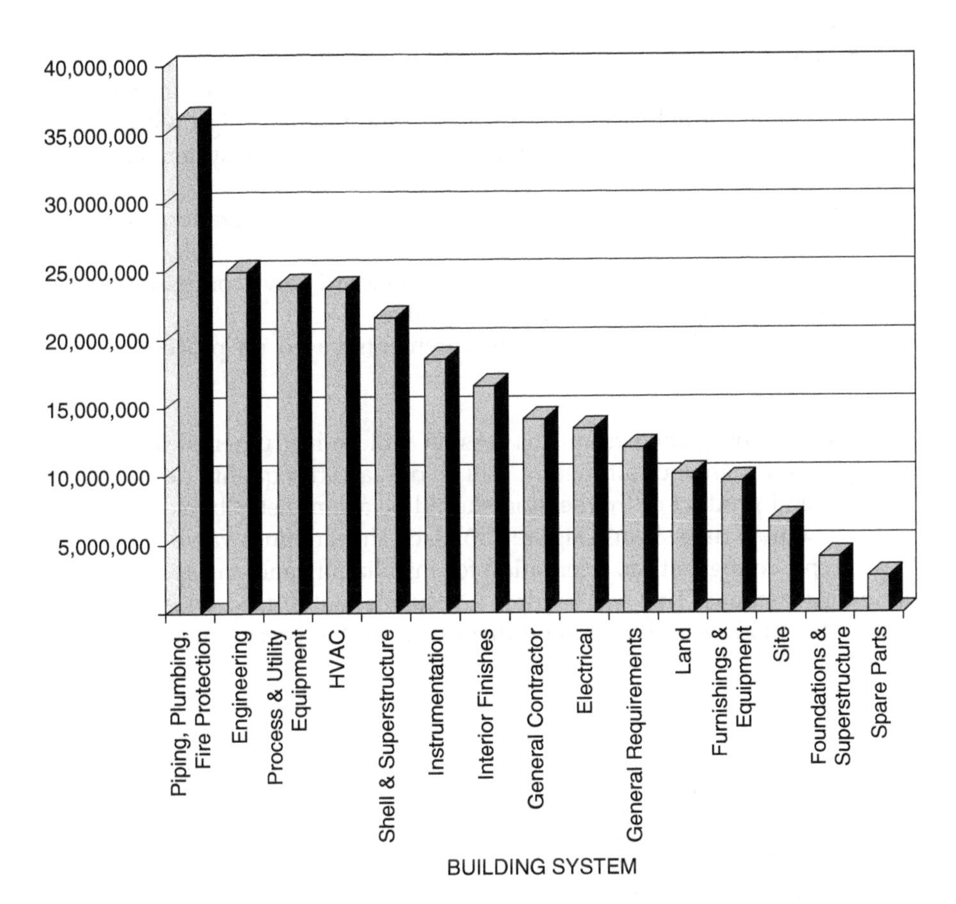

Figure 7.3 Costs for a typical biochemical manufacturing facility.

GMP. Depending on process requirements and overall design philosophy, there may be cases where configurations must be much more stringent than the 6-D rule identified by the FDA — one example being for cold systems. There may also be instances where, due to building configurations or equipment locations, routings may be less than desirable for locating valves. All of these situations must be "buildable" not only in terms of the manufacture of the valve, but also in system installation. Just because it looks good on the drawing board does not mean that it will be easy to install.

Once P&IDs are at or near an approved stage, it is good practice to have a formal review session to look at every sanitary diaphragm valve and define its orientation, configuration, and type based on an approved or preferred vendor product. The purpose of this session is to

- Define orientations in horizontal and vertical flow situations to minimize dead legs
- Identify where sterile access ports are required for CIP/SIP applications

- Identify situations requiring GMP-type valves
- Investigate the use of valve assemblies in order to minimize the number of field welds and simplify the piping design with a focus on improving overall system quality and reducing system cost (Figure 7.4)
- Ensure accessibility of all valves/assemblies for operation and maintenance
- Identify any situations where a nonstandard valve configuration will be required (a special or custom valve)
- Verify minimum dead-leg requirements based on process requirements

The information gained from this session will be instrumental in defining the system design and, ultimately, will lead to a better quality design effort.

The end product of this session should be a complete "book" of valve/valve assembly data sheets (Figure 7.5). Each sheet defines valve number(s), valve type, configuration, orientation for installation, and any special installation instructions. This will become a valuable tool during system installation in the field, as well as a good document for the procurement of valves.

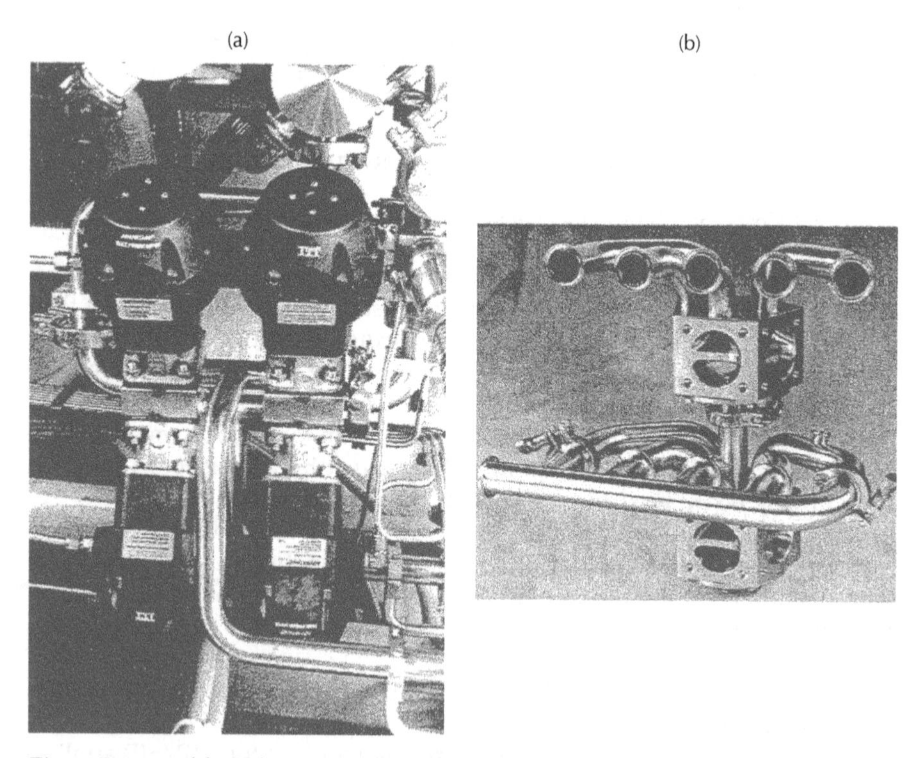

Figure 7.4 a and b. Valve assemblies. Photos courtesy of ITT Sherotec, Simi Valley, CA.

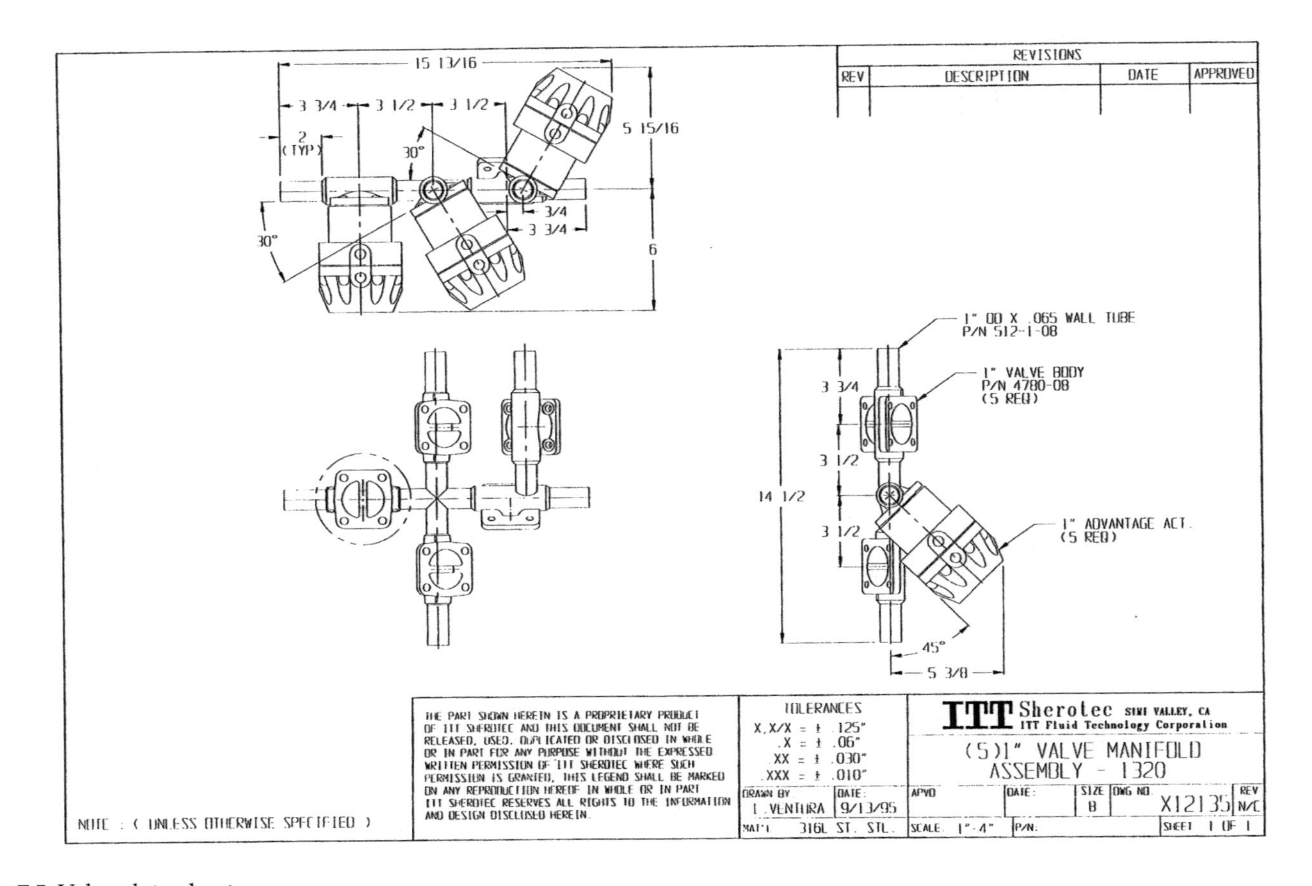

Figure 7.5 Valve data sheet.

A well-executed review program can save considerable time and money by allowing items to be corrected during design, not during construction when costs are increased by a factor of ten. It also provides user groups with a final opportunity to ask, "Is that what we really need?" If the answer is no, changes in design are, again, minimal in cost and schedule impact.

Another area where many projects fail to recognize potential benefit is the use of contractor resources during preconstruction design activities. The review of isometric piping details, generated from P&IDs and orthographic drawings, can yield valuable information about issues such as interferences, routing problems, support configurations, total number of welds to be made, or potential "blind weld" scenarios. Obtaining the services of experienced contractor personnel to review design details and assist in isometric drawing development can help eliminate many simple errors which can have significant schedule impacts or cause resources to be used in a less-than-effective manner. Companies that fabricate and install these systems on a regular basis can add tremendous value for very little cost by lending their fabrication expertise and assisting in the design effort as another set of eyes to review constructability issues regarding the systems.

All of these steps are good insurance and effective ways to decrease the risk component of system installation.

Material procurement

One area of system fabrication that deserves close scrutiny is in the procurement of materials. Contrary to what may be perceived, there are no real standards for construction materials; much of what is specified in design documents is based on preference.

Material finish can vary greatly between clients and projects. Surface finish can be defined at the minimum 180 grit all the way to a 320 grit electropolished finish for tubes, fittings, and component surfaces. Finish specification can impact material cost (the higher the polish criteria, the higher the cost), delivery times (longer manufacture time), and inventory availability (only common items held in stock), especially for some hard-to-get fittings.

Another issue is that specifications vary greatly for the dimensional sizing of fitting components used in these systems. The chart in Figure 7.6 shows how dimensions vary between manufacturers' standards and industry standards for the same fitting, a 90° elbow. During design, the A/E firm will select a particular base fitting type and use these dimensions as the basis for dimensioning and tolerances of installation. Consistency in buying fittings, not only during construction but also for spare parts inventory, will make installing and replacing fittings a much easier task.

In the design of sanitary process systems, there are many cases in which the final configuration of pipe routing leads to a situation of complex fitting-to-fitting arrangements, along with numerous in-line instruments and devices. This condition normally occurs at equipment connections and interfaces. This variation in dimensions of fittings can lead to problems in

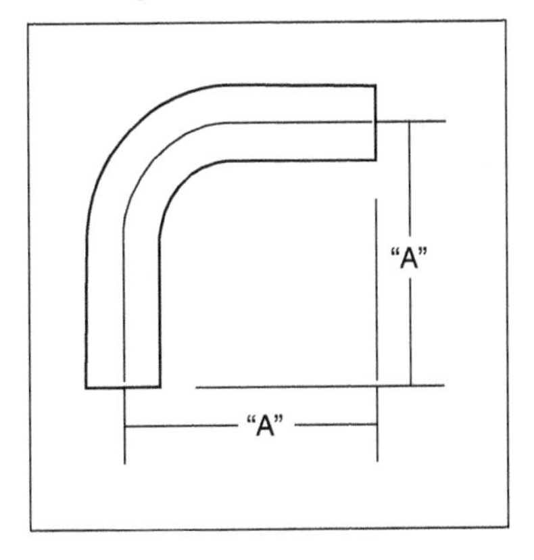

	Tri-Clover	Valex	Superior	ANSI B16.9	Tangent	BPE
Size	A	A	A	A	A	A
1/2	3.00	2.62	3.00	1.50	1.50	3.00
3/4	3.00	2.62	3.00	1.50	1.50	3.00
1	3.00	4.00	3.12	1.50	1.50	3.00
1 1/2	4.50	4.75	3.88	2.25	2.00	4.25
2	5.75	5.50	4.62	3.00	2.00	5.00
2 1/2	5.75	8.00	5.38	3.75	2.00	5.75
3	6.50	10.00	6.31	4.50	2.00	6.50
4	8.50	11.00	8.31	6.00	3.00	9.00

Figure 7.6 Industry and manufacturing standards for an automatic tube weld 90° elbow.

the makeup of line routings if consistency in specifying a particular brand or specification of fitting is required. If materials are purchased from different manufacturers, the dimensions may not correspond to those chosen by the design engineer.

Another issue that must be addressed is the tolerance allowed on material wall thickness. Variations of ±5 percent in the wall thickness can lead to a joint mismatch that will impact fit-up of the joint to be welded. Acceptable alignment cannot be achieved unless the components being welded have similar OD and ID dimensions. In Figure 7.7, a graphic of mismatch worst-case conditions is shown, along with the resulting condition from welding with improper alignment.

Improper joint alignment can cause a convex or concave weld bead that may interfere with the complete drainage of the piping system during cleaning. It also increases the amount of time required to prepare each joint for

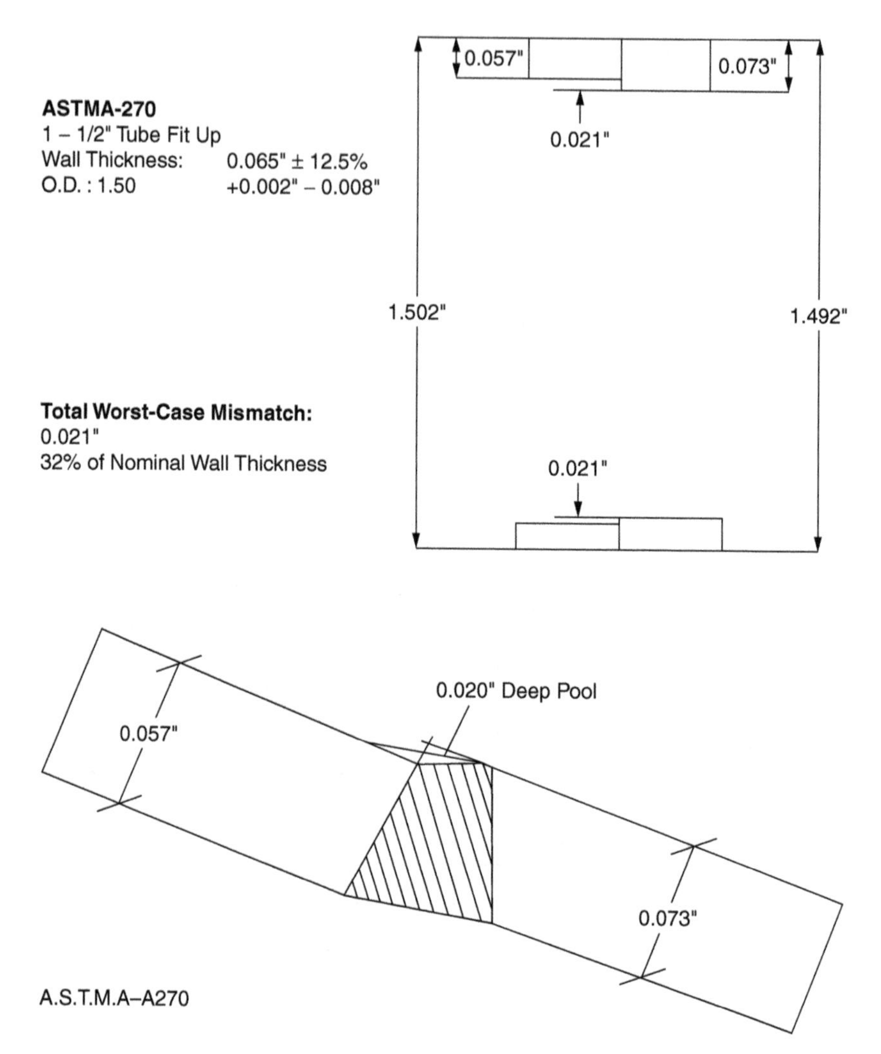

Figure 7.7 (a) Alignment mismatches. (b) A.S.T.M. A-A270.

welding. The graph in Figure 7.8 represents data on weld joint labor, showing that the fit-up of the joint is the most time-consuming operation under normal conditions. Any deviations in joint geometry will only increase the time required, thus increasing overall labor cost and creating potential schedule delays.

The performance of automatic welding equipment is also impacted. End-preparation specifications for automatic orbital welding are much more stringent than for a manually welded joint. Although it is possible to make acceptable welds with less-than-ideal joint fit-ups, there is always a quality issue if joints are not properly prepared.

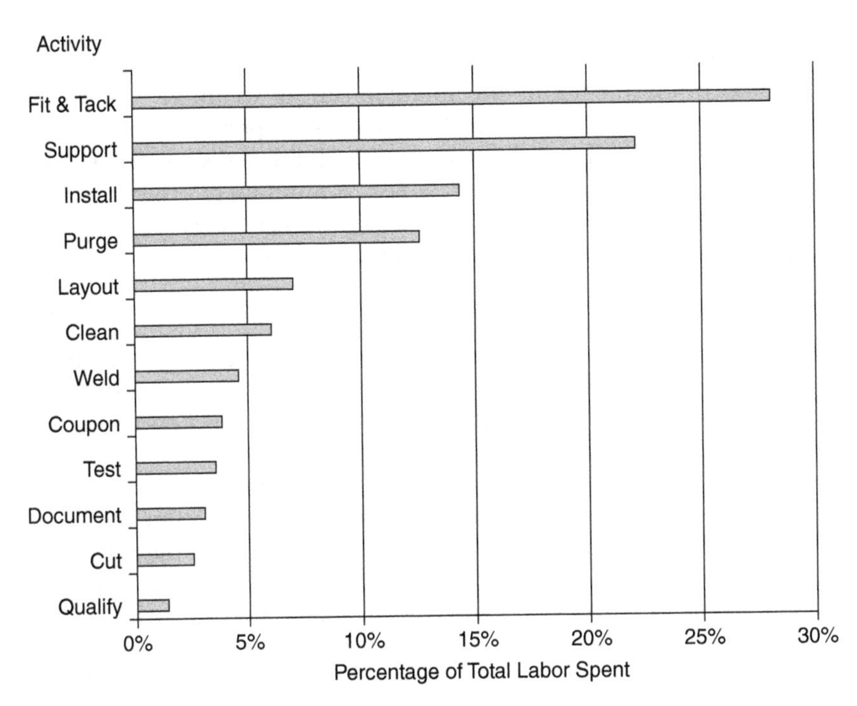

Figure 7.8 Sanitary tube installation at biopharmaceutical manufacturing and filling facilities, January 1991 to March 1994.

Cleanliness

Understanding the impact of maintaining cleanliness during system installation is a key to success. With most construction efforts occurring in less-than-ideal conditions in regard to a clean environment, knowing how to minimize material and system contamination during installation can go a long way in saving time, money, and potential rework due to quality issues.

The fabrication and installation specification will be the primary control document for dictating cleanliness in a fabrication/installation program. The specification must be very specific in describing the methods and procedures for the handling, storage, and working of stainless materials/components used in the system.

All sanitary process piping materials must be stored and handled to minimize contamination. Sealed ends, bagged fittings and valves, protected storage racks, and segregated storage areas should be the norm, not the exception. Measures to prevent carbon contamination of stainless materials would include no carbon contact from tools or fabrication aids/devices, dedicated tooling for cutting and polishing, etching tools for marking versus common pipe markers with normally high chloride content, and segregated work areas away from carbon steel grinding and welding.

Further protection should require that workers wear clean cotton or latex gloves during material handling to prevent soiling. Oil or dirt on workers' hands could result in chloride precipitation during welding.

During installation, all open ends should be capped as soon as possible. It is good practice to have daily routine walkthroughs by the installing contractor to verify that all ends are capped. A program of protection should also be developed for areas where other discipline activities, such as painting or fireproofing, could create potential contamination concerns.

Documentation

Another important aspect of sanitary process installation is the documentation that must be included in the IQ for facility validation. The first step is to define documentation requirements early in the project and ensure that the installing contractor understands not only what the documents are, but also when the completed documents must be submitted.

Too often, a good installation effort goes bad because documents that confirm successful completion of a required inspection or test are misplaced or not generated in the appropriate time frame. Other problems can occur if the proper information from the test or inspection is not obtained and verified by the proper personnel. These problems can be avoided by taking some simple steps.

A document matrix should be developed that defines all forms that will be required to verify design compliance (Figure 7.9). This matrix becomes a part of the system TOP for the IQ package at validation. Sample forms for all inspections should be approved early, whether they come from the contractor, the A/E firm, or the client. What is important in this approval is the information each form contains and the approvals required on the form. Samples of the document matrix and the inspection forms should be part of the contractual bid documents with the installing contractor, along with a table defining required submittal time frames. Thus, there can be no misunderstanding of what the requirements are and when they are expected to be met.

The documents required include material purchase orders, receiving records for materials and components, certified material test reports, welder qualification records, weld samples maintained for weld schedule verification, weld maps identifying each weld by unique number, weld logs that identify the details and time of each weld performed, purity certification for all weld gases, weld inspection reports, slope verification reports, hanger/support verification reports, cleaning and passivation records, pressure test reports, and videotaped records.

All of these documents must be collected and controlled to ensure their availability for audit and turnover at the end of the project. All documents should be originals, complete with all required approval signatures clearly legible and all information completed as required per the project document control procedures.

There is another aspect of the documentation process that merits discussion — as-built drawings. While as-built drawings are a specified require-

TOP NO. ________ TOP SYSTEM ________																				
SECTION 3-PIPING																				
3A Pipe Specification																				
3B Pipe Line Index																				
3C Cleaning Certification																				
3D Pipe Material Certification																				
3E Pressure Test Certification																				
3F Welding Certification																				
3G Passivation Certification																				
3H Line/Device Verification																				
3I Line/Device Checklist																				
3J Line Label Verification																				
3K Line Slope Verification																				
3L Line Support Verification																				
3M Isometric Drawing List																				
3N Orthographic Drawing List																				
3O Weld Inspection Records																				
3P Valve & Specialty Data Sheet																				
3Q Safety Valve Verification																				
3R Dye Leak Test Results																				

Figure 7.9 TOP matrix.

ment for IQ documentation, many projects seem to have difficulty deciding when to have these drawings produced, who should produce them, and in what format should they be submitted for archive.

In sanitary process systems, it is common to have conditions that are conceived by the design engineer one way but installed in the field in another, as seen in Figure 7.10. Both are acceptable from an engineering perspective, but what is most important is the condition of the installed system. Be sure that a clear understanding of the as-built drawing program is negotiated with the A/E firm and the installing contractor so that the proper as-built drawings are included with the TOP.

In many cases, companies use the same isometric drawings as for design and installation, with field markups or red lines to show as-built conditions.

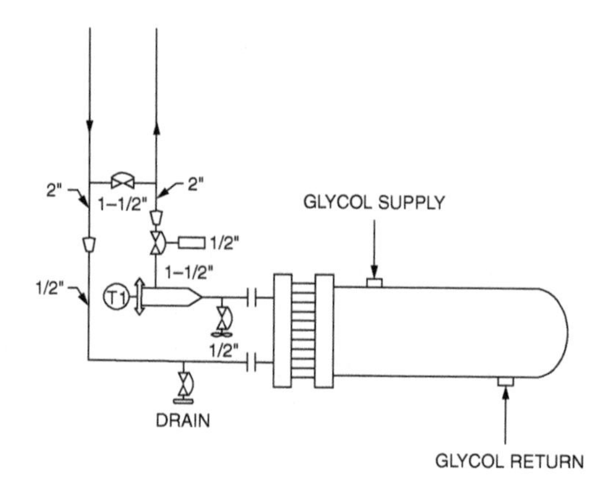

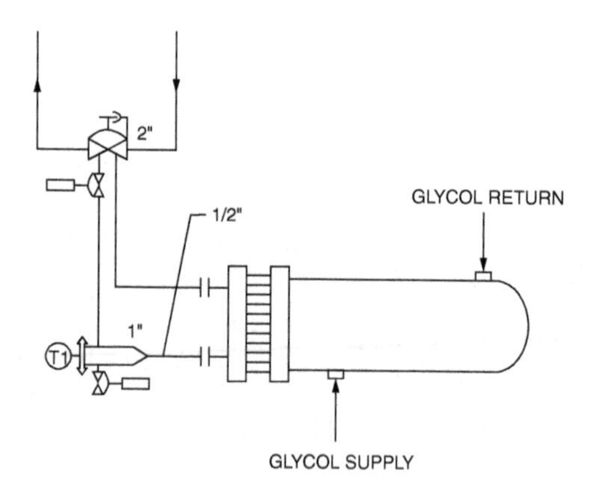

Figure 7.10 "As conceived" vs. "as built."

Whether this information is ever transferred to a final as-built drawing or simply remains on the isometric depends on the amount of money you want to spend.

Testing, cleaning, and passivation

With installation complete, the focus shifts to final testing, cleaning, and passivation of the systems. Systems should be flushed from low-point drains through temporary strainers to ensure particle removal. Testing of the systems should be included as part of the installation contractor's scope of work, so that any defects due to fabrication/installation errors can be corrected under the scope of the installation contract at no additional cost. Final pressure testing in accordance with design specifications using approved test media (deionized water, clean air, argon, nitrogen) must be documented for validation.

Stainless steel is naturally passive and resistant to corrosion. But during fabrication operations, defects will occur to the surface oxide film. Once the oxide film is damaged or weakened, the stainless steel may start to corrode at the location of the defect. Passivation does not remove oxide film; it is designed to do light cleaning by removing soluble matter adhered to the metal's surface, thus restoring the uniform finish of the material.

It is common practice in the industry for passivation to be performed by an outside firm that specializes in this type of operation. That is not to say that using an outside firm is mandatory; many installation contractors and some clients choose to perform their own passivation activities. Whatever the scenario, be sure that the personnel performing passivation operations are experienced with the medium chosen and the potential impacts on system components.

Passivation with nitric acid or other solutions requires special precautions. Be sure that all valves are flushed and cycled with sufficient flow through the valve. Diaphragm valves should be removed and any embedded materials cleaned from the diaphragms. Be sure to remove or isolate any components that are not 300 series stainless material. Also, be sure to check with local authorities to see what treatment measures are required before dumping passivation wastewater into the sewer system.

Modular fabrication concepts

With the trend in facilities moving toward very large projects, both in terms of scale of operation and capital cost, many companies are looking at a modular approach to fabrication and construction as a way to decrease costs, improve construction schedules, and maximize efficiency in the execution of equipment acceptance testing. While the concept of modularization is not new for process and mechanical equipment, it has become a growing trend for large capital pharmaceutical and biotech manufacturing projects.

To discuss modular concepts, you must first define what a module is. Many firms define it differently. For this discussion, we will define a module as any component of a facility that can be fabricated, constructed, and tested off-site, then shipped to a project site for installation in the finished facility. Using that definition, however, you could conclude that an equipment skid is a module, a group of similar equipment components assembled is a module, a prefabricated room with some equipment components is a module. And you would be correct.

There are two basic modular concepts (Figure 7.11):

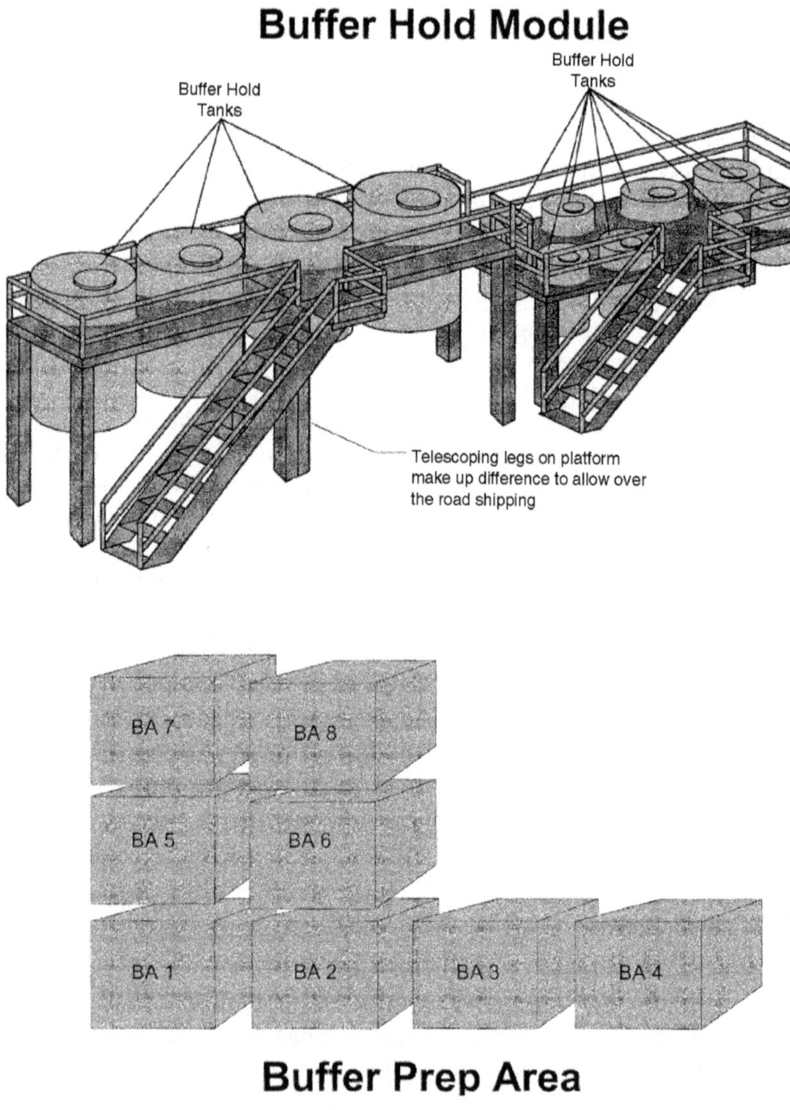

Figure 7.11 Basic modular construction approaches.

- **Process modules** — The building is constructed by conventional means. Modules will consist of internal system piping and equipment that are shop assembled off-site into assemblies, whose size is dictated by building access or shipping constraints.
- **Facility modules**_The entire building is divided into modular structures that include both building elements (walls, structure) and mechanical/process/electrical systems. Each module is preassembled off-site. The modules are then assembled at the site to construct the entire facility.

There are a number of advantages and disadvantages to the modular approach that should be investigated before implementing such a program:

- **Cost** — Modular construction will not decrease your facility capital cost. While it can decrease rework due to poor quality, the same number of man hours to fabricate and install piping and equipment will still remain. Cost savings occur when site resources are freed up during peak construction periods, reducing potential overtime situations, and in the improved time to have final system qualification and commissioning activities completed. On many projects, this can represent hundreds of thousands of dollars.
- **Schedule** — One advantage of implementing a modular approach is that system start-up and turnover of mechanical systems can generally be accelerated due to early completion of factory acceptance testing (FAT) at the vendor shop. Current industry claims are that up to six months can be gained from using this approach. But there is also a critical schedule element that must be identified and addressed very early in the project: engineering design data must be completed very early in the project to support the fabrication effort. To do this requires that critical decisions related to design scope, equipment selection, control philosophy, and system configuration be defined and "cast in stone" much earlier.
- **Quality** — There should be measurable improvements in quality due to fabrication/assembly work being completed in a shop environment. Generally, a shop environment is cleaner, has a more highly skilled and more stable workforce, and implements more stringent quality procedures on a daily basis. Another quality advantage can be found in the execution of FATs in a controlled shop environment. FAT execution in the shop allows for more refined evaluation of test results and will also allow for a more efficient investigation/repair approach when issues are found. Documentation is also better controlled in a shop environment than in field installation situations.

Potential problems — What can go wrong

Even with the best laid plans, problems can still occur. The problems that can be encountered on projects are varied — both in cause and impact. The following are examples of problems from case projects that resulted in quality issues impacting cost, schedule, or both. Each has been categorized in an attempt to define the source of the problem.

- **Design** — Prepurchased fittings did not match fittings used as a design basis; all fitting dimensions were incorrect. Material had to be repurchased, resulting in severe schedule delay to the project (26 weeks).
- **Design** — Design location of vessel nozzles did not correspond to as-built location of nozzles. Prefabricated spools had to be reworked, producing a 4-week schedule delay.
- **Design** — No specification was given on weld seam for tubing. Material had to be returned, impacting the schedule by 3 weeks.
- **Design** — Duplicate isometrics issued for the same system had different configurations. The fabrication of both configurations resulted in unnecessary material usage and welding time.
- **Materials** — Tubing and fittings were received with out-of-round end conditions. Rework cost of $6000 was incurred for facing end preps.
- **Materials** — Valves were received that did not meet minimum wall thickness requirements; valves had been overpolished. Replacement valves required 10 additional weeks for delivery.
- **Materials** — Cracks were found in base metal of tees during final testing of system. All similar fittings required reinspection; unacceptable fittings had to be cut out of the system and replaced. The schedule impact was 6 weeks.
- **Materials** — Contaminated welding gases were used. All welds had to be removed and reworked in system; the schedule impact was 6 weeks.
- **Fabrication** — Design dimensions were not verified in the field. Thirty-seven prefabricated spools had to be reworked.
- **Fabrication** — Tubing material was not identified by heat number; identification had been removed during polishing operation.
- **Fabrication** — Weld identification numbers were found in the wrong location; the latest drawing revision was not used.
- **Fabrication** — Material was received without certifications and installed in system. All tubing had to be removed and new material installed; the schedule impact was 6 weeks.

Inspection

One of the greatest areas of controversy regarding the installation of sanitary process systems is in the philosophy that governs the level of quality that goes into the system design and the inspections required to validate that level of quality. The idea that "more quality is better quality" or that "the more money spent on inspection, the better" can cause as many problems as an ill-defined policy on inspections.

Table 7.2 Philosophies of System Specifications

	Low Side	High Side
Inspection Requirements	Self-Police	100% Videotape
Surface Finish	180 Grit	320 EP/ESCA/SEM
Tube	Welded	Seamless
Weld Acceptance	ANSI/AWS	No Mismatch/Color
Fitting Tolerance	None	±5% wall, .0050 O.D.
Slope	No Pockets	1/40 per Foot
Material Joining	No Restrict	No Clamps, 100% Auto
Dead Legs	6D	3D
Cost Impact	$X	$2.6X

Example courtesy of Kinetic Systems Inc., Santa Clara, CA.

Many biotech and pharmaceutical companies react fearfully to FDA guidelines by stipulating very stringent system specification requirements and then requiring excessive inspection requirements to prove compliance. At the same time, there is no clear, consistent philosophy in the industry regarding how much is enough; each company interprets compliance differently. Table 7.2 illustrates two different philosophies regarding system specifications, each meeting the intent of GMP guidelines. The irony of this table is that both facilities were using a similar protein-derived manufacturing process and they were located in the same business park!

FDA criteria found in 21 CFR, Part 211.42, addresses the end results that must be achieved through the design and installation of the facility, including the sanitary process systems [2]. What it does not tell you is how to achieve these results. It is important to address the inspection philosophy for sanitary process systems early, so that design documents can reflect the desires of the client and so that budgets accurately reflect anticipated costs for implementing the inspection program. Standard industry practice and previous experience will play an important role in the development of this philosophy.

Table 7.3 gives a general overview of inspection approaches and the potential consequences of their implementation. Each approach carries with it some level of risk and cost, from little risk and high cost to high risk with a much lower cost. The challenge is finding the approach that makes sense.

A good example of this dilemma is the requirement for finished weld inspection. Current industry practice has finished welds receiving a 100 percent visual inspection of the exterior weld surface and a boroscopic inspection in order to validate interior weld surface finish compliance. Many companies require 100 percent boroscopic inspection of all sanitary process system interior welds. In addition, a trend has emerged in the past few years to archive videotaped records of all boroscopic inspections for future use by the FDA. Conversely, some companies take the approach of random inspections based on some predetermined percentage, say 20 percent, of completed welds. If defective conditions are found, then the frequency of inspections increases until some evidence of correction of the problem is developed.

Table 7.3 Inspection Approaches and Their Consequences

Inspection Approach	Potential Consequences
100% Inspection	Costly, inefficient, usually unnecessary, and sometimes physically impossible. Documentation more costly.
No Inspection	Low system integrity, possible contamination, difficult to maintain sanitary conditions. Lack of documentation to support validation.
In-House Inspection	Burden of proof on Owner. Inspectors usually poorly trained.
A/E, Contractor Inspection	Potential conflict of interest—problems result from what firm did or did not do.
Independent Inspection	Objective evaluation. Can be costly. Quality of inspection dependent of experience of firms personnel.

The cost to inspect an average orbital weld in a sanitary process system is approximately $70 to $75. Add the cost of videotape and archive, and the cost reaches $80 or higher. In systems that have thousands of welds, these costs can reach extravagant proportions. By defining an acceptable level of inspection, project costs can be managed while defining an acceptable level of risk.

Contracting — Finding the right company

Many companies have the resources and expertise to fabricate and install sanitary process systems for biotech and pharmaceutical facilities that will be validated before commercial use. Because of the risk associated with these systems and the extreme cost of having them fabricated, installed, and tested, there is always a concern about how to ensure the right firm is chosen to execute the work. While there are never any guarantees, some steps can be taken during the bid/evaluation phase of a project which can greatly increase the chances for success.

Obtaining good, relevant information regarding a company's resources and capabilities is important in the evaluation process. The development of a thorough questionnaire will provide much of the needed information to begin the evaluation process. Example 7.1 provides a sampling of typical questions that are important in evaluating contractor capabilities. Taking this information, an evaluation matrix can then be developed that provides a weighted factor of importance for each item (Example 7.2). Each potential contractor would receive a total score based on the sum of all the weighted factors, resulting in a ranking of contractors.

Another important evaluation tool is usually made a part of the actual bid document — information regarding the experience of the proposed staff that will be assigned to the project. The level of experience that the sanitary process piping contractor brings to the table should be on a level similar to that of the design firm and the construction organization. The critique of individuals should be no less; the emphasis on comparable work done in the biotech/pharmaceutical industry is just as important.

Example 7.1 Sanitary Process Piping Qualification Questionnaire

1. How many years have you been in business under current name?
2. What is your previous biotech/pharmaceutical experience? List projects completed over the past five years.
3. How many recent projects exceeded 10,000 linear feet of sanitary installation?
4. List any previous experience with ABC Biopharmaceuticals, Inc.
5. List any previous experience with Processing Engineering, Inc.
6. How many projects were executed in the state of North Carolina in the past five years? List all projects.
7. What is your firm's largest current project and its completion date?
8. Has your firm failed to complete a project in the past five years? If so, provide reasons.
9. What is your current boding capacity? Provide name of bonding company.
10. What is your firm's labor posture?
11. Describe your firm's understanding of philosophy regarding a design support arrangement with a firm such as Process Engineering, Inc.
12. Describe your shop fabrication facilities and capabilities in the Research Triangle Park area.
13. What is your current manpower (pipefitters and welders) availability for the immediate 6-to-12 month period? Provide a breakdown between shop and field labor.
14. What is your current shop capacity?
15. Describe your firm's orbital welding experience with stainless steel materials, including the number of machines and type of support equipment.
16. Describe your firm's CAD capabilities.
17. Does your company have a written safety program? Provide statistics for
    Lost workday cases (LWC)
    Restricted workday cases (RWC)
    Medical attention cases
    Fatalities
    Man-hours, previous year
    Substance Abuse Program (yes/no)
18. Interstate Experience/Modification rate (average 3 years)
19. Describe the project control tools utilized by your company.
20. Does your company have a written quality program?
21. Is your firm a licensed contractor in the state of North Carolina? Provide license number.
22. Provide the following financial information:
    Annual sales
    Current or total assets
    Current or total liabilities
    Net worth
    Net Quick Multiplier (NQM)
23. Provide three references for your company.

***Example* 7.2** Contractor Evaluation Matrix for Sanitary Process Piping

CRITERIA	WEIGHT	CONTRACTOR A	SCORE	CONTRACTOR B	SCORE	CONTRACTOR C	SCORE
YEARS IN BUSINESS (10+ = 4)	7	18	4	28	4	22	4
PARTNERSHIP EXPERIENCE	8	YES/SEE PROPOSAL	3	YES/SIGNIFICANT	4	YES/SEE PROPOSAL	4
PARTNERING AGREEMENTS	8	YES/SEE PROPOSAL	4	YES/SEE PROPOSAL	3	YES/SEE PROPOSAL	4
EXPERIENCE, ABC BIO (Y/N)	5	NO	1	YES	2	YES	1
EXPERIENCE, PROCESS, INC (Y/N)	5	YES/1 PROJECT	2	YES/8 PROJECTS	4	YES/6 PROJECTS	4
BIO/PHARM EXPERIENCE (Y/N)	10	YES/39 PROJECTS	4	YES/12 PROJECTS	4	YES/61 PROJECTS	4
SAFETY							
EMR (AVG 3 YRS) < 1 IS RISK	10	1.13	3	1.07	3	0.84	4
SUBSTANCE ABUSE (Y/N)	10	YES	4	YES	4	NO	0
FATALITIES		0	OK	0	OK	0	OK
LWC		7		7		23	
RWC		7		13		3	
CURRENT FIELD MANPOWER	3	399	3	200	2	1800	4
SHOP CAPACITY (# WELDERS)	4	60 PEOPLE/23 WELDERS	4	10 WELDERS	3	230 WELDERS	4
PROJECT CONTROL TOOLS	4	FINEST HOUR/P3/EXPED	4	PRIMAVERA/LOTUS	3	MS PROJECT/MC SQ	3
QUALITY PROGRAM	4	YES	4	YES	2	YES	4
BONDING CAPACITY							
SINGLE (MIN $5 MM)	GO/NO GO	15–20 MM	OK	14 MM	OK	21 MM	OK
AGGREGATE (MIN $10 MM)	GO/NO GO	30–40 MM	OK	25 MM	OK	30 MM	OK
ANNUAL SALES (MIN 15 MM)	GO/NO GO	22,600,000	OK	28,000,000	OK	94,800,000	OK
LABOR POSTURE	GO/NO GO	OPEN SHOP	OK	OPEN SHOP	OK	UNION	OK
SHOP FACILITIES (SF)		12,000 SF	OK	77,000 SF	OK	LOCAL 10K SF	OK
ORBITAL WELDING EXPERIENCE	GO/NO GO	EXTENSIVE	OK	EXTENSIVE	OK	EXTENSIVE	OK
REFERENCES		YES		YES		YES	
NC LICENSE (MUST OBTAIN)	GO/NO GO	YES	OK	YES	OK	YES	OK
TOTAL WEIGHTED SCORE			266		262		253

Finally, be sure to check references. Be open in asking questions regarding adherence to schedule, budget management and control, the number and size of change orders, resource utilization, quality and safety efforts, and the condition and timeliness of documentation submittals. Ask for examples of document control, inspection, and testing procedures. Ask also for copies of progress reports and progress monitoring data. The information contained in these documents and the format in which they are delivered will be important in your efforts to manage the contractor.

HVAC systems

Manufacturing facilities involved in the manufacture of human drugs are required by 21 CFR, Part 211.46 to have HVAC systems that

- Provide adequate ventilation
- Provide equipment for adequate control over air pressure, microorganisms, dust, humidity, and temperature when appropriate for the manufacture, processing, packing, or holding of a drug product
- Use air filtration systems, including prefilters and particulate matter air filters when appropriate, on air supplies to production areas to control dust and contaminants [3]

The basis of cleanroom design focuses on protecting the product from contamination sources, such as people. Developing expertise in the contamination control industry has led to the creation of standards for measuring space cleanliness in terms of the size and number of particles present. The current standards that provide the basis for cleanroom classifications used within the pharmaceutical industry are the ISO Cleanroom Standards. These are

- **14644-1 and 2** Cleanroom Standards (issued)
- **14644-3** Testing and Qualification (in final draft)
- **14644-4** Design and Construction (issued)
- **14644-5** Operations (in committee)

ICH Q7A classifies HVAC as a utility and defines system requirements in section 4.2.1:

> Adequate ventilation, air filtration, and exhaust systems should be provided where appropriate. These systems should be designed and constructed to minimize risks of contamination and cross contamination and should include equipment for control of air pressure, microorganisms, dust, humidity, and temperature, as appropriate to the stage of manufacture. Particular attention should be given to areas where APIs are exposed to the environment [4].

Table 7.4 Comparison of Clean Room Specs

Particles per m³ ? 0.5 μm	US 209E 1992	US 209E equivalent per cu. ft.	EU Annex 1997	France AFNOR 1989	Germany VDI 2082 1990	Britain BS 5295 1989	Japan JIS B 9920 1989	ISO 14644-1 & CEN 243
1								
3.5					0		2	2
10	M 1							
35.3	M 1.5	1			1		3	3
100	M 2							
353	M 2.5	10			2		4	4
1 000	M 3							
3 530	M 3.5	100	B-at rest A all times	4 000	3	E or F	5	5
10 000	M 4							
35 300	M 4.5	1 000			4	G or H	6	6
100 000	M 5							
353 000	M 5.5	10 000	C-at rest B-dynamic	400 000	5	J	7	7
1 000 000	M 6							
3 530 000	M 6.5	100 000	D-at rest C-dynamic	4 000 000	6	K	8	8
10 000 000	M 7							

Table 7.4 presents a comparison of cleanroom specifications and classification levels from various international regulatory agencies, including the now defunct U.S. Federal Standard 209E. These classifications, from class 100,000 to class 1, are based on limits measured in particles per cubic foot of size equal to or larger than the particle sizes shown. To put this into a more visible perspective, Figure 7.12 compares particle sizes to a more understandable benchmark.

It is important to understand that different regulatory agencies view classification criteria differently. Table 7.5 provides a comparison of U.S. FDA definition of class 10,000 against the E.U. view of an equivalent class 10,000 space.

It is also important to recognize that air classifications are dynamic in nature, based on the unit operations that may occur within the space at any given time during the manufacturing process. Table 7.6 provides a comparison of FDA guidance on in-operation conditions.

Facility requirements

Parenteral manufacturing facilities typically require class 100,000 cleanrooms for areas that involve a low risk for environmental contamination of product or components; class 10,000 for aseptic areas used to store or transfer product, components, or personnel that involve a substantial risk of environmental contamination of product or components; and class 100 for areas critical to

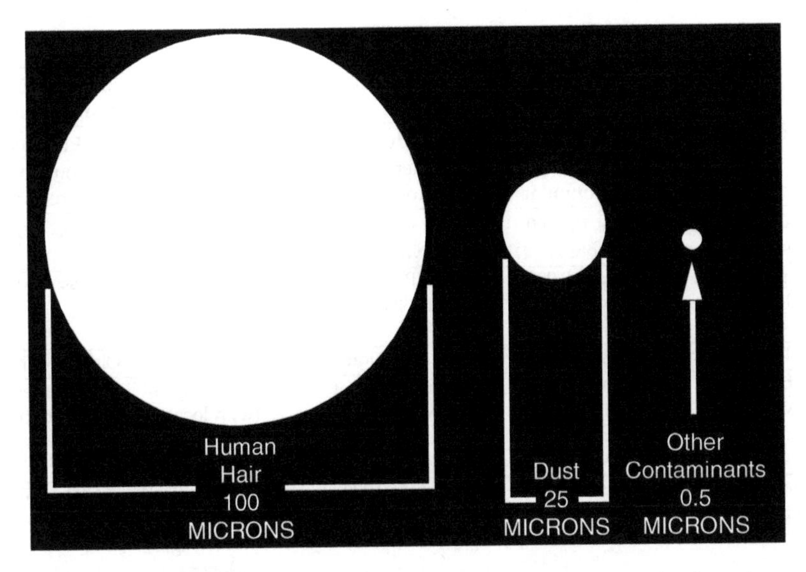

Figure 7.12 Sizing up the enemy. The basic unit of measurement is the micrometer (micron) — one millionth of a meter (1×10^{-6}) or 39 millionths of an inch (0.0000390). These diagrams wil give you a feel for just how small a submicron particle is. The human eye is capable of seeing particles down to approximately 25 micrometers.

Table 7.5 US vs. EU GMP

US FDA Aseptic Guideline	EU GMP Annex
No class 10,000	Grade B (10,000) Grade D (100K at rest)
20 air changes/hr	15–20 minute recovery
0.05″ DP (12.5 Pa)	10–15 Pa DP
90 ft/minute in class 100 hood	0.45 M/s (90 fm) "at working position
HEPA integrity tested 2/yr in class 100K	HEPA in Grades A, B, C
No gowning room	Gowning room is airlock
Isolator in Class 100K dynamic	Isolator in Grade D (100K at rest)

maintaining product sterility. Classifications 10 and 1 are normally found in the microelectronics industry, where smaller particles are more difficult to manage in the fabrication of small component geometries. While the biotech and pharmaceutical industries are less concerned with removing all particle matter, there is an emphasis on minimizing the number of viable organisms (living organisms that can reproduce) present.

The majority of sterile drug products today are parenteral drugs that are injected directly into the blood stream. By controlling particle count inside cleanrooms, the risks of product contamination are minimized, especially where the process involves open-system elements.

Table 7.6 Requirements by Unit Operation

Typical Process	Aseptic Processing (All classifications are in operation)		Terminal Sterilization (All classifications are in operation	
	Background Environment	Product/ Container/Closure Exposure	Background Environment	Product/ Container/ Closure Exposure
Raw material dispensing	Class 100,000 (Note 1)	Local Protection (Note 2)	Class 100,000	Class 100,000
Compoundig & (sterile) filtration feed	Class 100,000 (Note 1)	Local Protection (Notes 2 and 3)	Class 100,000	Class 100,000
(Sterile) filtration	Class 10,000	Class 100 (Note 7)	Class 100,000	Class 100 (Note 5)
Initial prep/ washing components	"Pharmaceutical" (with local monitoring) (Note 6)	"Pharmaceutical" (with local monitoring) (Note 6)	"Pharmaceutical" (with local monitoring) (Note 6)	"Pharmaceutical" (with local monitoring)
Final rinse of components	Class 100,000	Class 100,000 (Note 2)	"Pharmaceutical" (with local monitoring) (Note 6)	Class 100,000 (Note 2)
Sterilization/ depryogenation of components - loading	Class 100,000	Class 100,000 (Note 2)	"Pharmaceutical" (with local monitoring) (Note 6)	Class 100,000 (Note 2)
Sterilization/ depryogenation of components - unloading	Class 10,000	Class 100 (or wrapped/sealed)	Class 100,000	Class 100 (Note 5) (or wrapped/ sealed)
Fillering and Stoppering	Class 10,000	Class 100 (note 7)	Class 100,000	Class 100 (note 5)
Lyphilization Operation	—	Closed system	—	—
Lyphilization Transfer	Class 10,000	Class 100	—	—
Capping and Crimping	"Pharmaceutical" (with local monitoring) Notes 4 and 6)	Local Protection (Notes 2 and 4, and Fig. 2-4)	"Pharmaceutical"	Local Protection (Notes 2 and 4, and Fig. 2-4)
Terminal Sterilization	—	—	"Pharmaceutical"	N/A
Inspection	"Pharmaceutical"	N/A	"Pharmaceutical"	N/A
Labeling and Packing	"Pharmaceutical"	N/A	"Pharmaceutical"	N/A

Design issues

When designing cleanrooms for sterile product manufacturing applications, there are a number of issues that the project team must address. These are visually summarized in Figure 7.13 [5]. It is important to remember that each facility and application is different; design criteria must be defined in relation to the product or process requirements needed.

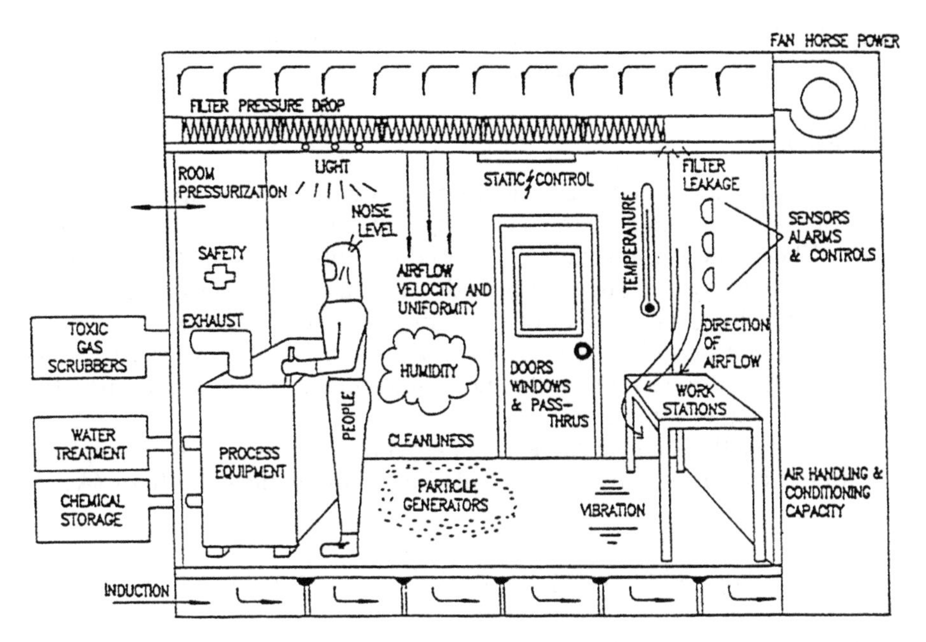

Figure 7.13 Regulatory production HVAC concerns: classroom design considerations.

Early in the programming of the facility design, decisions must be made regarding the classification of rooms and spaces. These decisions are affected by the following:

- The number of products in the facility
- Types of products
- The asepticity of the environment
- Process support requirements, such as sterilization of incoming and outgoing materials, gowning requirements, and pressurization levels
- Temperature and humidity requirements

For most production operations, companies will have defined criteria to meet their process needs in existing facilities. Issues regarding containment must also be addressed, since they will also have a tremendous impact on the overall HVAC system design. It is typical for fermentation operations, media and buffer preparation, and many cleaning operations to be carried out in class 100,000 environments; cell inoculation, bulk filling, and many purification operations will occur in class 10,000 areas; final filling operations will usually occur in class 100 environments.

It is important to define classifications based on process requirements, since this will be the basis for validation of the space. Companies often design spaces to higher classifications than are required, with the intent of not defining the more stringent classification as their basis for validation. This is a risky proposition, not only from the standpoint of questions from CBER and CDER, but also in terms of the costs associated with constructing and operating higher classified spaces.

Filtration

The heart of cleanroom design is in the airflow and filtration system. Particle control is accomplished by the use of high-efficiency particulate air (HEPA) filters. These filters are usually comprised of a fiberglass medium within a stainless steel housing. They can provide a rating of up to 99.99 percent clean air by controlling the number of times the air within the space is changed within a given period of time. The theory is simple: Remove particles via air movement through filters; the more times the air is moved through the filters, the cleaner it becomes.

What exactly is an air change? An air change is defined as the replacement of one room volume. The calculation of air changes per hour (ACH) is simple:

$$\text{ACH} = \text{cubic feet of supply air/hr (CFH)} \div \text{room volume (cubic feet)}$$

$$\text{CFH} = \text{CFM} \times 60$$

Depending on the classification of the space, the number of air changes per hour will greatly increase as the cleanliness level becomes more stringent, as shown in Table 7.7

So how much air is needed? Air supply must meet cooling load needs. It is also a good rule of practice to have at least 20 CFM of outside air per person working within the space. You must also have enough air supply to offset any exhaust or exfiltration from the room.

As the number of air changes increases, so does the cost of operating the system. For example, a good rule-of-thumb value for the cost of moving a cubic foot of air volume (CFM) in a class 100,000 space is within a range of $3.25 to $4.00 per cubic foot. While the CFM cost of air-handling units decreases with the size of the unit, the overall cost will increase as air change rates increase the number of CFMs.

The location and number of filters also play an important role in the filtration system. For class 100,000 applications, filters are usually located in-line (Figure 7.14). To achieve class 10,000 levels, the number of terminal filters is increased and, in many cases, the return air inlets are also filtered at low points within the space to provide better .circulation (Figure 7.15).

Temperature

The U.S. Pharmacopeia identifies certain temperature requirements for sterile products. These include

Table 7.7 Pharmaceutical Cleanroom Air Handling Criteria

Class	Air Changes/ Hr @ 10′ Ceiling Ht.[a]	HEPA Coverage	Nom. CFM per Filter	Room Air Velocity (fpm)
100,000	30	10–20%	350	5
10,000	60	20–40%	500	10
1,000	120	40–60%	600	20
100	540	80–100%	720	90

[a]A.C./hr. = (V_{room} × 60)/Height

Note: 99.99% on 0.3μm HEPA filters, low wall returns.

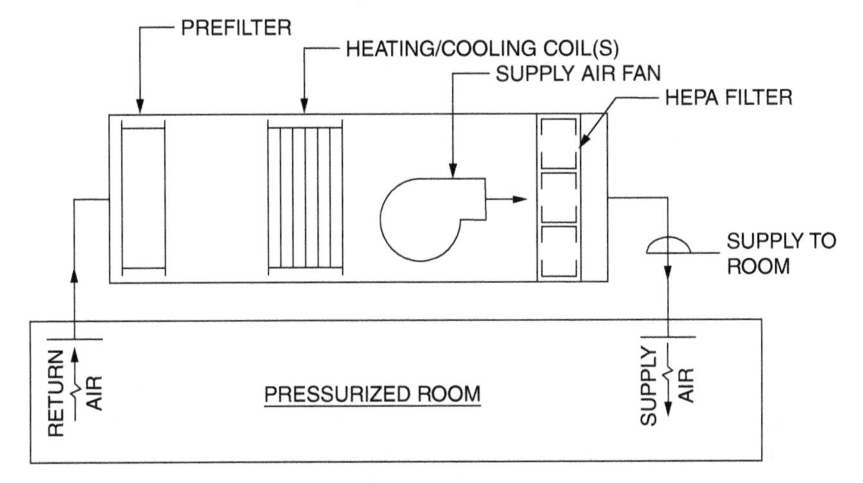

Figure 7.14 Filter requirements for class 100,000.

- Raw materials at 10–40°C
- API at 15–30°C with mean kinetic temperature < 25°C

Most bioprocess operations require the interface with personnel on a routine basis to perform the various manufacturing, quality, validation, and maintenance functions required to support the process. These people work in special gowns that are not necessarily meant for warm environments. Therefore, comfort also becomes an important issue.

As gowned personnel work in these areas, they tend to perspire. This will cause them to shed particles and bacteria from their skin that will increase the bioburden levels within the room.

The HVAC system must be designed to control temperature and humidity within the space to eliminate this situation as much as possible. Typical process spaces are designed to 66–68°F at 40 to 50 percent relative humidity (RH). Other work areas are normally 72°F at 50 percent RH or higher. It should also be noted that to maintain an even distribution of temperature

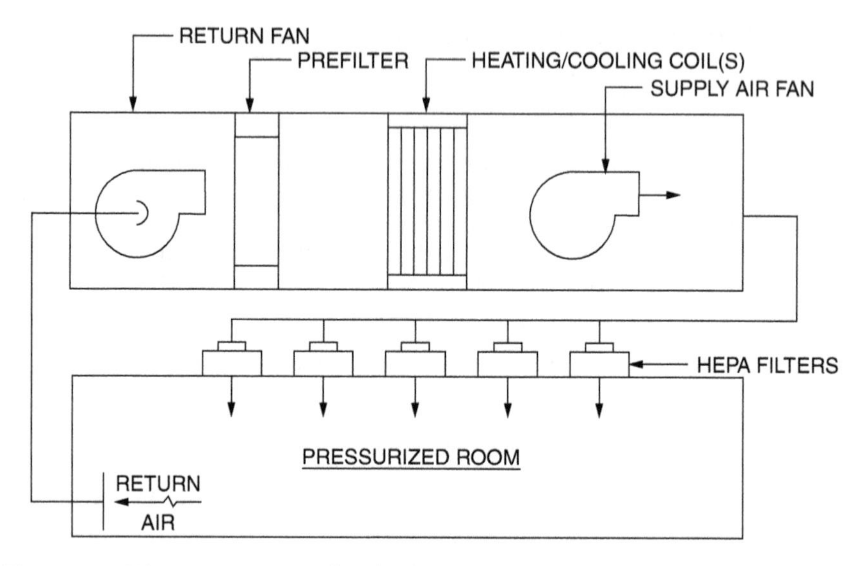

Figure 7.15 Filter requirements for class 10,000.

throughout the space, a higher room air change rate and greater filter coverage may be required.

Pressurization

Air pressure differentials are used to isolate or contain critical spaces within a process area. Typically, these are areas where the classification between rooms is different — 100,000 versus 10,000, for example — or in situations where containment is an issue, as discussed in Chapter 12.

In the United States, the traditional pressure differential between process rooms is .05″ ιν> wg. A very simple formula for calculating air balance and room pressure is

$$VP = (V/4005)^2$$

Where VP (velocity pressure) = $(CFM/4005A)^2$ and A is equal to a 1 foot square opening, 890 CFM will create 0.05 in. wg differential pressure.

A typical facility layout might see pressure differentials as shown in Figure 7.16, where positive pressures between critical spaces and support areas are maintained in order to control contamination between the spaces.

Pressure differentials are normally established by design in two ways:

- Passive control through the use of airflow balancing
- Active pressure control, in which pressure-sensing devices monitor conditions and allow for system recovery due to periodic disturbances, such as doors opening or operators entering spaces

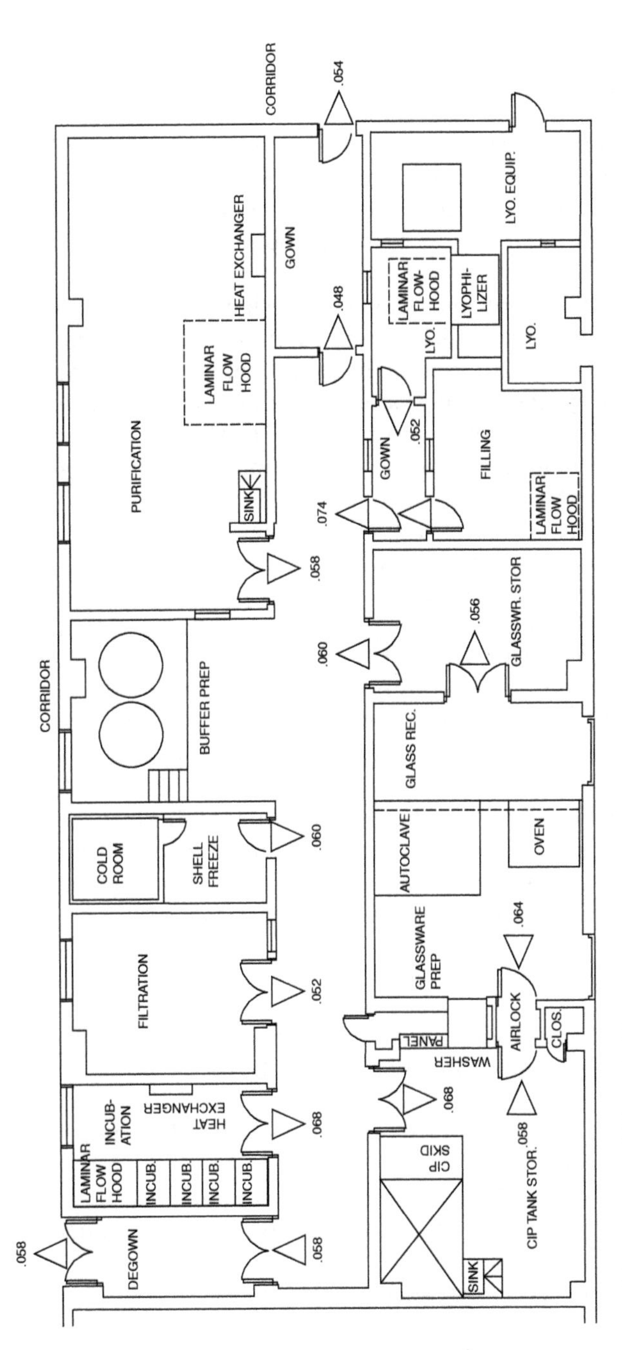

Figure 7.16 Air pressure differential layout.

Active control is the preferred method of control in the industry, but it has an additional cost factor for the instrumentation required to monitor and control airflow through the system.

Facility layout

Facility layout plays an important role in the cost and effectiveness of HVAC systems for a sterile product manufacturing facility. Because of the large volumes of air being moved throughout the facility, it is important to keep adjacency relationships in mind as conceptual layouts and flows are being developed.

Process and HVAC equipment areas should be designed as close together as possible for maximum system efficiency by minimizing duct runs and, therefore, decreasing power consumption. This is normally accomplished by utilizing large interstitial spaces above process areas or incorporating a penthouse on the building roof to house most of the HVAC equipment and supply ductwork. These spaces should be large enough to allow proper access for operations and maintenance activities.

Interstitial space, in the current trend, allows easy access to ductwork, filters, dampers, reheat coils, volume boxes, fans, utility piping, HVAC controls, and, possibly, air handlers themselves in a space located directly above the process areas. The advantage to this type of layout is that maintenance of system components can occur without disruption of the process area. Maintenance personnel have access to equipment without going through a gown/degown procedure and do not risk contamination of the process space through the performance of any repair or replacement activities.

Interstitial space is expensive, however. Estimated average costs for interstitial space can easily fall within the range of $100 to $150 per square foot (building cost only) for a typical sterile product manufacturing facility. Interstitial spaces should be as large as the construction budget will allow; finding the funds early will make for a more successful project.

Cost savings — Where to look

In the design of HVAC systems, there are a number of places to look when cost becomes an issue. Some have already been identified:

- Identification of classified space requirements only where necessary to support the process
- Design basis for temperature and humidity requirements
- Design basis for the number of air changes, especially for nonclassified areas
- Size and location of interstitial space or penthouse

There are other areas on which the project team should focus in reviewing cost:

- **Equipment sizing** Be sure that identified loads are met and an acceptable factor of safety applied in load calculations. Many projects end up with much more capacity than is needed to support current operations.
- **Sizing of ductwork and piping** Again, be sure that sizes meet your anticipated needs. If plans include future expandability in the budget, then oversizing can be acceptable.
- **Redundancy** The project philosophy on redundancy must be defined early. What equipment on both the wet and dry sides of the system should be included and at what level should the redundancy be approached — full system or only partial?
- **Contamination control** Should there be separate air-handling systems for each process application or separate systems only for specialized areas?
- **Duct construction** Use round ductwork instead of rectangular.
- **Equipment selection** Decide on standard versus custom or modular units.
- **Materials of construction** Use galvanized ductwork where application will allow instead of all stainless steel.
- **Environmental monitoring** The type of monitoring/control devices selected can have a tremendous impact on the cost of the system.
- **Cleanroom construction** Some process applications may be better suited for a modular approach to cleanroom construction than the traditional custom-built approach.

Construction and installation issues

With the primary focus of HVAC systems for parenteral facilities on controlling contamination, it should come as no surprise that the majority of issues in the construction and installation of these systems deals with maintaining system cleanliness and integrity. Construction budgets and schedules must account for these issues.

Cleanroom HVAC systems should be designed to minimize leakage from any source. Ductwork should be of welded stainless steel construction downstream from HEPA filters. The use of galvanized materials may be allowed for some systems. Spot-welded fittings should not be used. All duct joints should be flanged and gasketed or sealed in a manner to allow for zero leakage.

During fabrication, each piece should be cleaned and sealed individually before transportation to the field. Nesting of duct sections should not be allowed. At the point of installation, all sections should be recleaned and any open ends capped at the end of the workday to prevent any external contamination of the system interior surfaces.

Sealants used on duct should be nonhydrocarbon based, meeting established criteria outlined by the food processing and pharmaceutical industries for manufacturing facilities. All component interface points, such as dampers or VAV boxes, must also be properly sealed.

During final testing of the system, care must be taken to preserve the integrity of HEPA filters. For system blowdown, install temporary filters on both the supply and return. The method of leak testing should not allow the introduction of residue or contamination into the system. Only after final blowdown, testing, and rough balancing should HEPA filters be installed.

Operation costs — What you should know

Cleanrooms are expensive to build. They are also expensive to operate. Decisions made early in the design development process will continue to have an impact on company operating costs for many years in the future. The design of the system must not only be capital cost effective, but also energy efficient from an operations standpoint.

To achieve optimum cleanliness, cleanroom HVAC systems are kept running at full capacity 24 hours a day, every day of the year, to support multiple-shift operations. This means that the mechanical systems constantly run at or near peak capacity.

We have already seen that cleanroom air change rates greatly exceed those found in typical environmental rooms. With this high volume of air being forced through HEPA filters, which have a much higher pressure drop than normal diffusers, the resulting system is quite large in relation to the total square footage it serves. The same holds true for the air conditioning system. Typical office or industrial air conditioning loads range from 250 to 400 ft^2/ton; cleanroom air conditioning loads can be 30 to 100 ft^2/ton.

Table 7.8 provides some ranges of operating cost for different facility types. It is a worthwhile exercise to look at anticipated energy costs for the system design basis for a cleanroom. The worksheet shown in Example 7.3 provides one way to look at the cost of cleanroom operation based on some easily defined facility elements.

Table 7.8 Air Quantity Drivers

CFM Driver	Average Supply Air Range CFM/ft^2	Operating Cost Range \$/ft^2/yr	Outside Air %
Thermal Comfort	.75–1.5	1.0–2.0	10–15
Animal Suite	2.5–3.5	6–9	70–100
GLP	2–6	4–8	50-100
GMP Oral Dose	3–4	3–6	20–50
GMP Biotech	3–8	3–8	20–100
GMP Aseptic Filling	3–10	4–15	10–20
OSHA Factory	5–8	4–8	15–25
Dust Collection	.5–1	1–2	100
Incinerator with Recovery	1	10	100

Example 7.3 Worksheet for Calculating Energy Costs

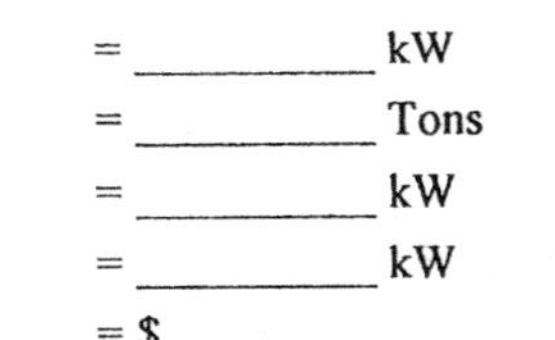

A. Determine that loading is for a consistent peak or average (e.g., average peak summer, etc.).

B. Input BHp kW = (BHp)(0.746 kW/BHp)(efficiency) = ________ kW

C. Process equipment KW = (RLA)(V)(✓ phase)(diversity)(0.001) = ________ kW

D. Lighting kW = (number of fixtures)(W/fixture)(0.001) = ________ kW

E. BHp Btu = (BHp)(efficiency)(2544 Btu/BHp) = ________ Btu

F. Process equipment heat rejection:

Btu = equipment kW from C - exhaust heat - cooling water = ________ Btu

F1. Exhaust heat Btu = (exhaust cfm)(1.09)(exhaust temp - room temp)

F2. Water cooling Btu = (cooling gal./min)(entering temp - leaving temp)(500)

G. Lighting Btu = (lighting kW)(3410 Btu/kW) = ________ Btu

H. Process cooling water Btu = (gal./min)(leaving temp - entering temp) = ________ Btu

I. Approximate building load Btu = (ft^2)(30 Btu/ft^2) = ________ Btu

J. Makeup air total load Btu = sensible + latent = ________ Btu

J1. Sensible Btu = (1.09)(cfm)(outside temp - room temp)

J2. Latent Btu = (0.68)(cfm)(outside grains - room grains)

J3. Makeup air for dehumidification = same as above, except substitute room dewpoint for room temp.

K. Total straight electrical load = B + C + D = ________ kW

L. Total AC tonnage = (E + F + G + H + I + J)/12,000 = ________ Tons

M. AC kW = (L)(AC efficiency) + any parasite kW (e.g., from pumps, etc.) = ________ kW

N. Total electrical consumption = K + M = ________ kW

0. Energy dollars: $ = (kW N)(rate $/k Wh)(hours of operation) = $________

References

1. *Code of Federal Regulations*, Title 21, Part 211.42. 1990. Washington, DC: U.S. Government Printing Office.
2. Ibid.
3. *Code of Federal Regulations*, Title 21, Part 211.46. 1990. Washington, DC: U.S. Government Printing Office.
4. FDA. 2000. Q7A Good manufacturing practice guide for active pharmaceutical ingredients. *Guidance for Industry* (November).
5. Weiss, M. 1992. Ongoing issues in the management and operation of biotech cleanrooms. *Genetic Engineering News* (November).

chapter 8

GMP compliance in architectural design and construction

The design and construction of a sterile product manufacturing facility requires strict adherence to the requirements outlined in the current Good Manufacturing Practice regulations, 21 CFR 211.42. The architectural aspects of GMP center on issues related to building size, layout, and materials conducive to extensive cleaning regimes.

Specifically, the GMP regulations state that "Any building used in the manufacture, processing, packing, or holding of a drug product shall be of suitable size, construction, and location to facilitate cleaning, maintenance, and proper operations" [1]. They go on to state that any such building layout shall be designed to allow products and components to flow through the building so as to prevent contamination or mix-ups [2]. General aseptic processing attributes are defined as "floors, walls, and ceilings of smooth, hard surfaces that are easily cleanable" and "a system for cleaning and disinfecting the room and equipment to produce aseptic conditions" [3].

Although these statements define requirements in very general terms, actual construction must comply with commonly accepted standards that are recognized in the industry and by the FDA. The challenge of developing an architectural system that complies with the intent of these guidelines is in designing and detailing a facility that incorporates highly technical operations using sophisticated engineering systems, involves numerous building codes and standards, and provides the required environment necessary to meet the needs of the client's manufacturing operations.

Layout

The layout of any sterile product facility must be developed around the needs of the facility, which are defined during facility programming. For new facilities, the ultimate premise is to design the building around the process. By understanding the process, the operations that support the process, and the operational philosophy, such as gowning and cleaning/changeover, the A/E firm can develop a step-by-step description of the facility from one space to another.

These relationships are normally defined through the use of basic logic diagrams (Figure 8.1) that provide a graphic tool for understanding the flow of process operations. Logic diagrams provide the first look at the flow of materials, personnel, product, and equipment through the facility, focusing on eliminating potential cross-contamination points and providing space for all the necessary process steps. The facility layout will be a translation of these flow diagrams, incorporating not only the operational requirements, but also integrating the anticipated mechanical systems required to support the spaces.

Segregation and flows

The design of a production facility for sterile products must implement the operating philosophy of the plant. The design's priorities must focus on the

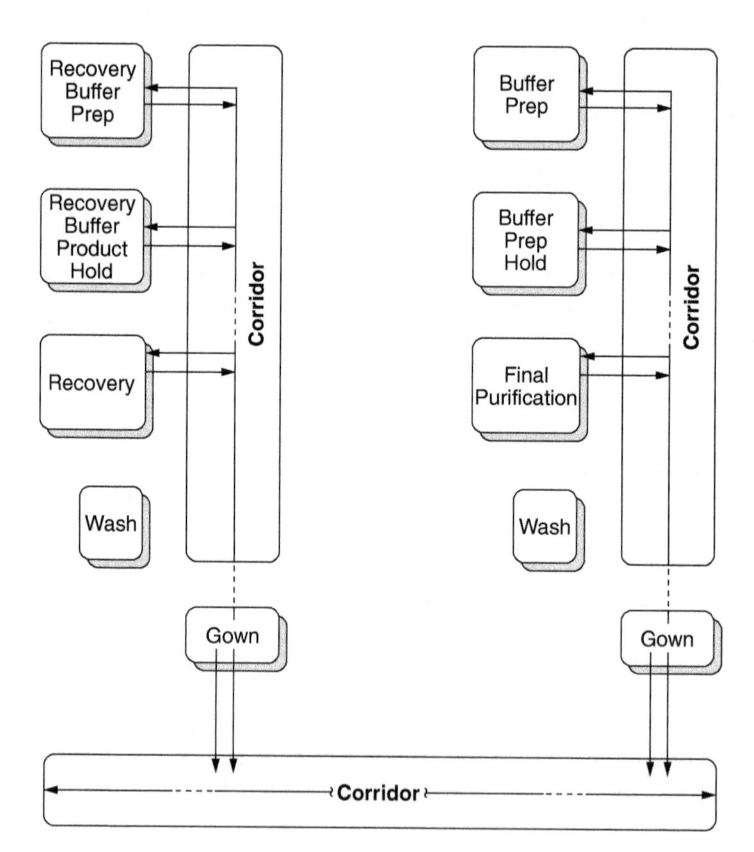

Figure 8.1 Logic diagram.

regulatory compliance aspect of functionality as it relates to protection of the product. To do this, engineers and architects define functional adjacency related to process unit operations and standard operating procedures.

The protection of the product from external contamination is a key GMP requirement. The concept of segregation is one basis to provide this assurance, and it is implemented in the facility design through a variety of avenues. These include physical separation, procedural implementation, environmental means, and chronological scheduling.

Segregation must be viewed from two concepts: primary and secondary.

Primary segregation is used to define the overall facility design and organization through environmentally controlled space envelopes around specific steps in the process (unit operations). Some examples of primary segregation include:

- Physical segregation between the production of two different products
- Segregation between different lots of the same product
- Physical segregation between upstream and downstream processing operations

Secondary segregation is applied to focus on procedural control of spaces, production activities, and personnel movement. It is normally applied in instances where equipment and components that support the production effort are "closed" and protected from the surrounding environment. Some examples of secondary segregation would be:

- Segregation of clean and dirty equipment
- Segregation of personnel flows via airlocks and implementation of gowning activities
- Chronological separation of activities via scheduling of process activities
- Segregation of stored materials for quarantine, released and finished product

Segregation by spatial means normally includes physically separated areas and dedicated paths of travel for personnel, materials, and equipment. Segregation by time (temporal segregation) includes sequencing the movement of clean and dirty materials and equipment through the same space, but at different times. Segregation by environmental control may include local protection of an open process operation by use of a classified area.

Open processing operations require segregation, and that is the primary driver behind flow. Flow considerations include:

- Direction of flow
- Type of flow
- Quantity/rate of flow

- Cleaning/decontamination
- Transfer requirements

The critical flows that are essential in the evaluation of a sterile production facility are

- Personnel flows
  - Manufacturing, support, and maintenance personnel
- Material flows
  - Media and buffer components, raw materials, finished goods
- Equipment flows
  - Clean or dirty components, portable equipment, product containers
- Product flows
  - In-process, intermediate, and final
- Waste flows
  - Solid, liquid, process, and contaminated

Figure 8.2 provides a look at functional adjacency and the relationships of flows and segregation.

Space considerations

The GMP regulations are clear in their emphasis on the "appropriate space" aspect of a facility. It is important that operations be performed within well defined areas of adequate size and that these operations create no potential contamination problems. The size of rooms, the size and placement of equipment, access to equipment, and personnel flow into and through rooms all have a direct impact on the ability to clean and maintain and, therefore, validate the facility.

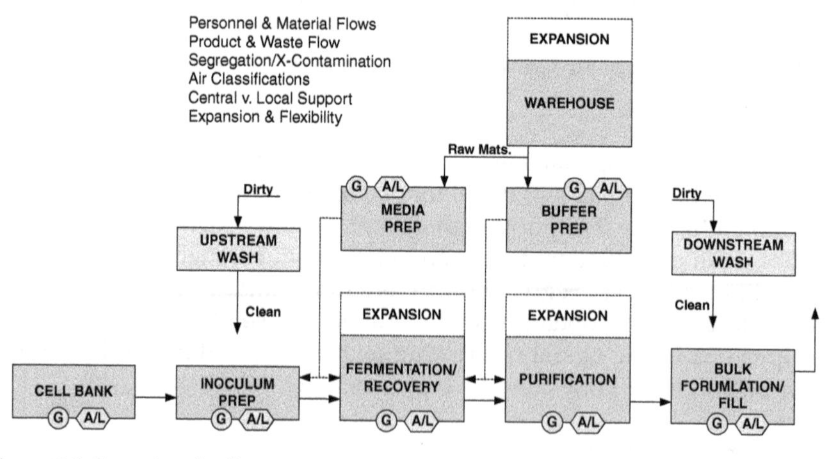

Figure 8.2 Functional adjacency.

To size the room properly, you must know everything that goes into it. Too often, the architect will place all permanently fixed equipment, platforms, and casework, for example, and omit space for items such as small portable vessels and carts that move in and out of rooms or spare items such as cleaning jumpers. A complete review of the equipment list, including those items anticipated as client-furnished, should be made to ensure the inclusion of all equipment.

In addition to process equipment, the layout must also provide adequate space for support items, such as carts, control panels, portable and spare equipment, and related items that take up space within any room. This is an area where most facilities fall short; when it comes time to cut costs, support spaces are usually the first to go.

The best method for defining room requirements in terms of needed space is room criteria sheets. In Example 8.1, a preliminary facility layout is provided with individual room designations. A more specific layout of individual rooms (Fermentation, Example 8.2) then provides the proposed equipment layout and support equipment taken from the preliminary equipment list for the facility.

The project team then meets to look at the functional operations that occur within each room and begins to define not only the design attributes of the space, but also every design component or system that will be present, by discipline. These are documented on room criteria sheets (Example 8.3). This is completed for every room within the facility.

This approach allows a clear understanding of operational requirements and provides a means for open communication among all user groups that have a stake in the operation of the facility. It also documents all decisions made and provides the A/E firm with a good tool for cross-checking design details against the concepts developed during facility programming.

Space considerations must also include the sizing of service areas for built-in items, such as autoclaves or lyophilizers; chase walls for concealing piping or ductwork from low wall returns; service corridors or gray space for valving, control devices, or support components; and interstitial spaces for primary HVAC and utility routings.

It is important that the A/E firm makes it clear to the entire project team that in defining space allocations, there are engineering systems present throughout sterile manufacturing facility spaces — above and below ceilings, running vertically and horizontally, exposed and concealed. Rarely does a facility have the luxury of excess space. The time spent in space allocation will have a tremendous impact on both facility cost and functionality of long-term operations and maintenance.

Service areas — Where do you put all this stuff?

An inherent problem in defining the layout of a parenteral facility is how to provide the necessary space for process and utility support components, such as valves, control devices, or heat exchangers, that require frequent

Example 8.1 Layout Overall Process Area

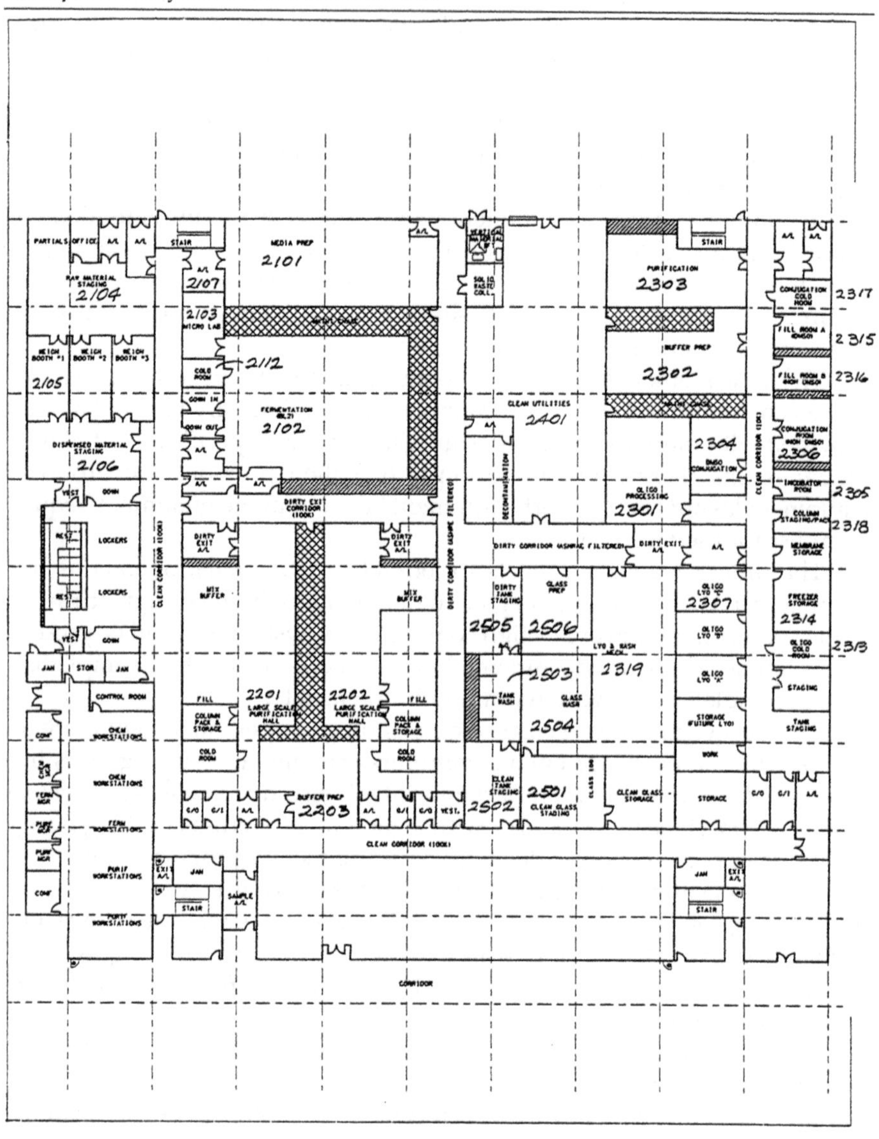

PROGRAMMING CRITERIA SHEETS

OVERALL PROCESS AREA

Example 8.2 Layout Fermentation Area

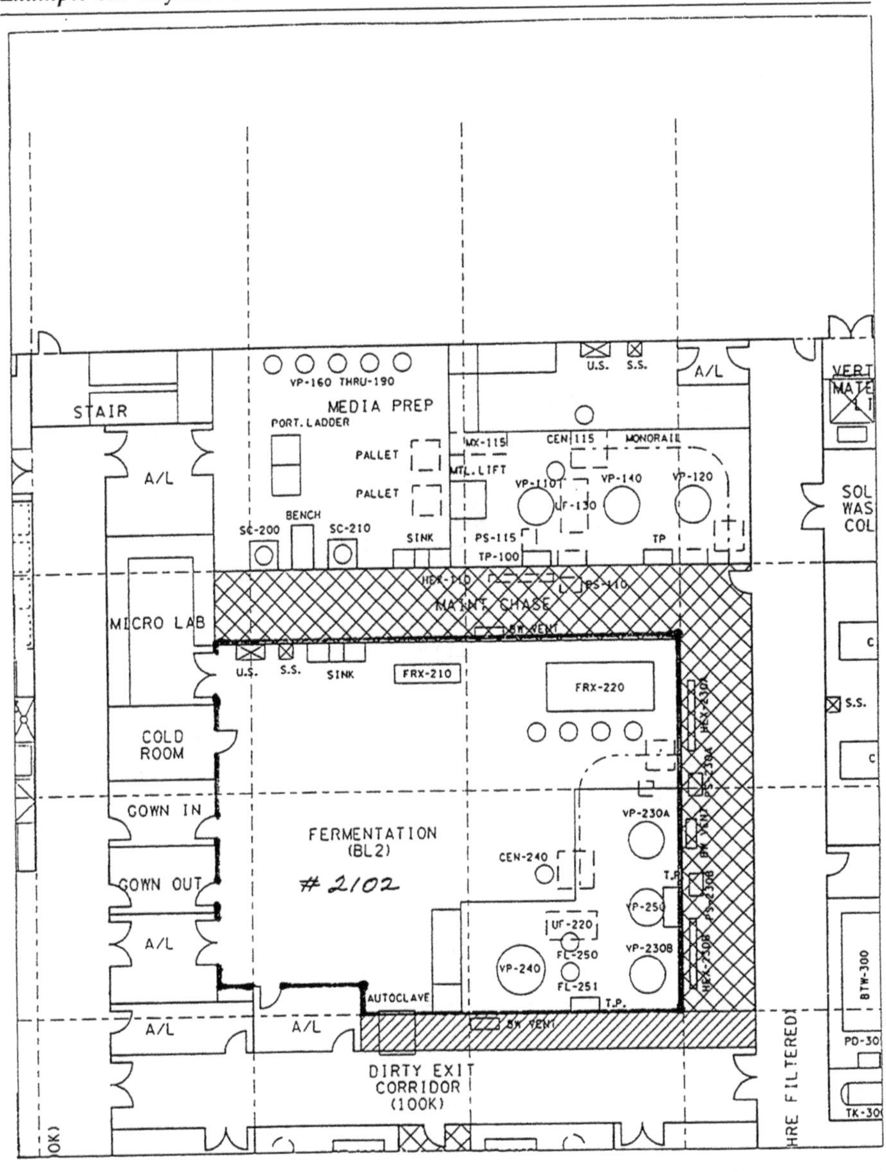

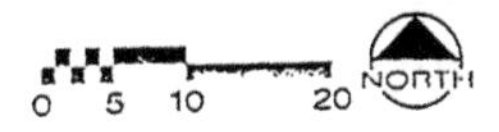

PROGRAMMING CRITERIA SHEETS

FERMENTATION

Example 8.3 Programming Criteria for a Biotech/Pharmaceutical Company (page 1)

PROGRAMMING CRITERIA FOR BIOTECH/PHARMACEUTICAL
REV.

SHEET 1

ROOM NAME	: FERMENTATION HALL	RECORDED BY	:
GENERAL AREA	: FERMENTATION	PROGRAMING TEAM	:
PROGRAM NUMBER	: 2102	DATE	:
ROOM NUMBER	:	REV.	:

FUNCTION	SUMMARY CRITERIA ITEM	DETAIL CRITERIA			
ROOM OPERATIONS (Architecture)	Room Function	FERMENTATION AND HARVEST			
	Gowning Level	Yes	No	Level	Requirements
				Street Level	
				Plant Uniform	
		✓		Gown Level 1	
				Gown Level 2	
				Gown Level 3	
				Gown Level 4	
		Other			
	Operating Schedule of Use	UNKOWN, TBD			
	Particulate & Micro Monitoring	NO			
	Cleaning	WIPE DOWN W/ WFI ; MOP			
	Projected Occupancy	4–8 PART-TIME			

ROOM CONSTRUCTION & FINISHES (Architecture)	Square Foot Allocation	2850				
	Room Dimensions	30×65 + 20×45				
	Ceiling Height	20 FEET AT FERMENTERS; 12 FEET REMAINING				
	Floor Finish	Yes	No	Selection	Area	Comment
				Dust Sealed Concrete		
				Waterproof Membrane Flooring		
				Epoxy Coating, Integral Base		
				Epoxy Terrazo, Integral Base		
		✓		Trowelled Epoxy, Integral Base		
				Vinyl Tile, Integral Base		
				Seamless Sheet Vinyl, Integral Base		
				Welded PVC Sheet, Integral Base		
				Carpet Tile/Carpet		
				Ceramic Tile		
				Metal Grating		
		✓		Checkerplate, S.S.		PLATFORMS
		Other				
		Other				
	Depressed Floor (Biocon./Spils)	Yes	No	Details		
		✓		WITH CURDS UNDER TANKAGE		
	Sloped Floors (Option)		✓	CIRCULAR DISH AROUND DRAINS		
	Wall Finish	Yes	No	Selection	Area	Comment
				Prefinished Metal Siding Panel		
				Exposed Structure Painted		
				Painted CMU		Latex/Epoxy
		✓		Painted Drywall W/PVC Wainscott		PVC LIMITED
				Painted Drywall		Latex/Epoxy
				High Build Epoxy Coating		
				Prefinished Insulated Panel, Metal		
				Ceramic Tile		
				Aluminium "Storefront"		
				Vinyl Wallcovering		
		Other				
	Sound Attenuation	Yes	No	Type	Comment	
			✓			
	Ceiling Finish	Yes	No	Selection	Area	Comment
				Exposed Structure Painted		
		✓		Seamless Drywall, Epoxy Painted		
				Seamless Drywall, Latex Painted		
				Suspended Mylar Faced Tile		
				Suspended Acousitcal Tile		
				Prefinished Insulated Panel, Epoxy		
				Prefinished Insulated Panel, S.S		
				PVC or FRP Panels, Lay-In. Tile		
		Other				
		Other				
		Yes	No	Details		
	Ceiling Access		✓			
	Walkable Ceiling		✓			

Example 8.3 Programming Criteria for a Biotech/Pharmaceutical Company (page 1A)

PROGRAMMING CRITERIA FOR BIOTECH/PHARMACEUTICAL

SHEET 1A

ROOM NAME	: FERMENTATION HALL	RECORDED BY	:
GENERAL AREA	: FERMENTATION	PROGRAMMING TEAM	:
PROGRAM NUMBER	: 2102	DATE	:
ROOM NUMBER	:	REV.	:

STRUCTURAL		Yes	No	Item		Comments
		✓		House-keeping Pads		
		✓		Equipment Access Platforms		
				Lifting Eyes		
				Hoist		
				Monorail		
			✓	Trench Drains		
		✓		Circular Depressions Around Floor Drains		
			✓	Floor Pits (Scales, Autoclaves, Etc.)		
		✓		Wall reinforcing for wall supported items		

OPERATIONAL ISSUES		Yes	No	Item		Comments
		✓		Portable Tanks		
				Material Lifts		
			✓	Wash Down		
			✓	Special Material Handling		
			✓	Forklift Access/movement		
			✓	Pneumatic System		
			✓	Trays of Bottles/Carts		

Miscellaneous		Yes	No	Item		Comments
			✓	Wall Bumpers		
			✓	Roof Hatches		
		✓		Armor Plate on Doors		
			✓	Work Surface		
			✓	Casework		
			✓	Storage Cabinets		
			✓	Shelving		

REMARKS	

Example 8.3 Programming Criteria for a Biotech/Pharmaceutical Company (page 2)

PROGRAMMING CRITERIA FOR BIOTECH/PHARMACEUTICAL

SHEET 2

ROOM NAME	: FERMENTATION HALL	RECORDED BY	:
GENERAL AREA	: FERMENTATION	PROGRAMMING TEAM	:
PROGRAM NUMBER	: 2102	DATE	:
ROOM NUMBER	:	REV.	:

Section	Item	Yes	No	Type	Number	Comment
Cont. ROOM CONSTRUCTION & FINISHES (Architectural)	Doors	✓		A		
				B		
				C		
		Other				
	Door Interlocks		✓			
	Door Seals	✓				
	External Windows		✓	A		
				B		
		Other				
	Internal Windows	✓		A		
				B		
		Other				
		Yes	No	Comments		
	Explosion Panels		✓			

Section	Item	Yes	No	Classification	Comments
ELECTRICAL NEMA CLASSIFICATION TO BE DETERMINED	Classification (Gases)		✓	D1	
			✓	C1,D2	
		Other			
	Classification (Dust)		✓	D1	
			✓	C2,D2	
		Other			
	Electrical Service	✓		120 V	MINIMUM
				220 V	
				440 V	
		Other		ALL RECEPTACLES TO BE FLUSH MOUNTED	
		Yes	No	Comments	
	Uninterrupted Power		✓		
	Emergency Power	✓			

Section	Item	Yes	No	Type	Number	Location
	Communications-Voice	✓		Intercom		
		✓		Telephones		
		Other				
	Communications-Data	✓		Computers		
		✓		Bar-Code Readers		
		Other				
	Security		✓	Card Readers		
	SAFETY	✓		Alarms		
		Other				
	Light Intensity (Foot Candles)	Provision				
	Light Type	FLUORESCENT, CLEAN ROOM TYPE				
	Operating Schedule	24 HRS				
	Equipment (watts/sf)	TBD				

Section	Item	Yes	No	Level	Comment
ENVIRONMENTAL/ PROCESS	Biological Classification			None	
				BL1	
				BL2	
				BL3	
				BL4	
				GLSP	
				BL1-LS	
		✓		BL2-LS	
				BL3-LS	
				BL4-LS	
		Other			
		Yes	No	Type	Location
	Noise or Vibration Control		✓	CHECK CENTRIFUGE REQMTS	
	Dust Producing Equipment		✓		
	Containment or Isolation of Hazardous Materials	✓		BIOWASTE CONTAINM T	FLOOR DRAINS
	Flammable or Solvent Storage		✓		

REMARKS	

Example 8.3 Programming Criteria for a Biotech/Pharmaceutical Company (page 3)

PROGRAMMING CRITERIA FOR BIOTECH/PHARMACEUTICAL

SHEET 3

ROOM NAME	: FERMENTATION HALL	RECORDED BY	:
GENERAL AREA	: FERMENTATION	PROGRAMMING TEAM	:
PROGRAM NUMBER	: 2102	DATE	:
ROOM NUMBER	:	REV.	:

HVAC CRITERIA		Data		Conditions	Comments
	Outside Conditions			Dry Bulb deg.F	
				Wet Bulb deg.F	
				Elevation	
	Inside Conditions	68		Dry Bulb deg.F	↑ 2ϒ
		50		% Relative Humidity	↑ 10%
	Filtration	Yes	No	Level	Comment
				30% ASHRAE	
				65% ASHRAE	
				95% ASHRAE	
		✓		100 K	
				10 K	
				100	
		Other			
		✓		In-Line HEPA	
			✓	Terminal HEPA	
		Other			
	Room Air Flow Rate			Per Sensible Load	
		✓		20 A.C./Hr (MIN)	
				60 A.C./Hr (MIN)	
				120 A.C./Hr (MIN)	
				90 Ft./Min. + /. 20%	
		Other			
	Relative Pressure Level			Neutral	
				Positive	
		✓		Negative	
	Supply Location	✓		Ceiling	
				Vertical/Wall Mounted	
				Low "Air Wall" at Floor	
	Return Location			Located at Ceiling	
				Vertical/Wall Mounted	
		✓		Low at Floor	
				No Return, all air exhausted	
		Other			
	Room Type			Cooler (Prefab.)	
				Freezer (Prefab.)	
				Warm Room	
		✓		Standard	
		Other			
	Special Exhaust			Dust	
				Vapour	
		✓		Bag-In/Bag-Out	TO CONFIRM
				Smoke	
				Toilet	
				Hydrogen	
				Halon	
		Other			
	Gauges	✓		Wall Mounted	
		Other			
	Operating Schedule			24 HRS; 7 DAY	

REMARKS	

Example 8.3 Programming Criteria for a Biotech/Pharmaceutical Company (page 4)

PROGRAMMING CRITERIA FOR BIOTECH/PHARMACEUTICAL

SHEET 4

ROOM NAME	: FERMENTATION HALL	RECORDED BY	:
GENERAL AREA	:	PROGRAMMING TEAM	:
PROGRAM NUMBER	: 2102	DATE	:
ROOM NUMBER	:	REV.	:

UTILITIES/ BUILDING SERVICES	Piped Services	Yes	No	Service	Abrev.	Qnt./Temp./Pros./Loc
				Clean Air	AC	
				Instrument Air	AI	
				Plant Air	AP	
				Argon	AR	
				Boiler Blowdown Drainage	BBD	
				Boiler Drain	BD	
				Boiler Feedwater	BFW	
				Biowaste (AG)	BW	
				Biowaste Condensate (UG)	BWC	
				Biowaste Drain (UG)	BWD	
				Biowaste Vent	BWV	
				Clean Condensate	CC	
				Clean Condensate Drain (UG)	CCD	
				Clean-In-Place Return	CIPR	
				Clean-In-Place Supply	CIPS	
				Carbon Dioxide	CO2	
				High Pressure Steam Condensate	CSH	
				Low Pressure Steam Condensate	CSL	
				Medium Pressure Steam Condensate	CSM	
				Ethanol	EA	
				Fuel Oil	FO	
				Sulfuric Acid, Condensate	HSOC	
				Sulfuric Acid, Dilute	HSOD	
				Helium	HE	
				Liquid Nitrogen	LN2	
				Nitrogen Gas	N2	
				Sodium Hydroxide	NAOH	
				Non-Production Waste Drainage	NPWD	
				Oxygen	O2	
				Process Buffer Transfer	PBT	
				Process Media Transfer	PMT	
				Product Transfer	PT	
				Process Waste (AG)	PW	
				Process Waste Drain (UG)	PWD	
				Production Waste Pressurized	PWP	
				Clean Steam	SC	
				Storm Drainage	SD	
				High Pressure Steam	SH	
				Medium Pressure Steam	SM	
				Sanitary Sewer	SS	
				Sanitary Sewer Vent	SSV	
				Steam Vent	STV	
				Vaccum	V	
				Process Vent	VP	
				City Water	WC	
				HVAC Chilled Water Return	WCVR	
				HVAC Chilled Water Supply	WCVS	
				Deionized Water	WDI	
				Water For Injection	WFI	
				Water For Injection Cold	WFIC	
				Water For Injection Hot	WFIH	
				Glycol Water Return	WGR	
				Glycol Water Supply	WGS	
				Hot Potable Water Supply	WHPS	
				Hot Potable Water Return	WHPR	
				HVAC Hot Water Return	WHVR	
				HVAC Hot Water Supply	WHVS	

REMARKS	

Example 8.3 Programming Criteria for a Biotech/Pharmaceutical Company (page 5)

PROGRAMMING CRITERIA FOR BIOTECH/PHARMACEUTICAL

SHEET 5

ROOM NAME	: FERMENTATION HALL	RECORDED BY	:
GENERAL AREA	:	PROGRAMMING TEAM	:
PROGRAM NUMBER	: 2102	DATE	:
ROOM NUMBER	:	REV.	:

UTILITIES/ BUILDING SERVICES (cont.)				Tempered Glycol Water Return	WKR	
				Tempered Glycol Water Supply	WKS	
				Plant Water	WP	
				Potable Water (Cold)	WCP	
				RO Water	WRO	
				Tower Water Return	WTR	
				Tower Water Supply	WTS	
				Softened Water	WS	
		Other				
	Pipe Penetrations	Yes	No	Location	Number	Comment
		✓		Walls		
		✓		Ceilings		
			✓	Floors		
	Utility Stations	✓			/	CONFIRM
	Floor Drains	Yes	No	Type	Number	Location/Comments
				Open		
		✓		Sealed		
			✓	Trenches		
	CONFIRM	?	?	Trapped		CHECK
			✓	Sumps		
			✓	Floor Sink		
				Tankwash Pit		PROCESS TO CONF.
		Other				
		Provision			Comments	
	Bio-Waste Containment	YES				
	CIP Provision	YES				
	SIP Provision	YES				
	Sinks	Yes	No	Type	Number	Material
				Single		
		✓		Double	/	SS
		✓		W/Drain Board	/	SS
		Other				
	Eyewash	Yes	No	Type	Comments	
		✓		Self Standing		
				Wall Mounted		
		✓		Combined with Emergency Shower		
				Deck Type		
				Fire Blanket		
		Other				
	Emergency Shower (SEE EYEWASH ABOVE)	✓		Self Standing		
				Wall Mounted		
				Ceiling Recessed		
				Multiple Nozzle		
		Other				
	Water Services		✓	Water Cooler		
			✓	Hot Water Heaters		
			✓	Water Filters/Polishers		
		Other				

FIRE PROTECTION		Yes	No	Provision/Type	Comments	
	Fire Resistance of Room					
	Spinkler System			Flush Sprinkler		
				Halon		
				PA or Preaction		
				Wet Sprinkler		
				Dry Sprinkler		
				Special Sprinkler		
		Other				
	Density					
	Design Areas					
	Equipment Provision	Yes	No	Location	Number	Comment
				Fire Hose Cabinet		
				Fire Extinguisher		
				Fire Alarm Manual		
				Smoke Detector		
		Other				
		Other				

Remarks	

Example 8.3 Programming Criteria for a Biotech/Pharmaceutical Company (page 6)

PROGRAMMING CRITERIA FOR BIOTECH/PHARMACEUTICAL

SHEET 6

ROOM NAME	: FERMENTATION HALL	RECORDED BY	:
GENERAL AREA	:	PROGRAMMING TEAM	:
PROGRAM NUMBER	: 2102	DATE	:
ROOM NUMBER	:	REV.	:

HOODS	Hood Type	Yes	No	Type	Number	Location
				Standard Fume Hood		
				Auxiliary Fume Hood		
				Bio-Hood (Class I, II, III)		
				Special Hood (Perchloric Acid)		
				Laminar Flow Hood (Vertical)		
				Laminar Flow Hood (Horizontal)		
				Radioisotope		
		Other				
	Hood Classification					
	Hood Data	Data		Information	Comments	
				Size		
				Face Velocity		
				Exhaust CFM		
				Static Pressure Loss		
				Other		
		Yes	No	Type	Number	Location
	Base Cabinets			Acid Storage with Vent.		
				Solvent Storage with Vent.		
	Working Surfaces	Yes	No	Type	Comments	
				Stainless Steel		
				Resin		
				Cup Sink		
				Faucet		
				Electrical Outlet		
				Utility Cutouts		
	Hood Services	Yes	No	Service	Quantity/Temp./Pressure/Loc.	
				Cold Water		
				Cup Sink		
				Compressed Air		
				Vacuum		
				Highvac Pump		
				Nitrogen		
				Cooling Water		
				Steam		
				Gas (Bio Hoods only)		
				Process Drain		
		Other				
	Electrical Service	Yes	No	Service	Comments	
				115/120 V Outlet		
				208 V Outlet		
				480 V Outlet		
				Uninterrupted Power		
				Emergency Power		
		Other				
	Blower	Yes	No	Details	Comments	
				Direct Drive		
				High Pressure		
				V-Belt Drive		
				Motor Single Drive 60 Hz		
				115 V		
				Coated Steel		
				PVC		
		Other				

Remarks	

maintenance, and still maintain the goal of a working space that allows for maximum cleaning and operational efficiency. If these items are located inside the designated clean process space, the amount of exposed horizontal surface area to hold contaminants is greatly increased, access protocols for maintenance operations must be addressed, and the risk of contamination due to component failure is increased.

A current trend that is gaining acceptance is the service chase or gray space concept. Facility layouts are designed to hold all support service components outside the clean space, in service corridors immediately adjacent to the process space (Figure 8.3). Gray space is defined as mechanical access areas that are environmentally unclassified, although controlled for humidity and supplied with filtered air. The majority of all valving, piping and conduit runs, supports, and utility equipment, such as heat exchangers and pumps, is located inside this space, in effect on the other side of the wall from the process space. Maintenance personnel have separate access to these areas, allowing for less stringent gowning requirements and providing a way to perform maintenance without disruption or shutdown of process areas.

Construction materials in gray spaces may be of a lesser grade while satisfying all required performance criteria. Since high-maintenance mechanical items are located close to the operational areas they serve, yet are in low-cost space, enhanced cost reductions may result.

Figure 8.4 provides an example of how far this approach can be taken, where the service corridor becomes as large as some of the actual process areas, even housing such items as transfer panel connections and HVAC units. As you might expect, with this approach comes added cost to the facility. Service corridor space can cost in the range of \$200 to \$300/ft^2, depending on the overall size of the area and the configuration of the building. But even at these numbers, gray space deserves a close look as a way to provide assurance of cleanroom maintainability over the life of the facility.

Accessibility

Another important aspect of layout comes in providing proper operation and maintenance access to equipment and components. Parenteral facilities are very mechanically intensive; thousands of feet of ductwork and piping, miles of control wiring, and hundreds or thousands of components are packed into relatively small spaces. Even when service corridors and interstitial spaces are provided for utilities, the access to valves, dampers, instruments, and connections must be thoroughly reviewed. All the open space in the world does no good if the operations or maintenance personnel cannot gain access to a needed device.

If design attributes such as catwalks or walkable ceilings are used, be sure to review access to all maintenance points. Provide proper design to comply with all state and federal safety standards for guardrails, head height, and fall protection.

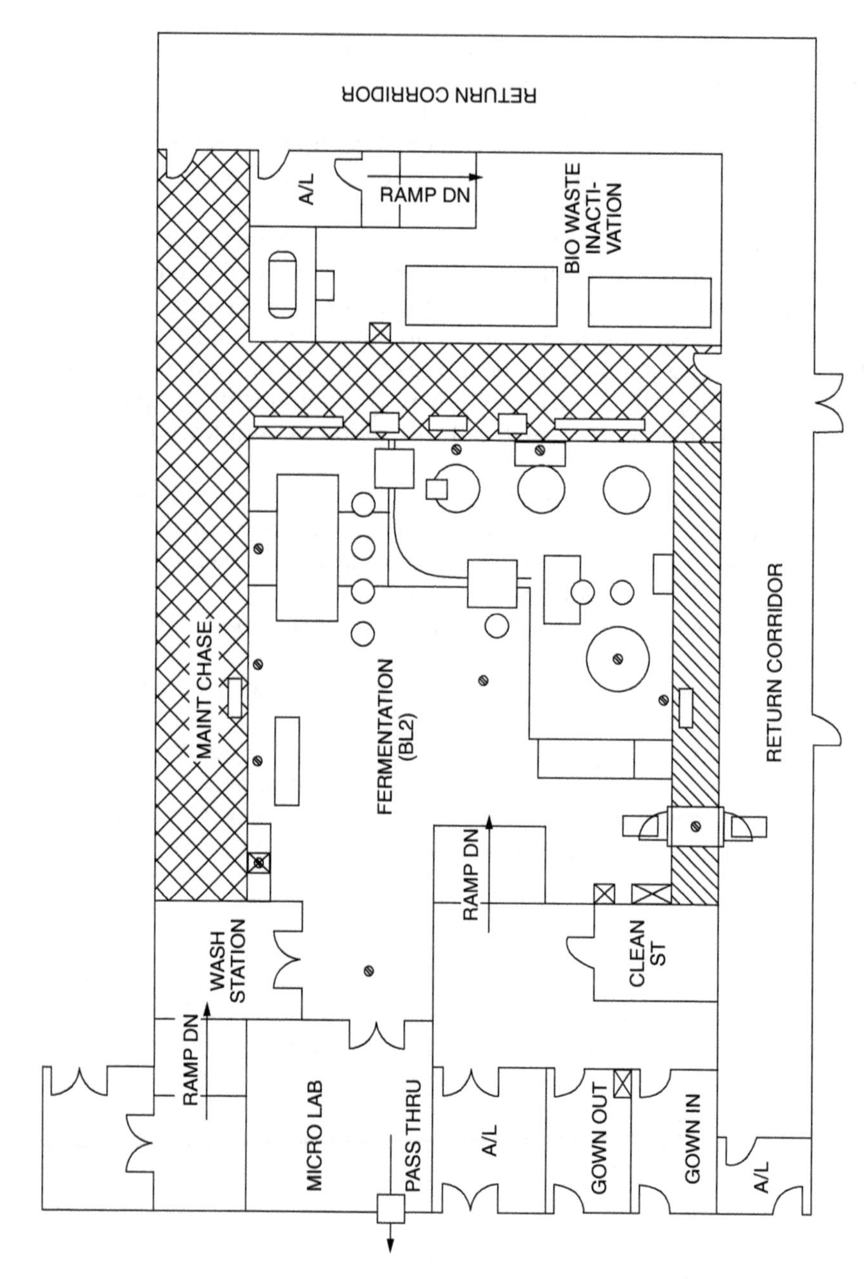

Figure 8.3 Facility layout showing "gray" areas.

Figure 8.4 CAD drawing of a service corridor. (CAD drawing courtesy of Jacobs Engineering, Cincinnati, OH.)

Access from within a cleanroom environment for servicing mechanical equipment should also be avoided. If process operations require aseptic conditions, the practice must be avoided. Room layouts and mechanical system routings in the facility should be carefully coordinated to eliminate access doors from cleanroom spaces. When service corridors or interstitial spaces are utilized, such coordination is not difficult to accomplish. But in cases where space constraints do not allow such flexibility, great care should be taken when determining locations for access doors to dampers, light fixtures, or control devices from processing areas.

Cleanliness and cleanability

GMP regulations require "smooth, hard surfaces that are easily cleanable" and a defined system for cleaning these surfaces to produce the desired conditions. This means that the construction materials selected must facilitate cleaning from rigorous daily cleaning regimes, be impervious to water and the selected cleaning solutions, and be resistant to chipping, flaking, oxidizing, and other deterioration.

The architectural design must also employ what is known in the industry as "crack and crevice free" construction; the attention to detail that focuses on eliminating areas of contamination ingress and egress, eliminating areas promoting stagnation or particle buildup, and minimizing to the greatest extent possible the availability of surfaces that will attract or collect particles.

Table 8.1 Negative Pressures and Corresponding Velocities through Crack Openings

Negative Pressure, Inches Water Gage	Velocity, fpm	Negative Pressure, Inches Water Gage	Velocity, fpm
0.004	150	0.060	590
0.008	215	0.080	680
0.010	240	0.100	760
0.014	285	0.120	828
0.016	300	0.150	930
0.018	320	0.200	1030
0.020	340	0.250	1200
0.025	380	0.300	1310
0.030	415	0.400	1510
0.040	480	0.500	1690
0.050	540	0.600	1860

(Calculated with air at room temperature, standard atmospheric pressure, Cu = 0.6)

With most process areas under strict cleanliness control and pressurization, even the slightest crack can offer a pathway for contamination flow that will be surprisingly effective in pushing particles into clean areas (Table 8.1).Therefore, it is important to provide seamless construction in door and window frames by requiring welded jambs and sealable hardware cutouts. Require electric boxes without knockout plugs. Be sure that walls are sealed above the ceiling to prevent the propagation between the walls of dust and debris that could potentially escape through a small crack or an opening such as an electrical outlet. And be sure that the integrity of fixtures is maintained by having gasketed seals in every one.

Material selection

GMP requirements for materials are very general. Operational needs, user group preferences, previous project experience, and manufacturer's performance data provide important input into the decision on material selection. But there are numerous materials and finishes that will fit performance requirements — all have different costs for installation, different life cycles, and different availability.

A large part of material selection will also focus on the finished product of the installation effort and the compatibility of the floor, wall, and ceiling materials:

- Do the materials join snugly?
- Do they resist thermal expansion and contraction?
- Are they resistant to changes in environmental conditions?
- Are they smooth and compatible to clean?
- Are they of dust-free design?

- What about the actual installation? Are special tools required?
- Are there hazardous conditions that must be addressed?
- Does the installation require defined environmental conditions for application?

What about the cost? Everyone wants the best, but what can the project budget absorb? Cost analysis must include not only the cost of material, but also the cost of installation, repairs, and the material life cycle.

Flooring systems

The selection of a flooring system reflects aesthetic, durability, and cleanability concerns. Many types of cleanroom flooring systems are available, each with its own impact on the construction effort. Some of the more familiar types include the following:

- **Epoxy terrazzo** A hard, durable, long-lasting surface that has good chemical resistance properties. Installation costs are usually higher, and a skilled installer is mandatory if the installation is to meet industry standards. An integral cove base is available, either trowel or precast. It is used most often in manufacturing spaces.
- **Epoxy resin** Popular in the industry due to the development of more chemically resistant resins. This flooring system is easier to install, with a more moderate price. Concerns come from overall durability and potential for discoloration of the topcoat due to exposure to chemicals.
- **Welded sheet PVC** A chemically resistant material often used where compatibility with adjacent wall finishes is preferred, as in laboratory environments. Durability becomes an issue in high-traffic areas.

The selection of a flooring system takes into consideration the presence of wet processes, wear characteristics, chemical resistance properties, and durability under projected traffic loads. Application time can be affected by the environmental conditions of the room, the safety of fumes generated during installation, the condition of the subfloor prior to installation, and the time required for the system to cure before being placed in service. All of these factors can increase the time and cost of installation and must be incorporated into developing work budgets and schedules for the project.

Walls

The two basic types of wall construction for cleanrooms are gypsum wallboard (GWB) on metal studs and concrete block. To the extent possible, GWB should be selected over concrete block to avoid any potential problems due to cracking of the block and the resulting cracks in the finish. Even with proper design of control joints to mitigate cracking, the control joints themselves are an issue on the exposed surface. Gypsum wallboard provides a low-cost and very flexible option for construction, as well as ease of maintenance and potential renovation.

The preferred wall treatments used in the industry today are epoxy paint coatings and flexible sheet vinyl panels. Epoxy paints provide a low-cost, hard, durable surface that is easily cleaned and repaired. Wall treatments using sheet vinyl material are normally found in high-traffic areas, such as equipment corridors or areas where the classification level is more stringent (i.e., where there is a need for a smoother surface for cleaning). Wall panel treatments can range from corner guards in high-traffic areas, to wainscoting to protect walls from equipment movement, all the way to floor-to-ceiling coverage to produce a more monolithic surface for cleaning.

There are other wall finish alternatives, such as seamless PVC coatings and troweled-on cementitious materials, which provide the same level of protection but are more costly to install and repair.

Transition points between wall surfaces must be smooth and uniform. At floors, ceilings, door and window frames, penetrations, and built-in components, the ability to seal the joint in a manner that facilitates cleaning and preserves the integrity of the clean environment is important. Silicone caulking materials are normally used for this purpose.

Ceilings

The most common material used in cleanroom ceiling construction is GWB, with a finish treatment of epoxy paint. As with GWB walls, flexibility and ease of installation are attractive attributes. It is also provides the most monolithic of surfaces before going to a more costly vinyl sheet coved transition at wall–ceiling joints.

The main construction concern with GWB ceilings is the extensive coordination required among all discipline trades during installation. Because of the many penetrations through the ceiling surface — HVAC registers, HEPA filter housings, light fixtures, and piping penetrations — it is important to emphasize coordination to minimize rework, even if repairs are relatively simple. The fragile nature of GWB is also a concern; walking or dropping heavy objects on the material can easily distort or destroy it.

Lay-in grid systems are also used in less stringent areas where cleaning is less of a problem. Gasketed cleanroom tiles are available which provide ease of installation and access to above-ceiling services for minimal cost. Again, the concern lies in the seams and joints in the system that provide potential breeding grounds for particle accumulation.

Construction issues

Details

It is the task of architectural design to translate the general requirements of GMP into specific details that will produce a finished facility that promotes cleanliness and cleanability while maintaining the clean environment. Simple in theory, this is the true measure of how well the architectural design is executed. Are the details easily understood, technically sound, and buildable?

The focus on details of construction for cleanroom drawings should center on penetrations and intersections or transitions. Penetrations include items such as door and window frames, piping, pass-through items, lighting fixtures, and air terminals (e.g., HEPA filter housings). Transitions are defined as any intersection between materials and surfaces, such as between floor and wall, wall and ceiling, or dissimilar materials. These points are where the potential for breakdown of the clean environment is greatest. They are also where substantial time and money can be spent during construction if poor details require substantial rework to correct.

Figures 8.5a–g give examples of some common details for cleanroom applications. It can be seen that great emphasis is placed on sealing all open areas, minimizing horizontal surfaces that can collect dust, and providing materials that can withstand the dailyrigors of cleaning and traffic within the space.

For facilities that do not have the luxury of service corridors for routing mechanical services, the architect must focus on the coordination of mechanical system routings in wall spaces and the sizing of the walls. Since common practice in the industry is to provide concrete curbing to control fluids present due to cleaning or spills at floor level, the width of curbs and the thickness of walls must accommodate the space needs of the piping and ductwork routed within the wall space.

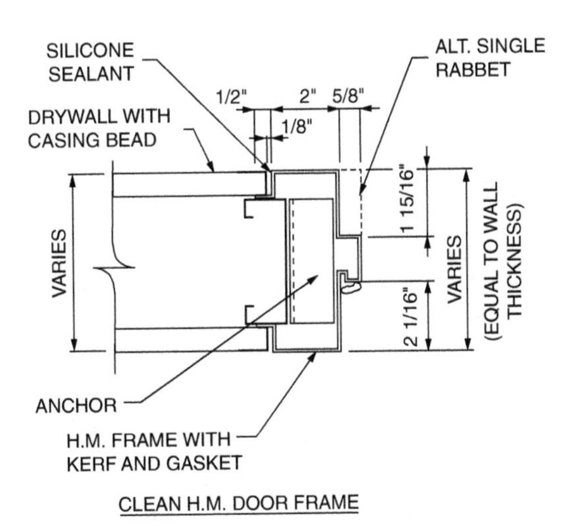

Figure 8.5a CAD drawing of clean H.M. door.

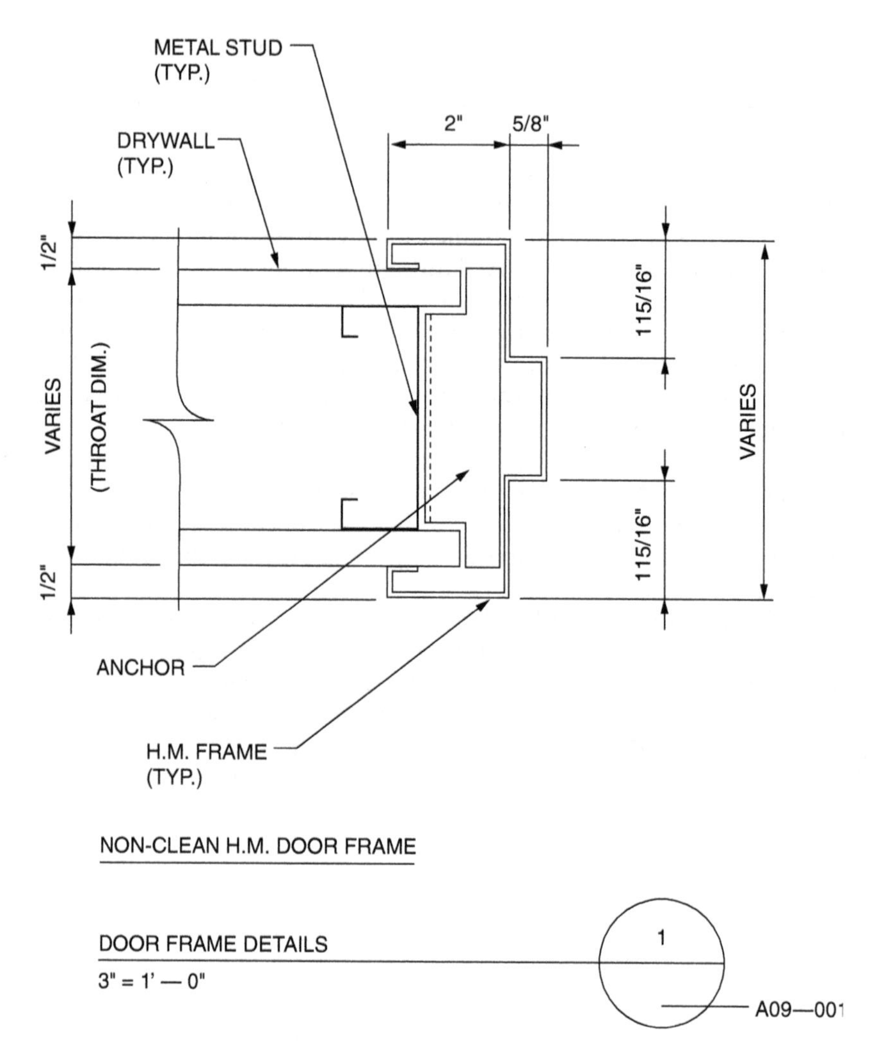

Figure 8.5b CAD drawing of nonclean H.M. door frame.

Air leakage through openings is another issue that must be addressed. There are many potential locations for leakage, some not so obvious. Selecting the proper seal and, sometimes, the proper component is important. For example, light fixtures used in cleanrooms must have gasketed seals around the lens and the fixture housing to provide an airtight seal. Many fixtures that claim to be sealed only provide seals around the housing. Another example involves door frames and the seals between the door and frame in the closed position. Air leakage through the door requires a seal material at the active rabbet. Many times, this material is either improperly installed at the factory and will not hold in place through frequent use or is a material that cannot withstand the daily cleaning regime required by protocol, deteriorating soon after start-up and allowing air leakage.

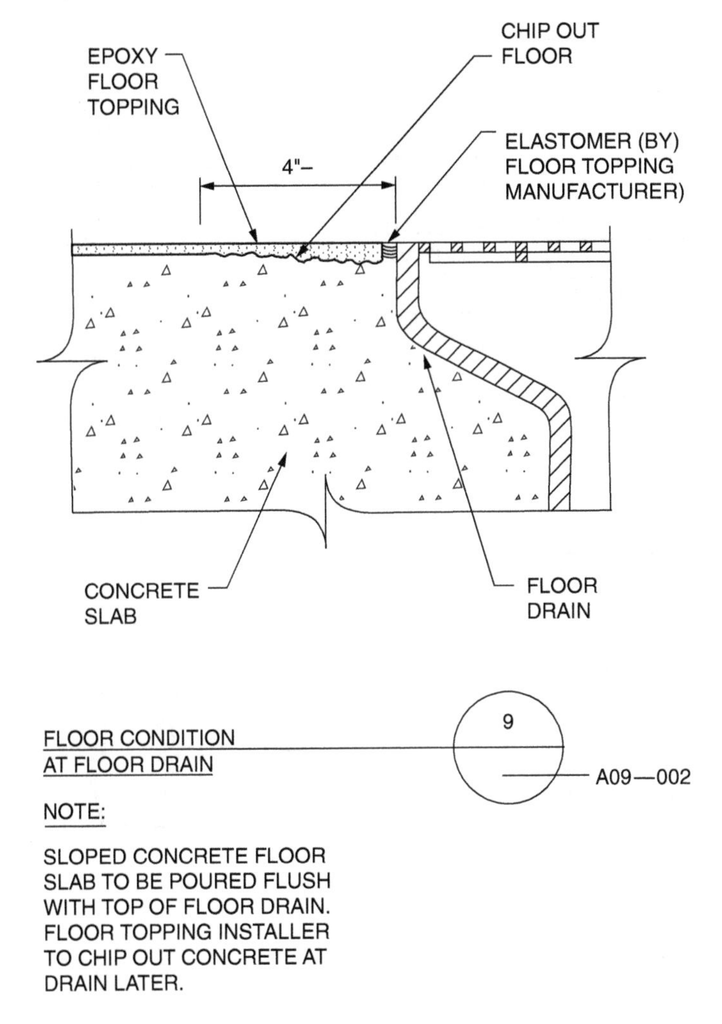

Figure 8.5c CAD drawing of floor condition at drain.

Another area of focus in maintaining clean or aseptic conditions should be on pass-through equipment, such as autoclaves and glass washers. These items must have bioseals within the framed walls to prevent air passage from dirty to clean environments. The type of seal material, the design of the gasket support and its spacing, and the method of securing the gasket material are very important. The cover plate that protects the seal and provides the aesthetic quality to the installation must fit snugly with the surface and receive a caulk seal at the transition. Surface continuity of the GWB wall surface is important when making this seal.

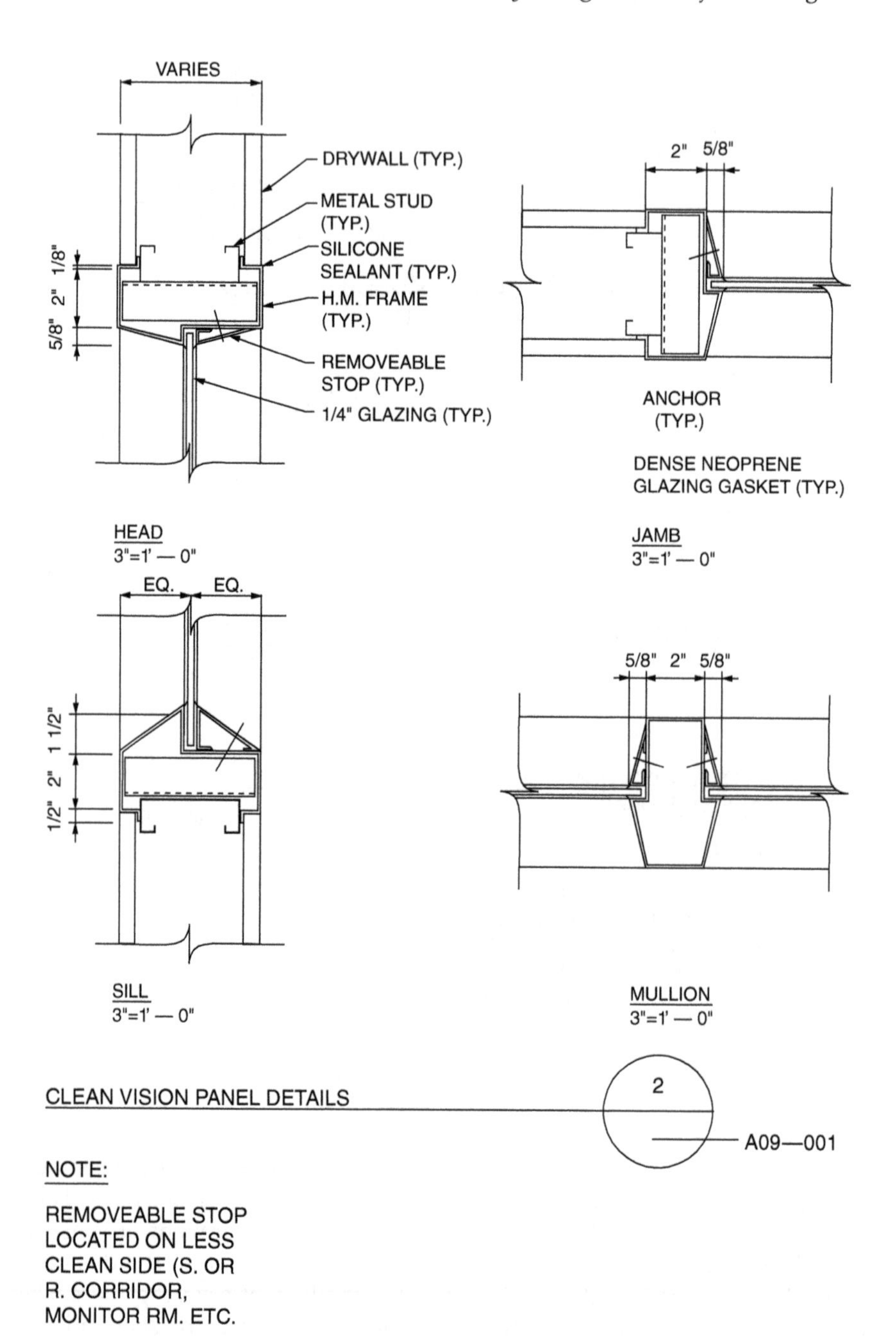

Figure 8.5d CAD drawing of clean vision panel details of head, jamb, sill, and mullion.

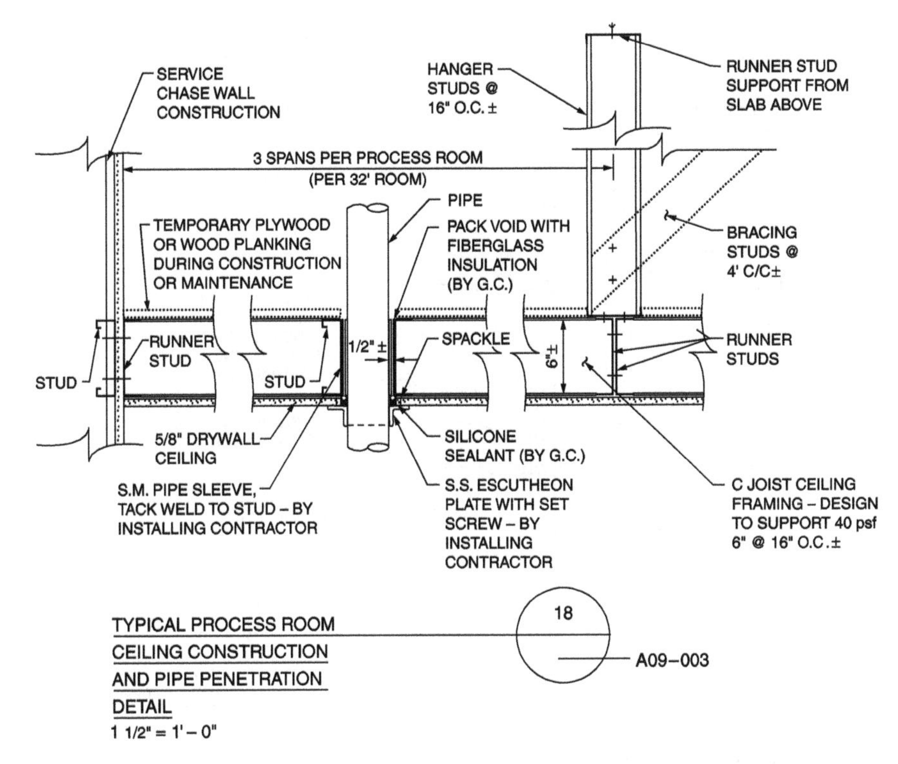

Figure 8.5e CAD drawing of typical process room ceiling construction and pipe penetration detail.

Costs

The challenge for the architect is to design a facility that balances minimal risk and optimal cost. As with any building project, large sums of money can be spent in providing the most expensive, elaborate, and durable materials and putting them into a facility that better serves as a palace than a functional building. Factors of safety can be added like "belts on suspenders" until risk of contamination is negligible, but at an extremely high cost for little benefit. Just look at the semiconductor industry, where contamination control is more stringent in terms of cleanroom requirements.

Figure 8.6 provides an overview of building cost in relation to the other project discipline costs. Many different approaches can be taken to meeting GMP requirements — different combinations of materials and finishes that all have an impact on project cost and facility life cycle. By evaluating life cycle costs, Table 8.2 gives a simple comparison of how combinations of alternatives can impact cost.

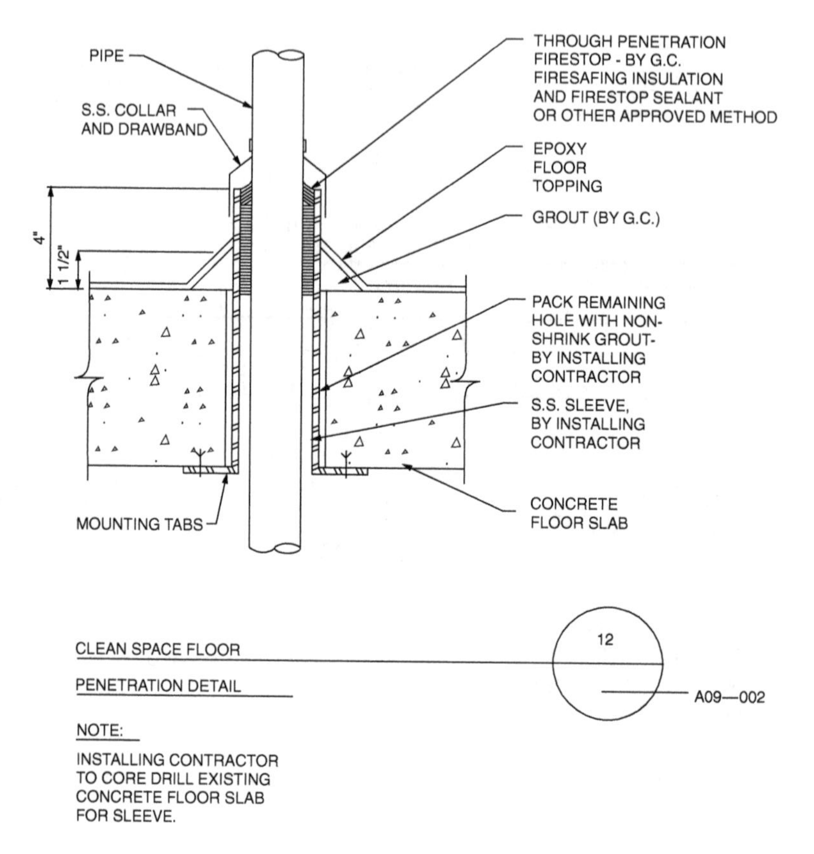

Figure 8.5f CAD drawing of clean space floor penetration detail.

Optional methods

One alternative to the traditional custom-built cleanroom that we have discussed is modular construction in which prefabricated panel systems are used to construct a cleanroom space that meets GMP and is configured to meet the process needs of the company.

Modular cleanrooms are attractive in a number of ways. They allow a tremendous amount of flexibility as compared to custom-built facilities. Modular cleanrooms provide a preengineered, self-contained space that can be designed to provide environments up to class 100. They can be located as a box-within-a-box (Figure 8.7), or they can be designed to fit integrally into a defined space within a facility layout. The panel systems can be reconfigured to meet process needs by adding or deleting panels and rerouting services that support the space.

The finishes provided by such panel systems are of a high quality and very durable. The interiors provide smooth surfaces that are easily cleaned, sealed openings for penetrations through the panels, and effective layout in locating lighting and air termination points (Figure 8.8).

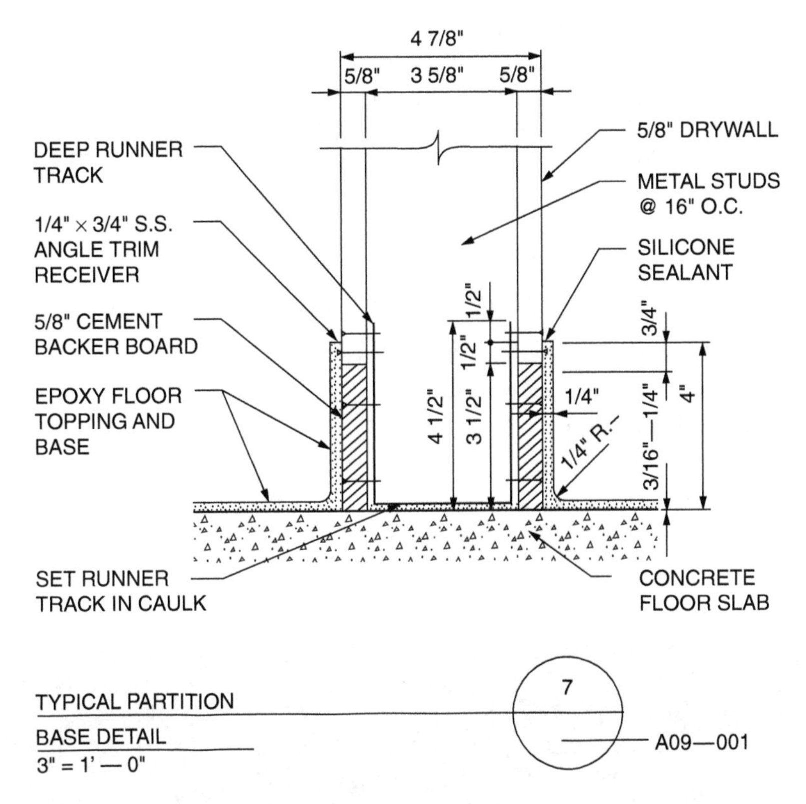

Figure 8.5g CAD drawing of typical partition base detail.

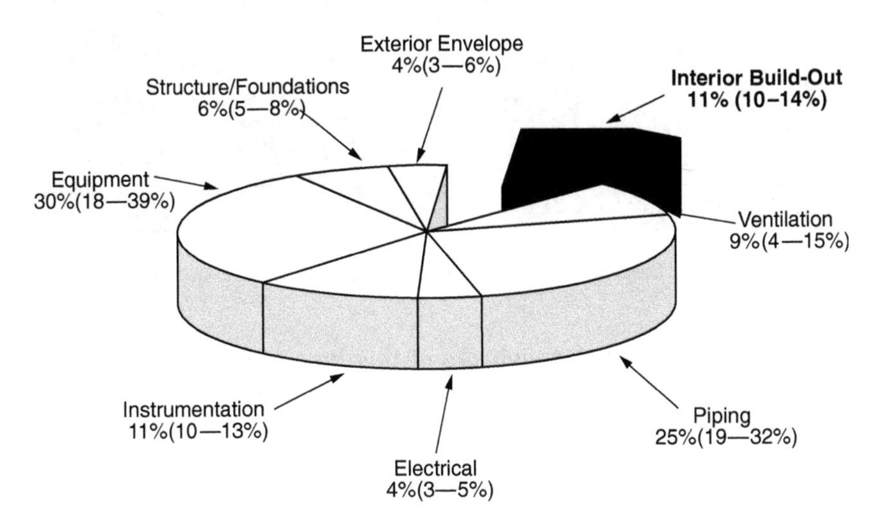

Figure 8.6 Architectural finishes costs.

Table 8.2 Architectural Alternatives

Areas		A147	A157	A247	A257	A346	A356
Ceilings	700 ft^2	$4,929	$4,929	$4,929	$4,939	$14,090	$14,090
Walls	1800 ft^2	16,840	16,840	15,832	15,832	34,660	34,660
Floor	700 ft^2	3475	6800	3475	6800	3475	6800
Base	180 ft^2	851	2482	851	2482	851	2482
Other (duct, lights, frames, cove corners)		16,977	16,977	16,977	16,977		
Total		27,112	45,568	42,064	47,030	53,076	58,032

Alternative A1: GWB partition w/vinyl sheet surface

A2: Block partition w/plaster epoxy finish

A3: Prefabricated wall system

A4: Vinyl sheet flooring

A5: Epoxy terrazzo floor

A6: Prefabricated ceiling system

A7: Plaster ceiling w/epoxy finish

Figure 8.7 Modular cleanroom facility. (Photo courtesy of Clestra Cleanroom, Inc., Syracuse, NY.)

One of the perceived drawbacks of this type of system involves cost. Modular systems are more expensive than traditional stick-built construction, as would be expected. However, depending on long-term operation planning and manufacturing philosophy, an analysis of initial capital cost versus future renovation cost could prove otherwise.

Another issue involves size. For large-scale manufacturing operations, modular systems are not practical. They are best suited to single-story

Figure 8.8 Modular cleanroom interior. (Photo courtesy of Clestra Cleanroom, Inc., Syracuse, NY.)

operations requiring spaces less than 10,000 ft^2. Individual suites are often constructed using modules in conjunction with custom-built facilities. Areas for testing and filling are good examples. For many clinical operations, where process applications are constantly changing, this method of constructing cleanroom space provides a tremendous advantage over constructing or renovating new facilities for new clinical products. Since most of these operations can occur in spaces less than 10,000 ft^2 and employ process equipment on a much smaller scale, the value is obvious.

Renovation

Many start-up companies are faced with an early decision whether to build a new facility or renovate an existing building. With no previous experience on which to draw, they are left to decide the best course of action, which could prove to be more difficult than expected.

On the surface, it may look simple: "Look at all the money we'll save by not having to build a new building." They begin by focusing on a "great deal" on existing property or perceived inherent advantages to an existing facility. They then begin to look for ways to adapt the facility to meet process needs, risking compromise where it may not be so apparent.

Unfortunately, it is not always safe to assume that renovation is the least expensive construction choice. Few existing facilities can accommodate the equipment and building systems needed for sterile production. In many cases, fitting the process to the building becomes an almost impossible task when trying to focus on contamination issues and maintain environments that meet GMP.

The steps used to define needs for project formation and facility programming are the same for a renovation project as for new construction. The challenge comes from analyzing the available data to make sound engineering judgments regarding space utilization, layout, and costs.

Issues

Some general categories of issues impact the decision to renovate for every project. While not comprehensive for every situation, they are areas that should be reviewed by the project team:

- **Sewage/discharge permits** Permitting is becoming more of an issue not only in liquid waste discharge, but also in air emissions and hazardous waste. The intended use of the renovated facility may cause more problems than local officials indicate, especially if neighbors see the facility as creating a real or imagined detriment to their business, employees, or property. The overall business climate of state and local governments can also impact the issue of these critical permits. Permitting agencies that are not biotech friendly can cause severe delays in permit issue, which can have tremendous impacts on schedule and cost to the project.
- **Existing utility services** Sterile product manufacturing operations require substantial amounts of water, electricity, natural gas, and sewage treatment, much more than would be adequate for R&D or light manufacturing operations. There may also be an issue regarding the quality of these utilities in terms of supporting parenteral manufacturing. The best example involves municipal water supplies. Municipal drinking water supplies vary from city to city, mainly due to the source of the water (deep or shallow well, river, lake), the type of natural mineral formations in the area, silt, organic potential, and the clarification methods employed by the municipality. Extensive pretreatment equipment (Table 8.3) may be required to remove or convert undesirable contaminants. The cost impact of such a situation is outlined in Example 8.4.
- **Site access** During construction there are critical issues related to storage, laydown areas, staging areas, temporary access, and construction office space that will impact renovation projects more than new site construction. Thousands of square feet of space may be required for storing equipment and materials, housing temporary offices, and providing the preassembly and staging areas required for the numerous trade contracts. Measures to provide this space can become very costly if off-site facilities are required or local zoning ordinances prohibit this size and type of temporary construction.

Table 8.3 Pretreatment Equipment for Municipal Water

Equipment	Intended Purpose
Heat Exchanger	Temper the water to 75°F for maximum reverse osmosis production.
Water Softener	Remove calcium and magnesium ions that cause hard water and scaling.
Activated Carbon	Remove high molecular-weight organics and chlorine.
Multimedia	Remove particles .20 mm.
Organic Scavenger	Remove organic matter.
Colloidal Scavenger	Remove particles <1 μm
Sodium Bisulfite	Remove chlorine.
Acid Injection	Lower water pH, keep participates in solution.
Caustic Injection	Raise pH, keep silica in solution.
Microfiltration	Remove particles down to filter rating size.
Greensand Filter	Remove soluble iron.
Ultrafiltration	99% removal of bacteria, particles, organics, pyrogens, and colloids .10,000 molecular weight.

Data courtesy of U.S. Filter Corp.

Example 8.4 Pretreatment Equipment Cost Impact

Equipment	Purpose	Cost (1994 $)
HCl Injection	Lower pH	2,102.00
Collodial Scavenger	Remove particles <1 μm	6,897.00
Prefilter Housing with 1 mm filters	90% removal of particles	876.00
Ultrafilter (10,000 MW)	99% bacteria removal	27,938.00

Data courtesy of U.S. Filter Corp.

- **Loading** Existing structures that were not designed for a comparable manufacturing operation will not be able to support the floor load due to equipment and services. In many instances, the proposed building will undergo a major functional change in conversion to manufacturing space. Additional structural steel may be required to support added interstitial space (Figure 8.9), equipment mezzanines, pipe racks, and access platforms. Figure 8.10 gives some recent data comparing four similar projects for biotech manufacturing that involved renovations or new construction on a clean site.

Figure 8.9 Structural steel erection for equipment mezzanine. (Photo courtesy of Gilbane Building Co.)

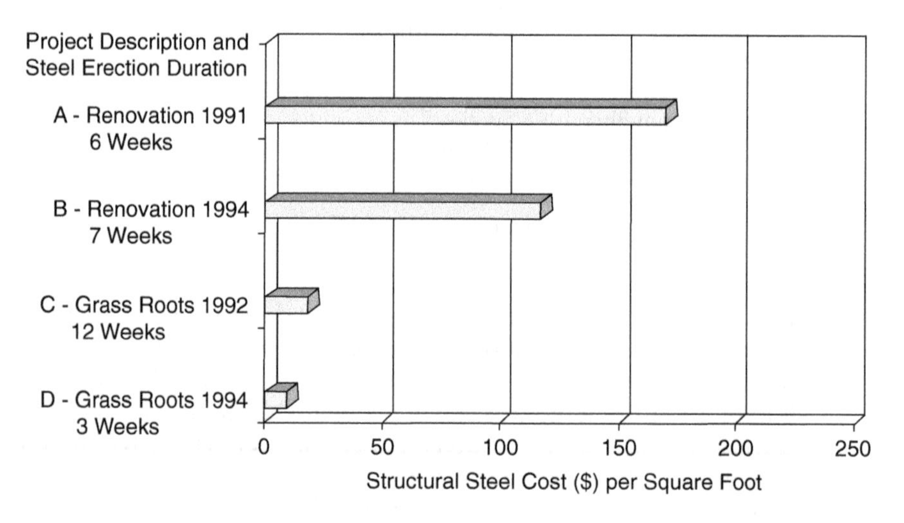

Figure 8.10 Interstitial space renovation is expensive.

Many state and local governments are providing attractive incentives to lure young biotech companies to their areas. These may take the form of tax incentives, free services, construction support funding, or the use of existing shell space for little or no long-term lease. If renovation of space is required to obtain these incentives, be sure to compare realistic construction and operating costs against the value of the incentive. What sounds too good to pass up at first may not look so good over the long haul.

References

1. *Code of Federal Regulations,* Title 21, Part 211.42. 1990. Washington, DC: U.S. Government Printing Office.
2. Ibid.
3. Ibid.

chapter 9

Commissioning

The completion of the construction phase brings to the forefront the most critical, and in many cases the most stressful, phase of project execution: The effort to complete a facility and execute an effective start-up requires a central shift in the execution philosophy of the project from a building approach to a systems approach. The activity referred to as commissioning focuses on maximizing the efficiency and effectiveness of the start-up effort, providing support for the validation effort through integrated testing and documentation; it also provides evidence to support compliance with GMP.

What is commissioning?

As children, many of us were involved in projects in which we painstakingly put together the pieces of a toy, carefully following the directions provided by the toy manufacturer. We came to the moment of triumph when we proudly looked at our accomplishment, turned on the switch — and watched helplessly as nothing happened.

Sterile product manufacturing facilities are a series of complex systems — complex process equipment, advanced process control automation, intelligent building management systems, and sophisticated environmental control systems. Waiting until the end to troubleshoot problems impacting the functionality of the facility could take years. That is why it is so important to know what commissioning involves and to plan for its execution.

Webster defines the term *commissioning* as "an act of bringing together." For our purposes, we will view commissioning as the process of ensuring that all building and process systems are designed, installed, functionally tested, and capable of operation in conformance with the *design* intent.

We will focus on commissioning as comprising those activities necessary to ready a facility or system for service — getting it ready to use or preparing to "push the button." This effort will focus on every piece of equipment, every system and its components, and every test and document necessary to prove compliance with design and functional requirements. Commissioning activities must be integrated in order to prove functionality of the entire facility as a tool for producing a product that meets GMP.

Commissioning and project execution

If any facility has as its goal the production of product to meet GMP, as sure as death and taxes, it will require a commissioning phase. How that phase of work is planned and executed will be a direct reflection of the overall project execution effort. Regardless of how far under budget or ahead of schedule your construction effort has been up to this point, failure during commissioning will be the legacy of the project — the item everyone will always remember.

The key elements of a sound commissioning program include the following:

- Organizing and planning activities
- Factory testing
- Static testing
- Operator training
- Walk-downs for tagging verification
- Equipment start-up
- Functional testing
- System punch listing, turnover, and acceptance
- Documentation
- Calibration

Commissioning should be viewed as the closing step of the execution phase — the final opportunity to verify that all systems are complete, safe, and functional. To focus on constructing a facility for months, only to come to the end of the project and frantically try to tie up the loose ends, is the equivalent of committing project management suicide. Costs will increase by an order of magnitude much greater than the scope of activities; schedules will extend beyond targeted goals, resulting in potential lost profits; and quality will be compromised as people rush to complete work under extreme pressure.

And the level of effort needed to complete the commissioning effort will increase dramatically, as seen in Figure 9.1.

Qualification

Commissioning and qualification activities are the foundation upon which validation is built. In this context, qualification is viewed as an examination of results against the stated requirements. The ISPE's *Commissioning and Qualification Baseline Guide* focuses on the concept of design qualification as the primary means of this examination, and espouses the concept of enhanced design review as the tool to verify conformance to operational and regulatory expectations. The *Baseline Guide* views this as a structured review of the facility design against the expectations of the end user, as defined in the user requirement specifications for each system.

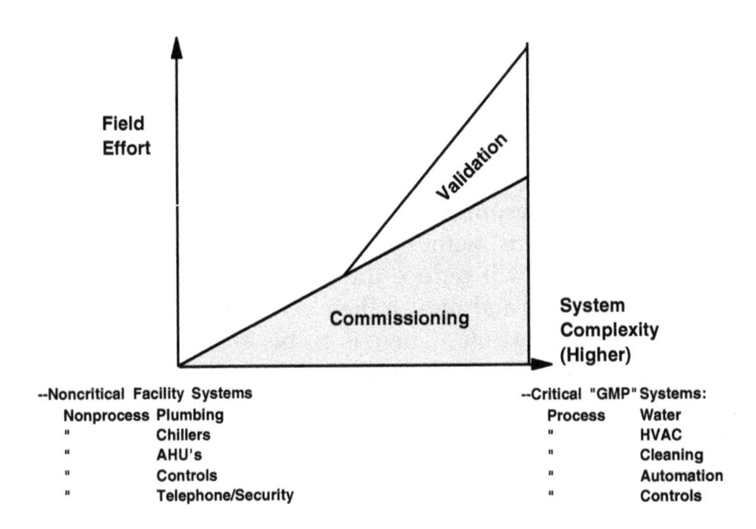

Figure 9.1 Commissioning effort profile.

This concept is being embraced by many regulatory agencies worldwide. Both the European Community Working Party on Control of Medicines and ICH Q7A have specifically addressed this concept, and the FDA has indicated it will also have an expectation for qualification activities on API (and bulk biological) facilities. This concept will be discussed in more detail in Chapter 13.

Commissioning to support validation

Even if the facility is complete, if it cannot be validated, it is useless. Understanding this point is important when developing a commissioning effort that will focus on an integrated commissioning/validation effort. But it is also important to realize that commissioning does not equal validation; it is a complement to validation. While all systems will be commissioned, only those systems deemed direct impact systems will ultimately be validated.

Validation, in most projects, is a very costly and time-consuming effort. Project budgets will typically show validation costs in a range of 5–15 percent of TIC; the duration for validation typically requires 3–10 months at the end of the project. Any effort that will help streamline the validation process and reduce costs is important. That is where a focused commissioning effort will provide value.

The testing and documentation necessary to support validation, specifically IQ and OQ activities, should be coordinated to take advantage of the commissioning effort. A well-planned commissioning effort will provide much of the data required of the validation team. The execution of commissioning activities should occur in the same time frame as the early validation activities. The two should be synergistic.

Benefits of commissioning

A well-defined and executed commissioning effort will provide a number of benefits to the project:

- **Effective use of resources** Coordinating the efforts associated with system walk-downs, witness testing, inspection, training, and document submittals will reduce the number of personnel required to support the effort and reduce their expended person-hours.
- **Eliminate duplication** There is no benefit to producing the same document twice or requiring the same test or procedure be executed more than once. Time will be constrained enough without the repetition of effort. From a quality standpoint, duplicate documentation floating around may also pose an audit problem, especially if the documents, for some unknown reason, contain conflicting data.
- **Maximize schedule** Conducting commissioning and validation activities in parallel will reduce schedule time dramatically and allow for an earlier start of OQ and PQ activities.
- **Reduce costs** By completing the commissioning activities in an efficient manner, the time required of construction support personnel on the project will be reduced. The sooner these individuals are off the project, the more the savings. This also would apply to any technical assistance provided by third-party agencies or vendor support groups. Proper management of the effort will also reduce the total number of person-hours expended to completion.
- **Improve quality** Effective planning and execution of commissioning activities, such as calibration, loop checks, or run-ins, and a focus on document accuracy and completeness will provide a higher-quality finished product at facility turnover.

These benefits will be realized by implementing a sound commissioning strategy that focuses on the qualification of all direct impact systems via a well-defined commissioning plan.

The commissioning plan identifies all systems to be commissioned. It details the activities necessary for commissioning and the sequence in which they should be executed. It also defines the responsible personnel or organizations that will execute the activities, and the documentation to be produced. A typical table of contents for a sound commissioning plan would include the following:

- Statement of Scope
- Definitions
- Roles and Responsibilities

- Practices and Procedures
- Testing and Start-Up
- Acceptance
- Technical Documentation
- Schedule
- Training Requirements
- Final Report

The commissioning strategy should be integrated with the validation program and provide a precursor to formal qualification of systems, especially the direct impact systems. The commissioning plan and the Validation Master Plan should, in fact, cross-reference each other.

The commissioning team

Commissioning is not the sole responsibility of the construction firm, the design firm, or the client user groups. It requires an integrated effort among individuals from very different backgrounds and professions. Recognition of this fact is the key to organizing a successful commissioning team and defining who the stakeholders really are, as shown in Figure 9.2.

While many projects attempt to execute commissioning activities by committee participation, the most effective form of team organization is the traditional hierarchy of a designated team leader and support staff, similar to the overall project organization (Figure 9.3).

Stakeholder Group	System Impact		
	Direct	Indirect	None
Engineering	X	X	X
Manufacturing	X	X	X
Quality Assurance	X		
Validation	X		
Contractors	X	X	X
Suppliers	X	X	X

Figure 9.2 Commissioning stakeholders.

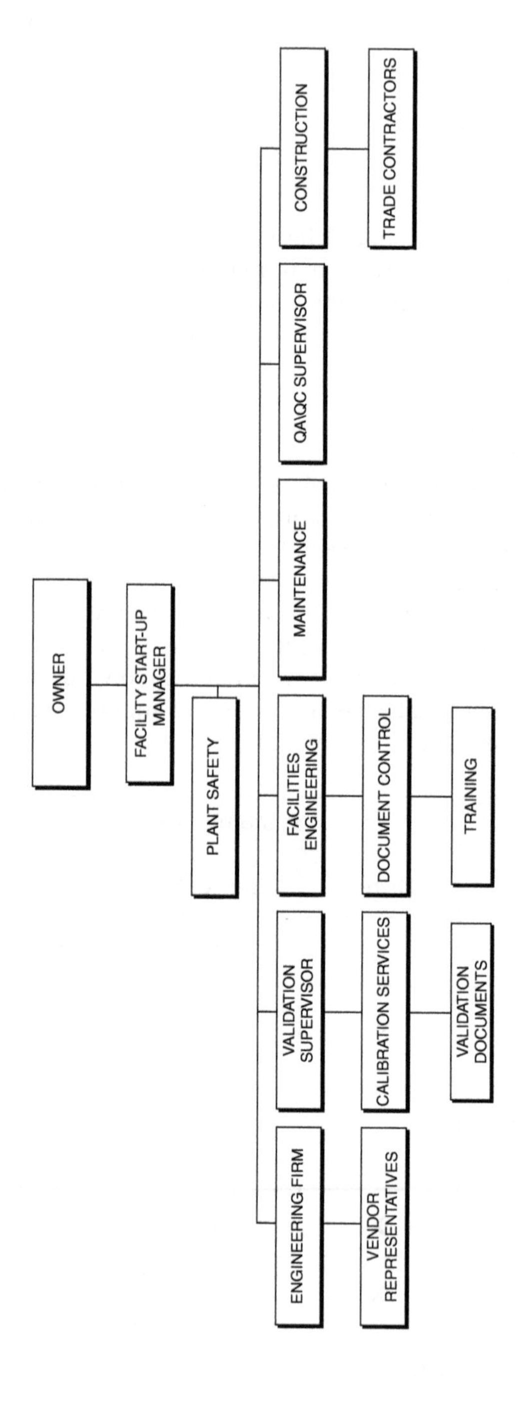

Figure 9.3 Project organization.

The makeup of the team will tap resources from a number of key groups and functions — design, construction, validation, quality assurance, operations, and engineering. Administrative, technical, and support personnel will be assigned to full-time participation for the duration of their assignment, even if they are on loan from other groups or projects. The true benefit is realized when personnel who will ultimately be assigned responsibilities in the new facility, such as manufacturing supervision, facility engineering, or technical services, are given the opportunity to become members of the team and transfer the knowledge gained during commissioning to their daily project operational roles.

As with any project team, empowerment to make decisions and to control work activities is critical. Especially for the commissioning team leader, this overall authority of execution is critical. Clear lines of communication and reporting responsibility should be developed and issued as part of an overall commissioning plan or procedure so that all project team members are fully aware of who is doing what.

Roles and responsibilities

Defining the roles and responsibilities of the commissioning team requires an understanding of the tasks involved, the strengths of the team members, their experience, and a focus on producing a quality effort. The first step in this effort is to recognize that commissioning cannot be performed in a "black box" environment. This means that operations personnel should not focus only on generating protocols, construction personnel should not be responsible only for tracking down documents, and engineers should not focus only on punchlisting systems. One of the easiest ways to define roles and responsibilities is through the development of a matrix, as shown in Figure 9.4.

The most effective team will integrate the experience of the project team participants and focus on maximizing results. Not only will operations personnel be involved in developing protocols, they will also participate in system walk-downs, equipment start-up, and documentation review. The validation team will participate in equipment start-up activities to gain experience, review documents for compliance, and assist in protocol development. Construction personnel will be involved in the development and management of commissioning schedules and start-up assistance.

The problem with most commissioning efforts is that everyone wants to assume that whatever is needed, it is someone else's problem. Calibration is a construction activity; cleanroom certification is the owner's problem; the validation group should worry about documentation. If this approach is taken, errors discovered during commissioning and validation will cause both schedule delays and increased project cost.

Legend: W = Write and maintain R = Review A = Approve E = Execute	Commissioning Leader	QA/Validation*	System Owner	Project Engineer	Contractor(s)/Vendor
Commissioning Plan	W & E	R	R	A	n/a
Commissioning Schedule	W	R	R	A	n/a
Commissioning Budget	W	R	R	A	n/a
Overall Test Plan	W & E	R	R	A	n/a
Pre-delivery Inspection (PDI) Plan	R	A	A	W	W & E
Pre-delivery Inspection (PDI) Report	R	A	A	W	W
Factory Acceptance Test (FAT) Plan	R	A	A	W	W & E
Factory Acceptance Test (FAT) Report	R	A	A	W	W
Inspection Plan	R	A	A	W	W & E
Inspection Report	R	A	A	W	W
Functional Test Plan	R	A	A	W	W & E
Functional Test Report	R	A	A	W	W
System Test Summary Report(s)	A	A	A	W	n/a
Final Commissioning Summary Report	W	A	A	A	n/a
* "Direct Impact" systems only					

Figure 9.4 Typical commissioning deliverables responsibilities matrix.

Scheduling the effort

"Commissioning involves start-up; that happens late in the project. Let's worry about it later." This is the wrong approach if you have any desire to succeed. Commissioning involves many activities — development of procedures, testing and calibration, training, drawing and document development, resource management, and reporting, to name a few. It will be necessary to understand each of these activities, their resource needs, and their anticipated durations, in order to develop a commissioning plan that has potential for success.

When to start planning

When is the best time to start to focus on commissioning activities? The simple answer is *early*. Plan very early in the life cycle of the project, not two thirds of the way through construction. The reasons are simple when reviewed individually.

- **Resources** It takes many people to execute a successful commissioning effort. The sooner resource requirements are discussed and developed, the better the odds that the right amount of resources will be available and that project budgets will accurately reflect the level required. Even for new biotech and pharmaceutical companies with no previous experience to go on, it is important to develop estimates for these needs early so that personnel can either be hired or obtained through outside sources to support the effort.

- **Utilization of time** The number of activities that will occur during commissioning will be in the thousands. Members of the commissioning team will have many commitments placed on their time throughout the process. But people cannot be in two or more places at the same time, nor can they do 20 things at once. Scheduling the effort to maximize time utilization is paramount to success.
- **Coordination of deliverables** Commissioning will produce a vast array of deliverables — test documentation, calibration records, training sessions, procedures, validation documentation. Identifying who is responsible for what early will avoid confusion later. It will also allow documentation requirements to be well defined as a part of contractual documents during equipment procurement, contracting, and third-party inspection and testing activities (Figure 9.5).
- **Optimization of schedule** One of the goals of a successful commissioning effort is to reach completion in an expeditious manner, providing a quality end product. The commissioning schedule should

Deliverable	Maintenance Status	
	"Living"	"Historical"
Commissioning Plan	X	
Commissioning Schedule	X	
Commissioning Budget	X	
Pre-delivery Inspection (PDI) Plan	X	
Pre-delivery Inspection (PDI) Report		X
Factory Acceptance Test (FAT) Plan	X	
Factory Acceptance Test (FAT) Report		X
Inspection Plan	X	
Inspection Report		X
Functional Test Plan This addresses: • Setting to work (including calibration) • Regulation and Adjustment • Testing and Performance Testing	X	
Functional Test Report(s)		X
System Test Summary Report(s)		X
Commissioning Plan Summary Report		X

Figure 9.5 Deliverables.

be a separate and distinct schedule of activities that is integrated with the turnover schedule of the facility. The sooner this schedule is developed, even if it only involves a rough sequence of events in its early stages, the more effective the detailed development of the actual schedule will be in terms of identifying all necessary activities and their predecessors. Figure 9.6 provides an example of a preliminary summary turnover schedule. Each of the identified tasks is then expanded to provide the sequence of activities required to complete the commissioning effort (Figure 9.7).

- **Management of cost** The philosophy of preproject planning as a means of effective cost management applies to the commissioning effort, as it does to the entire project. Knowing estimated resource requirements and anticipated durations will help in cost management. For example, if personnel requirements indicate the need for outside resources, these must be accounted for in the budget. If special training requirements or start-up assistance above what is normally provided by vendors is deemed necessary, identify these costs in the estimate.
- **Transfer of responsibility** At some point, the life of the project will turn from a focus on construction to a focus on system function and operation. Transfer of care and custody of systems and equipment from the construction organization to the client is important. Defining this point in terms of time and physical work is also important from a contractual standpoint; knowing when construction resources will no longer be available becomes more of a concern the closer the "drop-dead" date comes.

One useful tool in starting the planning effort is an interactive planning session, similar to the effort executed for the overall project. By bringing together representatives from all the required groups and organizations to look at some common start-up requirements, the commissioning team and the user groups can obtain clear picture of all the activities involved. Figure 9.8 gives an example of the results from one such interactive planning session.

The most important tool in the planning effort is a well-developed start-up plan or start-up and acceptance manual. The commissioning team should develop the plan with a focus on performing a full functional check of all instrumentation and equipment. The classification of systems, the strategy for facility turnover, and documentation requirements must be defined. The plan must incorporate criteria for performance tests that will challenge the operating characteristics of all equipment and instrumentation items within those ranges specified by vendors. It must also define the necessary safeguards for ensuring that personnel are not subject to dangers from on-line equipment.

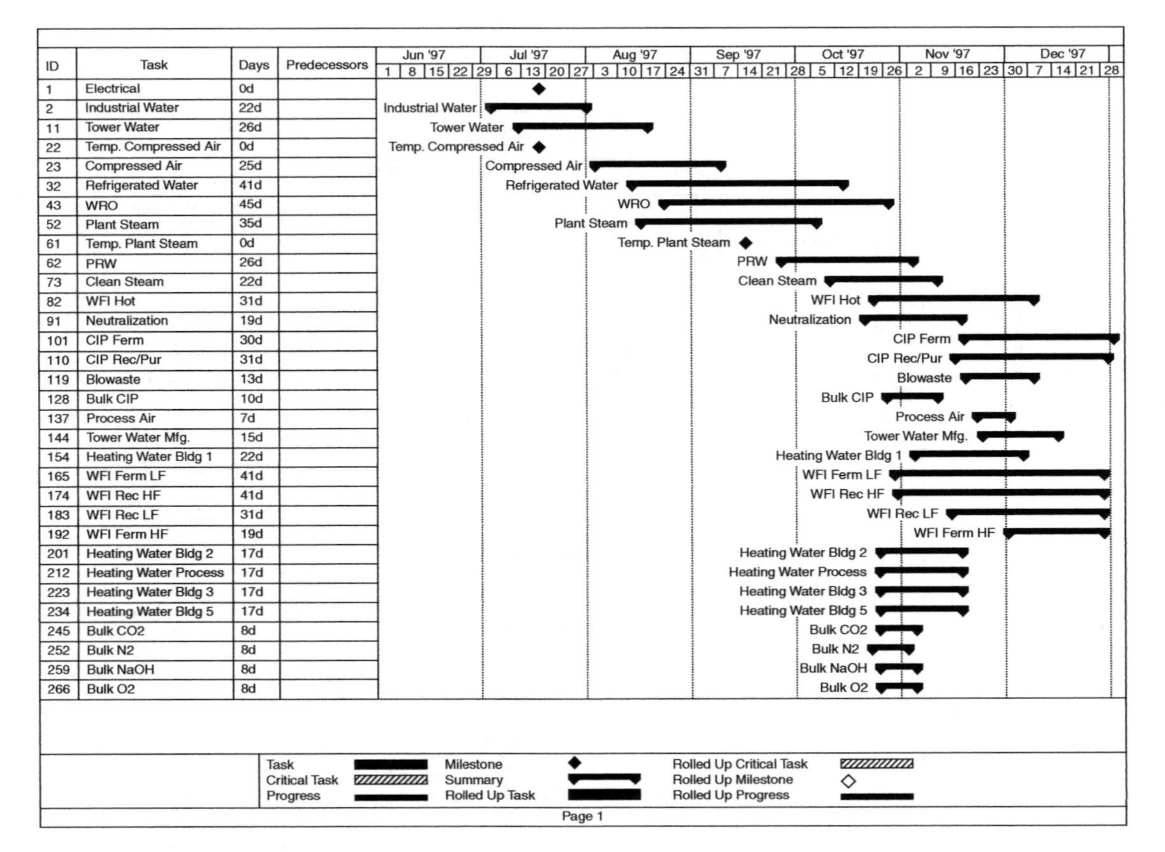

Figure 9.6 Preliminary summary turnover schedule.

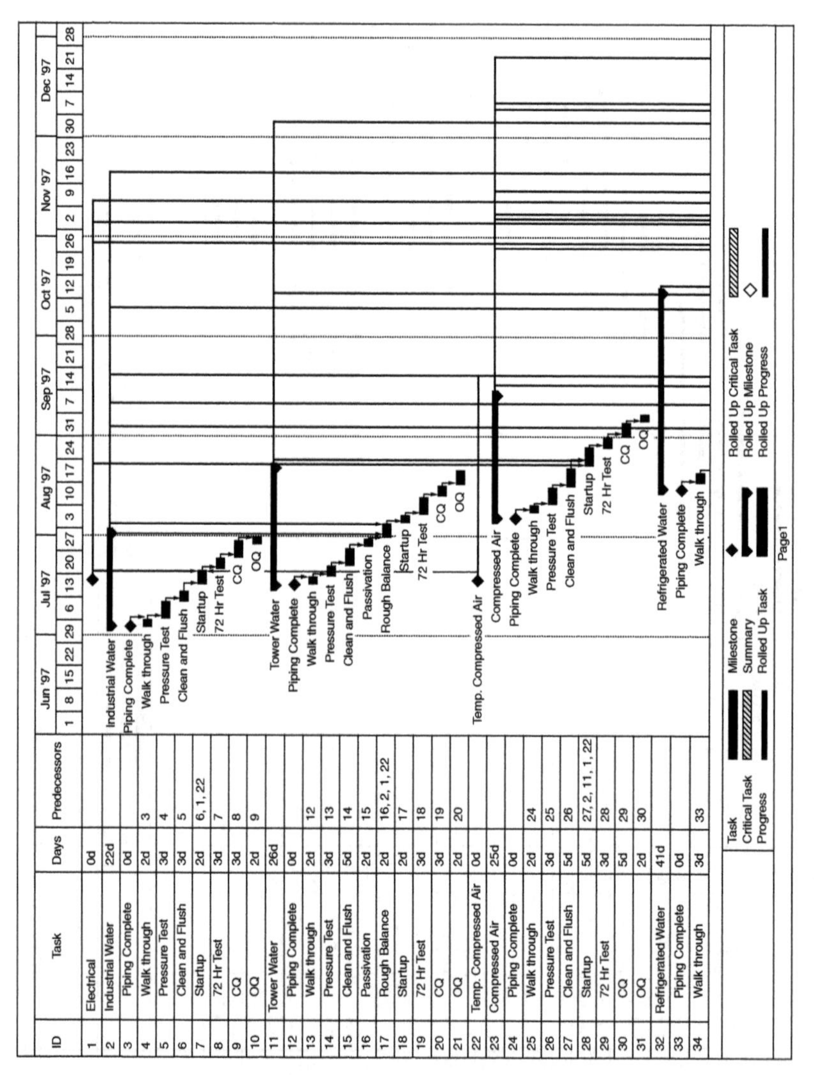

ID	Task	Days	Predecessors
1	Electrical	0d	
2	Industrial Water	22d	
3	Piping Complete	0d	
4	Walk through	2d	3
5	Pressure Test	3d	4
6	Clean and Flush	3d	5
7	Startup	2d	6, 1, 22
8	72 Hr Test	3d	7
9	CQ	3d	8
10	OQ	2d	9
11	Tower Water	26d	
12	Piping Complete	0d	
13	Walk through	2d	12
14	Pressure Test	3d	13
15	Clean and Flush	5d	14
16	Passivation	2d	15
17	Rough Balance	2d	16, 2, 1, 22
18	Startup	2d	17
19	72 Hr Test	3d	18
20	CQ	3d	19
21	OQ	2d	20
22	Temp. Compressed Air	0d	
23	Compressed Air	25d	
24	Piping Complete	0d	
25	Walk through	2d	24
26	Pressure Test	3d	25
27	Clean and Flush	5d	26
28	Startup	5d	27, 2, 11, 1, 22
29	72 Hr Test	3d	28
30	CQ	5d	29
31	OQ	2d	30
32	Refrigerated Water	41d	
33	Piping Complete	0d	
34	Walk through	3d	33

Figure 9.7 Expansion of turnover schedule in Figure 9.6.

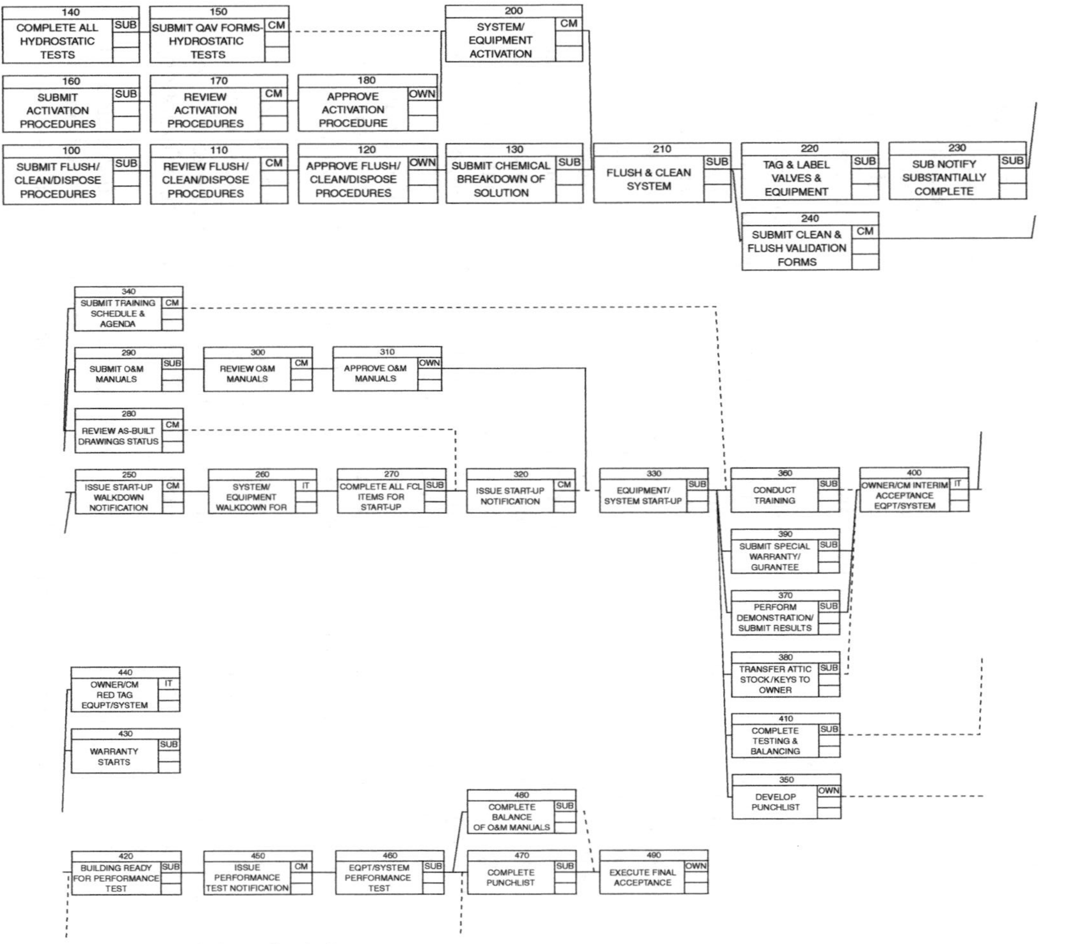

Figure 9.8 Start-up commissioning activity schedule.

When to start execution

When does commissioning really start? At the start of walkdowns? At the start of static testing? At the end of mechanical completion? At turnover? The truth is that many of the activities critical to successful commissioning start early in the project — and many may not actually be recognized as commissioning activities.

One of the first commissioning activities begins with the development of the schedule. Each system within the facility must have clearly defined boundaries and have its own schedule developed, defining clear turnover dates, along with procedures and TOPs to support validation. Because of the large number of systems involved, it is important to allow sufficient time for the development and review of these documents by the commissioning team. These procedures include inspection and documentation requirements.

During the development of equipment procurement documents, it is important to look at factory testing requirements as an essential element of the commissioning program. All mechanical and electrical equipment, all digital control systems, and all software should be factory tested prior to shipment. The debugging of errors in the factory is much more cost-effective and reliable than field correction. Specific test plans should be a submittal requirement from all vendors where factory testing is deemed essential.

Instrument calibration is another important commissioning activity that must be planned and executed early. Coordinating specified tolerances in design specifications with protocol requirements will help maximize the efficiency of the commissioning and validation efforts. Too often, a protocol is written without knowing what tolerances are stipulated in the design specifications. This situation leads either to a required recalibration to ensure that the tolerance can be met or to a revision to the protocol so that the capability of the instrument can be addressed. Either situation causes extra expenditures of time and money.

Unfortunately, many projects start the real focus of commissioning when system walkdowns begin for punchlisting. During walk-downs, every foot of pipe, every valve and control device, every component of the system is visually inspected for proper identification, location, orientation, and support.

Too often, punchlist activities do not start until systems are deemed 90 percent complete or more. This leads to a crisis management situation; too little time is left to perform the substantial number of work items necessary to complete a system. The only alternatives are to extend the schedule, add more personnel, or work overtime hours, none of which is usually acceptable. Begin system completion verification early by focusing on the *systems* rather than the building. Don't wait until 90 percent to review system completeness.

Another often overlooked aspect of the commissioning effort is training. During start-up, client personnel should be given sufficient opportunities to become familiar with the equipment installed in the facility so that less time is spent during OQ execution learning what the equipment can do. It is important to establish training requirements so that their impact to cost and schedule can be properly factored into the project. The most effective way to accomplish this is to develop training matrices that identify types of equipment and descriptions of the necessary training. Table 9.1 provides a sample of the kind of information that would appear in a training matrix.

Functional and performance testing

Performance testing is the process by which independent system performance is demonstrated within a specific set of operational tolerances. The results of these tests are demonstrated in delivering a required capacity or duty. Testing performed during commissioning should be designed to ensure that the designated systems have been constructed to perform in accordance with predetermined design criteria. Such tests include the following:

- Alarm tests
- Emergency power tests
- Fluid flow
- Valve tests for stroke, let-by, and authority

Functional and performance testing performed during commissioning may be used to support validation IQ, OQ, and PQ if the system is classified as direct impact. In this case, the inspection process should be managed and performed within the context of established system qualification practices.

Documentation

Planning and successful management of the commissioning effort requires a significant amount of documentation — procedures, lists, test results. These documents will focus on every piece of equipment and every system in terms of identification, installation, calibration, testing, and performance. The commissioning effort will have two levels of documentation. The higher level will include all of the necessary documents required for successful management and execution of the effort.

The second level of documentation utilized in the commissioning effort will be developed to meet the requirements defined in the TOPs for IQ package development for each system. Thus, it is important for the commissioning team to review document requirements carefully against the TOP requirements to avoid unnecessary duplication or documentation. This also highlights another advantage of integrating the commissioning and validation efforts — creating a paper trail that provides continuity between the two efforts.

Table 9.1 Training Requirements

Spec Section	Spec Name	Description of Training
15323	Air Compressors & Accessories	Provide a technical representative to instruct personnel on operation & maintenance breathing air equipment for one day (8 hr).
15330	Central Vacuum Cleaning System	The manufacturer shall provide instructions on the operation & maintenance at the vacuum cleaning equipment for one day (8 hr).
15380	Reverse Osmosis and Deionized Water System	Operation at system equipment shall be thoroughly demonstrated to owner's satisfaction. At least 3 sessions on-site. An additional session on-site to explain material in O&M Manuals.
15381	Deionized & USP Purified Water System for PS	Operation of system equipment thoroughly demonstrated to owner's satisfaction. At least 2 sessions on-site. An additional session on-site to explain material in O&M Manuals.
15383	Water for Injection (WFI) System	Operation of WFI sill shall be thoroughly demonstration to owner's satisfaction. At least 2 sessions on-site. An additional session on-site to explain material in O&M Manuals.
15402	Domestic Water Pressure Booster System	Provide instruction to owner's personnel on operation and maintenance at pumps for minimum of one day (8 hr).
15409	Neutralization System	Provide technical representative to instruct owner's personnel on O&M neutralization system equipment for one day (8 hr).
15409.1	Neutralization System for Solvent Storage Area	Provide technical representative to instruct owner's personnel on O&M of neutralization system equipment for one day (8 hr).
15531	Combined Wet Standpipe & Automatic Water Sprinkler Fire Extinguishing System	Provide 10 man-days (80 hours) installation and start-up assistance on all fire protection controls.
15541	Diesel Engine-Driven Centrifugal Fire Pump	Provide operation training to each shift of owner's personnel. Each session shall include a complete demonstration of pump. At least 3 training sessions, one per shift.

What types of documents?

There are numerous types of documents involved in the commissioning process. These can be broken down into some well-defined categories.

Lists

There will be numerous lists associated with a successful commissioning effort. These lists provide a wealth of data to the commissioning team regarding the systems and the components that make up these systems. Lists are also an important planning tool in defining the scope of the commissioning effort and a benchmark for progress measurement of the overall commissioning effort. Some common lists include the following:

- Design drawings and specifications
- Vendor drawings
- Equipment
- Instrumentation
- Pipelines
- Valves
- Lubricants
- Spare parts

Plans and procedures

Plans and procedures provide a methodology for the execution of the commissioning effort and a basis for functional performance. They must contain enough detail so that questions related to OQ issues can be addressed. These documents include the following:

- Standard operating procedures
- Start-up plan
- Start-up procedures
- Operation and maintenance manuals
- Validation protocols

Installation and inspection

Installation and inspection documents provide the final validity of the commissioning effort. The content of these documents and the accuracy of the data contained in them is critical to supporting the validation effort. These documents include the following:

- As-built drawings
- Calibration records
- Loop checks
- Static tests
- Factory tests
- Dynamic tests

Submittal requirements

The submittal requirements for commissioning documents must be well defined in contractual documents. This is true for all submittal categories. Not only must the specific type of document be defined, such as an inspection report or a certified drawing, but also the number of copies required and the specific format requirements of the document (e.g., originals, blue lines, or reproducibles). Having the right number of copies available for review and distribution is important in maximizing the efficiency of the commissioning effort.

The general terms and conditions of contracts should also include a submittal schedule that defines when documents must be submitted to the project. By knowing when documents should be available for review, the commissioning team can do some preplanning to avoid the last-minute deluge of review time that is so common at the end of a project.

Closeout

The final act of the commissioning effort must involve the closeout of the project and the transfer of custodial care of all systems and facilities to the owner. Not until this activity is complete should the commissioning team, particularly the team leader, be released from the project.

Because of the contractual implications of the closeout process, it is important to define the intent of the commissioning effort and the scope of commissioning activities in each and every trade contract. A clear summary of functions and responsibilities (Example 9.1) for each of the project participants must be defined, and a detailed description of the closeout process must be identified (Figure 9.9) in the project closeout procedure.

Example 9.1 Functions and Responsibilities for Closeout

A summary of the functions and responsibilities of each party involved in the closeout process follows. Detailed steps performed by each party are illustrated in Figure 9.5.

Construction Manager	The main focus of the construction manager in this process is to ensure conformance by the trade contractor to all contractual requirements and to coordinate the necessary closeout activities of the Architect/Engineer, owner, and trade contractor.
Trade Contractor	The responsibility of the trade contractor is to conform to all contractual requirements. This includes completion of all contract and punchlist work, submission of all required documentation, and in some instances, development of equipment start-up procedures for approval.
Owner	Involvement of the owner includes review and approval of various procedures, input to punchlists, coordination of internal departments for training and start-up, and partial or final acceptance of all work.
Architect/Engineer	Architect/Engineer responsibilities include input to punchlists and review of documentation submitted by the trade contractors.

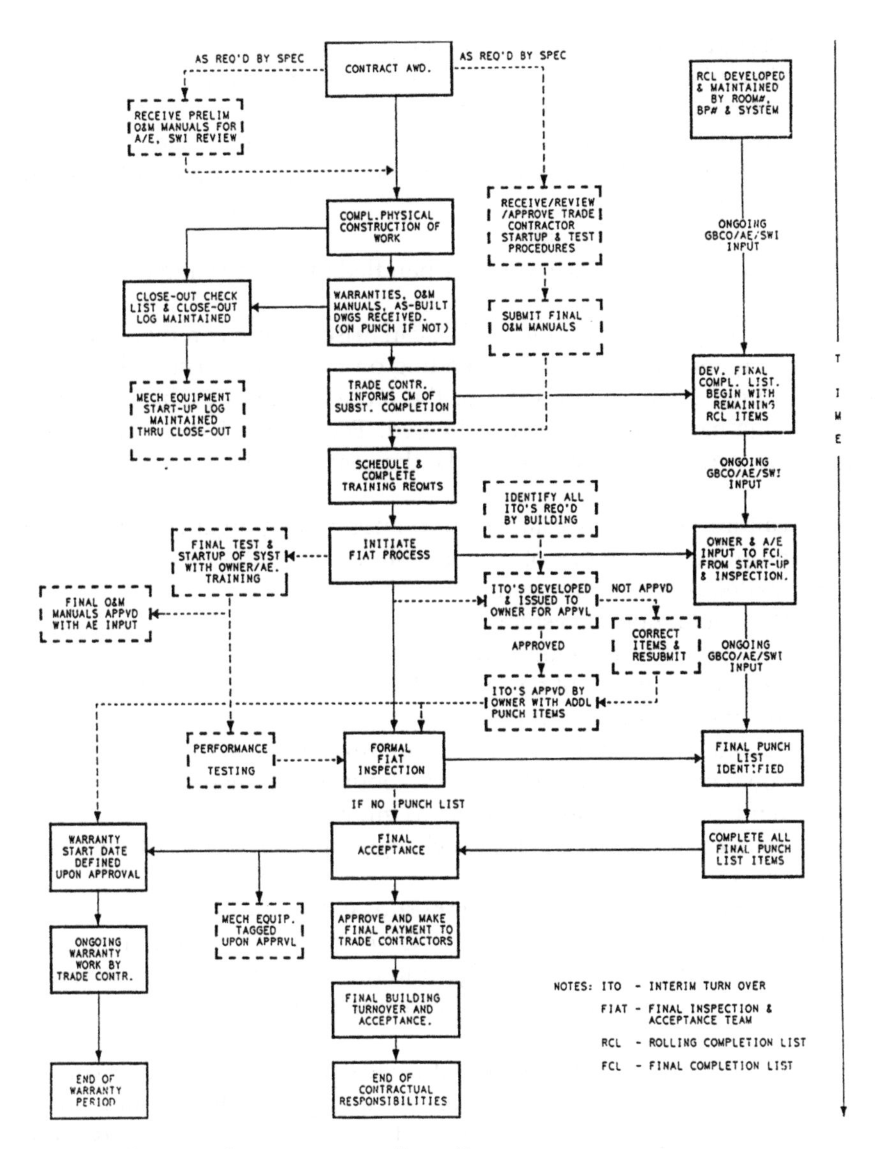

Figure 9.9 Contract closeout process flow diagram.

Commissioning problem sources

When a commissioning program fails, it is usually due to one or more of ten basic reasons:

- Ill-defined scope; the owner thought they bought the service from the subcontractor
- Frequent and undocumented field changes
- Field acceptance system never clarified or implemented

- Design specifications unclear about supplier expectations
- Commissioning not properly budgeted in the project
- System boundaries not defined for testing
- Commissioning plan never developed or implemented
- Assignment of owner personnel to commissioning effort too little, too late
- Field progress overstated
- Belief that the service would be handled under the validation contract

The list of potential problems that can arise during commissioning can also be long and very diverse, depending on the project. The following list is only a brief summary of some of the more common problems that can occur; some have already been addressed. The value of such a list is this: Make sure your project does not make the same mistake.

- Failure to properly establish a commissioning organization with appropriate definitions of responsibilities and authority
- Inadequate staffing resources
- Poor definition of systems
- Poor schedule development
- Lack of continuity of personnel throughout the commissioning effort
- Insufficient training of operators
- Inadequate funding to support commissioning activities
- Poor communication to vendors about requirements
- Tendency to look for shortcuts to improve schedule
- Poor coordination of design changes that impact multiple design areas
- Poor planning for quality support of witness testing
- Inadequate coordination between engineering, procurement, and construction functions in regard to documentation requirements
- Poor communication among project team members

Finally, another problem area that occurs on many projects involves a failure to focus on the building as an integral part of the commissioning effort. Many times, the emphasis placed on start-up and commissioning activities target systems — what is known on P&IDs. Items such as occupancy permits for buildings, plant access, security, materials management, or QC support facilities are not thought about until it is too late to avoid impacting the commissioning schedule. Be sure to look at the entire facility, not just the process, mechanical, electrical, and instrumentation systems.

References

1. The ISPE Commissioning and Qualification (C&Q) Baseline Guide.
2. The FDA has indicated that it will have an expectation for design qualifications to be performed on API (and bulk biotech) facilities.

chapter 10

Quality management to meet regulatory requirements

A cornerstone of the pharmaceutical industry is the understanding that quality cannot be tested into a product. It must instead be designed into facilities and systems that will ultimately produce finished drug products. The ultimate use of biotech and pharmaceutical manufacturing facilities demands that quality standards be met in facility design and construction, not just in operations. This is demanded by the industry, by regulatory agencies, and by consumers.

If quality is defined as "conformance to standards," then quality control is "a system for ensuring the maintenance of proper standards." [1] The FDA's current GMP regulations address certain requirements governing processes, procedures, and characteristics for pharmaceutical production facilities. It is well known that design quality in a biotech facility centers on compliance with these regulations. Over the years, the industry has adopted standards to define the general terms of FDA/CBER/CDER regulations. These standards are commonly incorporated into an engineering firm's design guidelines and details.

Engineering quality and compliance

Quality overkill

The idea that "more quality is better quality" or that "the more money spent to design in quality the better" can cause as many problems as the lack of quality in a facility design. Organizations that attempt to ensure quality by selecting or specifying the highest grade of materials, state-of-the-art equipment, complex automation, or the most stringent operating classification criteria are missing the point.

A more reasonable approach is to examine specific requirements for individual operations and areas within the facility, establish realistic and achievable criteria, and design the facility to meet these specific needs. Spending money for unnecessary technology or more stringent criteria will not impress the FDA, nor will it guarantee the elimination of errors.

GMP reviews

Implementing sound design policies and procedures benefiting from experience in the industry enables design firms to "design in" quality through adherence to GMP guidelines. The formal review process to implement such a system involves the development of review meetings and design checklists for all aspects of the design.

Projects should establish a review committee to review the facility design for regulatory compliance. While this committee should not have any design responsibility, it will provide a review of the design based on applicable regulations and guidelines. Representatives from each of the primary engineering disciplines, under the coordination of an independent committee chairperson, will review documents and provide technical expertise and interpretations of GMP and other applicable regulations based on their area of specialization.

For each area of the facility design, a guideline (Example 10.1) is developed to identify attributes critical to providing a quality system design that will meet GMP. Each responsible discipline lead engineer will see that a checklist is filed for each document checked and that each item on the checklist is checked or identified as "not applicable." The checklist is also provided with a sign-off sheet (Example 10.2) to verify review and approval and provide for any comments.

Specification language

Good engineering practice dictates the use of written specifications to establish requirements for materials, systems, and performance criteria. Conformance to these requirements must be measurable.

A major contributor to poor design quality is vague specification language. If specific, well-defined acceptance criteria are not provided or detailed performance criteria established, interpretations can occur. Interpretation can lead to quality problems.

Written specifications should include every requirement that is not covered on the drawings. Specification language must clearly address the GMP attributes of the design. The specifications must give clear direction on expected performance capabilities and define execution and testing issues clearly. Specifications must also clearly address workmanship standards. Phrases such as "in accordance with standard industry practice" or "per manufacturer's recommended procedures" should be avoided.

Example 10.1 Guideline for HVAC Design

1.0 Introduction

The HVAC DPE assigned to the project should use the following guidelines to control the quality of the design from a cGMP perspective. The checklist that follows should be completed and used to control the quality of working drawings, revisions, and related documents.

2.0 Guidelines for HVAC Design

As a predecessor to detailed design, the HVAC DPE should meet with project management and the client to review basic design parameters for the facility on an area-by-area basis. Meeting results should be documented and form the basis for design criteria. The approval of these criteria by the client must be obtained prior to proceeding with design.

The purpose of the design criteria is

- To establish a basis for facility HVAC design.
- To provide the Owner with a document for FDA facility review.

The design criteria sheet should address both outside and inside design criteria. Outside design criteria should be obtained from ASHRAE (American Society for Heating, Refrigeration, and Air Conditioning Engineers) data. Inside design criteria should include, but are not limited to, temperature, relative humidity, filtration level, minimum air change rate, pressurization requirements, exhaust requirements, and cleanliness level. Both temperature and humidity should be listed as design set points/minus tolerances. A listing of temperature and humidity ranges is subject to interpretation and should be avoided. Filtration should address final filter efficiency and location. Air change rates may be required for cooling load, or as required for a given area classification. Pressurization should be noted on the drawing in the form of actual room pressure levels from a common reference point. The reference point should be noted on the design documents, along with any special exhaust requirements. The location of laminar flow hoods, biosafety hoods, and fume exhaust hoods should also be noted on these documents only if the room is to be validated as a clean room or containment area. Many clients require cleanroom design without claiming an actual classification in their FDA submittals.

2.2 Calculations

A. Airflow Leakage Rate Calculations

In general, the calculation of the airflow leakage rate for each room should be based on the pressure differential established on the design criteria sheet. Assumed leakage rates based on a percentage of supply air are unacceptable. For leakage calculations, each wall should be considered separately regardless of whether the room is interior or exterior. In addition, it should be noted whether leakage is into or out of the room. Leakage calculation should contain a safety factor to compensate for less than ideal construction, deterioration of gasket sealing, and so on.

B. Cooling Load Calculations

Cooling load calculations should be done on a room-by-room basis. Careful attention should be given to process equipment loads because these can be significant heat generators. In calculating this load, consideration should be given to motor load, convection, and radiation from heated vessels and thinly insulated process and utility piping. Attention should also be given to potential heat gain from the air handling unit supply fan due to the high air change rates and multistage filtration. The actual operation of process equipment should be understood so that realistic assumptions on heat gain factors and diversity can be made. Process equipment heat loads with basis should be included in design criteria and reviewed and approved by the owner.

C. Heating Calculations

Heating is generally accomplished by a "reheat," therefore, the heating calculation should include the energy required to raise the supply air temperature from its summertime design point to room temperature plus sufficient energy to offset winter heat loss.

— continued

D. Fan Static Pressure Calculation

Fan static pressure should always be calculated, rather than relying on "rules of thumb."

2.3 Airflow Diagrams

Airflow diagrams should show the air handling unit components arranged in their proper sequence, flow measuring stations, reheat coils, location of each level of filtration, humidifiers, exhaust, and any other system components. Airflow diagrams should show supply, return, exhaust, infiltrations, and exfiltration airflows from each room. Obtaining client approval of airflow diagrams is required prior to starting ductwork layout.

2.4 HVAC P&IDs

P&IDs should be developed for each individual air handling system. The HVAC discipline should generate P&IDs for chilled water and hot water. Each of these systems should be independent from similar process-related systems and should be the responsibility of HVAC engineers, not process engineers. The HVAC department may also be required to produce P&IDs for other general utilities.

2.5 HVAC Ductwork and Equipment Arrangement Drawings

Detailed HVAC drawings are an absolute necessity for a facility with the complexity common to pharmaceutical facilities. Single line drawings are unacceptable. All ductwork should be drawn as double-lined, fully dimensioned duct (including centerline elevation for round ducts and bottom of duct elevation for rectangular ducts). The location should be shown from the column lines. All ductwork should be coordinated with piping and electrical disciplines. Density may require multiple levels of duct drawings. Sections are required at any points where the plans do not completely and clearly define the design. A meeting must be convened to preplan special requirements for duct/pipe/electrical prior to commencing any design. Ductwork plans must clearly indicate duct material, insulation type, and pressure classification using duct construction tags.

When sizing return air ducts, keep in mind that calculated room leakage rates are empirical and will vary depending on construction tolerances; in turn, R.A. quantities may vary from theoretical design. Return ducts should be sized to handle at least 25 percent more than design airflow. Return air plenum designs should be avoided unless there is no potential for contamination problems. The velocity in Class 100 R.A. walls should be <600 FPM.

2.6 Control Diagrams

Control diagrams should be drawn in P&ID format and should show all coil piping, including all valves, line numbers, and so on. A separate P&ID should be developed for each air-handling unit and should show all control devices individually. Control diagrams should be supplemented with sequences of operation either on the drawing or as part of the specifications. Control diagrams should include a control valve schedule that supplies the size and specifications for each valve.

2.7 Air Handling Units (AHUs)

A. Insulation

Air handling units should not contain internal insulation exposed to the airstream. Double wall construction sandwiching insulation between two metal panels or single wall construction with external insulation is a requirement.

B. Filters

Air handling units should be provided with prefilters, two inches thick at minimum, Farr 30/30s or equal, and 80–85 percent efficient filters. If HEPA filters are located within the air handling unit, they should be the last component in the direction of the airflow into the room.

C. Fans

Air handling unit fans should contain a provision for modulating airflow to compensate for increased static pressure losses as the loading of filters increases. This is usually accomplished through the use of inlet vanes; other methods would be discharge dampers and variable speed drives.

Fan motors should be sized at 25 percent above the brake horsepower. Fans should be belt driven with variable pitch type drives up to and including 25 HP and fixed pitch drives at about 25 HP.

D. Coils

Air handling unit coils should have no more than 8 fins per inch and be no more than 6 rows in depth to facilitate coil cleaning. Where more than 4 rows are required to obtain the cooling capacity, 2 coils should be placed in series. Spacing between the coils should be a minimum of 24 inches to permit cleaning of the coil face. All coils should be piped so that counterflow of air and water is achieved. Fins should be a continuous, flat (noncorrugated) type. Note: On large air handling unit systems, consideration should be given to bypassing the cooling coil with part of the return air to minimize the amount of reheat required. With coil bypass, pretreating the outside air for dehumidification may be required.

E. Notes

The HVAC specifications should state that before the activation of any air handling unit, all construction debris should be cleared away and the unit thoroughly vacuumed. Units should be vacuumed again before the installation of cartridge/bag filters. Prefilters should be installed before initial unit start-up. Cartridge/bag filters should be stored in a clean/dry place. Cartridge/bag filters should be installed after room finishes are complete. If HEPA filters are contained within the air handler, these should also be installed at this time. Comprehensive cleaning guidelines for duct and equipment must be provided as part of the construction specifications.

2.8 Reheat Coils

Reheat coils should be electric or hot water. Use of steam does not allow for stable control. Electric reheat coils should be multistage or SCR controlled. Maximum fin spacing on reheat coils should not exceed eight fins per inch. Reheat coils rarely require more than two rows. Small reheat coils should not be sized for a large water-sided temperature difference, which could result in lower water flow rate, and, at part load conditions, would result in control stability problems. Reheat water should be operated at as low a temperature as is reasonable to improve controllability.

2.9 Ductwork

A. Materials

Ductwork may be constructed of lock-forming quality aluminum, galvanized steel, or stainless steel. The use of galvanized steel ductwork may be controversial because of the potential for flaking of the galvanized coating within the duct stream. This potential is minimal and current FDA opinion allows for the use of galvanized steel. A similar problem can exist with aluminum duct because it can oxidize and release particles into the airstream. Use of stainless steel eliminates these potential problems; however, its use is generally cost prohibitive and benefits are generally judged to be minimal.

Where space allows, round or oval ductwork is preferred to rectangular ductwork to reduce leakage and pressure drop and to facilitate cleaning. Provide adequate access doors for cleaning and inspection.

B. Specifications

Ductwork specifications should be comprehensive and detailed; simple reference to SMACNA (Sheet Metal Associated Contractors of North America) tables affords the contractor excessive latitude. Specific tables should be generated for each pressure classification that will be encountered, and the pressure class for each duct system should be defined clearly on the design drawings.

C. Flexible Ductwork

Round, flexible duct should contain an aluminum foil liner in lieu of a vinyl liner. The helix of flexible ductwork should be formed on the outside of the duct's surface to minimize crevices that could cause contamination. All flexible ductwork should comply with NFPA (National Fire Protection Association) 90A and 90B. Flexible ductwork should not penetrate any wall or exceed a maximum length of 5 feet. Connections to flexible ductwork from sheet metal work should be made using conical taps wherever possible. Straight spin-ins with or without scoops should not be permitted.

D. Elbows

Wherever possible, elbows should be of the radius type and without turning vanes. Where necessary, use square mitered elbows and equip with single-thickness, extended-edge, turning vanes. Double-thickness turning vanes should not be allowed because they have internal cavities that could become problematic (i.e., cleanliness problem). All vaned elbows should have an access door located nearby to provide for cleaning.

E. Dampers

Manual balancing dampers should be installed where required and should be shown clearly on the design drawings. Balancing dampers should be provided with a self-locking regulator suitable for securing the damper at the desired setting and making future adjustments as required.

F. Access Doors

Access doors should be located at all balancing dampers to provide for cleaning the damper surface. Access doors should also be located at reheat coils and any other surface that could collect material within the duct. Access doors should be clearly shown on the drawings and should be large enough to allow a person to clean obstructions from the interior of the duct. Access doors should be provided with extension collars and sash locks. The use of hinged access doors is discouraged; leakage rates around the hinges are too high to be considered acceptable, and there is no means for adjusting the door at the hinge to provide a positive seal.

G. Hangers and Supports

Hangers and supports for ductwork within concealed spaces may be the same as those used in commercial and industrial applications. Exposed ductwork within process areas should be provided with smooth rod hangers threaded only at the end to allow for a connection.

H. Sealing

All ductwork should be sealed in accordance with SMACNA Class A rating, which requires all seams, joints, fasteners, penetrations, and connections to be sealed. Sealant should be FDA acceptable for the application, and nonhydrocarbon based. Leakage rates as low as 1 percent total airflow are not uncommon. All ducts passing through a clean room wall or floor should be provided with stainless steel sheet metal collars and sealed at the opening. Details of sealing methods should be provided on the design documents.

I. Leak Testing

Before ductwork is insulated, the installing contractor should leak test each duct system at 125 percent of the operating pressure. Leak testing should be witnessed and signed off to signify approval. Acceptable leakage rates and leak testing procedures and reports should be based upon the *SMACNA HVAC Duct Leakage Test Manual*, 1st ed. 1985.

2.10 Insulation for HVAC Ductwork

All insulation should be in accordance with the "flame-spread" and "smoke-develop" ratings of NFPA Standard 255. Ductwork should be externally insulated. The use of internal duct liner is not acceptable. Some criteria will not allow the use of fiberglass insulation even on the exterior of the ductwork. Whenever ductwork is exposed in clean spaces, it should be insulated with a rigid board-type insulation and jacketed with either a washable metallic or PVC-coated jacket. Jacketed ductwork should be of sufficient density to minimize the dimpling effect when the jacket is applied. Flexible blanket insulation is acceptable in concealed spaces. No insulation should be applied on the ductwork until the leak test has been performed, witnessed, and approved. Rigid duct insulation should also be used in mechanical rooms.

— continued

2.11 Sound Attenuators
Sound attenuators should not be used in systems requiring sanitizing because the perforated face (interior) of the sound traps can collect dust and microorganisms.

2.12 Humidifier

Humidifiers serving clean areas/process areas should use only clean steam for humidification. Carbon steel piping and headers are not acceptable; 316L grade stainless steel should be used. All humidifier components (main body, valves, piping, manifold, etc.) should be made of 316L stainless steel.

2.13 Air Distribution

A. Air Supply
Standard type (turbulent flow) diffusers should be used in Class 100,000 areas. Terminal HEPA filters should be used in the ceilings of the areas that are Class 10,000 or cleaner.

B. Return/Exhaust Air
Ceiling return/exhaust is acceptable in Class 100,000 areas. Low wall returns should be used in Class 10,000 or cleaner areas. All returns/exhausts must be a louvered, removable types.

Specifications should also clearly define acceptance criteria and the basis for rejecting unsatisfactory work. They must clearly address attributes such as pressure tests, weld finishes, surface smoothness, temperature ranges, response times, chemical resistance, and pressure differentials.

Specification language must also clearly address GMP attributes such as cleanability of finished surfaces, drainability of systems, and maintenance of system cleanliness. For a sanitary process piping system, this could read as follows:

- "All openings in tubing, fittings, components, or equipment shall be covered at all times to keep foreign particles out of the system. A plastic cap should be used for temporary protection. Rags or paper bags should not be used due to the potential for depositing lint or material in the opening."
- "Slope requirements of 1/8 inch/foot minimum are required."

Detail

Quality design requires drawing details that are specific, precise, and informative. Details should never be open to interpretation, especially in classified spaces and manufacturing areas. The philosophy of good detailing should apply to all areas, systems, and disciplines. The execution of good detailing will bring the old adage, "a picture is worth a thousand words," to life by giving contractors a clear understanding of what is expected in the fabrication and/or installation of a represented facility component.

Example 10.2 cGMP Conformance HVAC Design Checklist

Document Number: ____________ Revision: ____________

Item **Comments**

General

1. Verify that formal design criteria have been developed and approved by the client and that the document being checked conforms in all respects to the approved design criteria.

Design Guidelines

1. Air intakes are located as far away as possible from any upwind air exhaust.
2. Ductwork materials are in accordance with design criteria.
3. Access doors are provided at all in-line devices requiring access (coils, vanes, etc.)
4. Outside intake and exhaust points are filtered with bird screens.
5. Duct hangers and supports, in clean areas, are designed to prevent dust collection.
6. Low-level return air ducts are located in room walls for all Class 10,000 or cleaner areas.
7. Design conforms with FDA Guidelines on Sterile Products Produced by Aseptic Processing, where required.
8. Design conforms to Federal Standard 209E.
9. Outside design criteria is from ASHRAE or designated by Client.
10. Inside design criteria include, but are not limited to, temperature, relative humidity, filtration level, minimum air change rate, and pressurization requirements.
11. Temperature and humidity are listed as design set points with plus and minus tolerances.
12. Filtration addresses final filter efficiency and location.
13. Pressurization is noted in the form of actual room pressure levels from a common reference point. Reference point is not outdoors (affected by wind).
14. Directional airflow based on pressure differential is noted along with special exhaust requirements.
15. Locations of laminar flow hoods, biosafety hoods, and exhaust hoods are noted.
16. Cleanroom classification is noted on design documents only if room is to be validated as a clean room.
17. Process equipment heat loads have been established and reviewed by the Owner.

— *continued*

Example 10.2 (continued) cGMP Conformance HVAC Design Checklist

Calculations

1. The calculation for the airflow leakage rate for each room is based on the pressure differential established on the design criteria sheet and not on a percentage of supply air. ________________
2. Cooling load calculations have been performed on a room-by-room basis. Considerations include motor load, radiation from heated vessels, and radiation from thinly insulated process and utility piping and equipment, electrical loads, AHU supply fan motors for air handling units. ________________
3. Heating calculations include heat required to raise supply air temperature from summer design point to room temperature plus sufficient heat to offset winter heat loss. ________________
4. Fan static pressure is calculated. ________________

Airflow Diagrams

1. Airflow diagrams show for air handling units components, in proper sequence, flow measuring stations, reheat coils, location of each filtration level, humidifiers, exhaust fans, and other system components. ________________
2. Supply, return, exhaust, infiltration, and exfiltration airflows from each room are shown on airflow diagrams. ________________

HVAC P&IDs

1. HVAC P&IDs, chilled water and hot water P&IDs are generated independently of process-related systems and have been checked using a P&ID checklist. ________________

HVAC Ductwork and Equipment Arrangement Drawings

1. HVAC drawings are detailed. ________________
2. Ductwork is drawn as double-lined and fully dimensioned duct (including center-line elevation for round duct and bottom-of-duct elevation for rectangular duct). ________________
3. Ductwork is coordinated with other disciplines, especially piping and electrical. ________________
4. Sections are provided at any design point not clearly defined. ________________
5. Duct centerline locations are shown from column lines. ________________
6. All ducts identify material of construction, insulation type, and pressure classification. ________________

— continued

Example 10.2 (continued) cGMP Conformance HVAC Design Checklist

Control P&IDs

1. Control diagrams show coil piping, including valves and line numbers. ______
2. Separate control diagrams have been generated for each air handling unit, showing all control devices individually. ______
3. Control diagrams have been supplemented with sequences of operation and a control valve schedule showing the size and specification for each valve. ______

Air Handling Units

1. Insulation ______
   a. Insulation has not been placed on the inside of the air supply ducts. ______
   b. Double-wall construction sandwiching insulation between two metal panels or single-wall construction with external insulation is used. ______
2. Filters ______

Air handling unit has been provided with prefilters two inches thick at a minimum and medium efficiency cartridge/bag filters. ______

3. Fans
   a. Air handling unit fans are able to modulate airflow by using inlet vanes, discharge dampers, or variable speed drives ______
   b. Fan motors are sized 25 percent above brake horsepower. ______
   c. Fans are belt drive with variable-pitch drives for 25 HP or less and fixed-pitch drives for drives above 25 HP. ______
4. Coils
   a. Coils are properly oriented. ______
   b. Air handling unit coils have no more than 8 fins per inch and are no more than 6 rows in depth. ______
   c. If cooling capacity requires more than 6 rows, 2 coils are to be placed in series. ______
   d. Coils are spaced a minimum of 24 inches apart. ______
   e. Coils are piped to achieve air/water counterflow. ______
   f. Fins are continuous and flat (noncorrugated). ______

Reheat Coils

1. Reheat coils are electric (multistage or SCR controlled or hot water). ______
2. Maximum fin spacing on reheat coils does not exceed eight fins per inch (more than two rows are rarely required). ______
3. Smaller reheat coils are not sized for a large water-side temperature difference. ______

— continued

Example 10.2 (continued) cGMP Conformance HVAC Design Checklist

Ductwork

1. Materials

Ductwork constructed of lock-forming quality aluminum, galvanized steel, or stainless steel. ______

2. Specifications

Specific tables are generated for each pressure classification and each duct system pressure class is clearly defined. ______

3.Flexible Ductwork

a. Round, flexible duct contains aluminum foil liner in lieu of a vinyl liner. ______
b. The helix of flexible ductwork has formed on the outside of duct's surface. ______
c. Flexible ductwork complies with NFPA 90A and 90B. ______

4. Elbows

a. Either elbows of the radius type and without turning vanes are used or square mitered elbows equipped with single-thickness, extended-edge turning vanes. ______
b. All vaned elbows have an excess door located nearby for cleaning the vanes. ______

5. Dampers

a.Manual balancing dampers are installed where required and shown clearly on design drawings. ______
b. Balancing dampers have a self-locking regulator suitable for securing the damper at the desired setting and making adjustments. ______

6. Access Doors

a. Access doors are located at reheat coils and any other surface that could collect material within the duct. ______
b. Access doors are clearly indicated on the drawings. ______
c. Access doors are large enough for a person to clean obstructions from the interior of the duct and are provided with extension collars and sash locks. ______
d. Hinged access doors have not been used. ______

7. Hangers and Supports

Exposed ductwork has smooth rod hangers threaded only at the end. ______

Insulation for HVAC Ductwork ______

1. No internal duct insulation has been used. ______
2. Exposed ductwork in clean spaces is insulated with rigid board-type insulation and jacketed with a washable metallic or PVC jacket. ______
3. Jacketed ductwork is dense enough to minimize dimpling. ______
4. Rigid duct insulation is used in mechanical rooms. ______

— continued

Example 10.2 (continued) cGMP Conformance HVAC Design Checklist

Sound Attenuators

1. Sound attenuators are not used in systems requiring periodic sanitizing. ____________

Humidifiers

1. Humidifiers use clean steam for humidification. ____________
2. Stainless steel (316L grade) has been used for piping, headers, and all humidifier components. ____________

Air Distribution

1. Supply Air Class 10,000 or cleaner areas have terminal HEPA filters in the ceiling. ____________
2. Return/Exhaust Air ____________
   a. Class 100,000 areas have ceiling returns. ____________
   b. Class 10,000 or cleaner areas have low wall returns. ____________
   c. All returns/exhaust are the louvered, removable core type. ____________

Drawings or Document Number: ____________
HVAC Design Engineer: ____________
HVAC DPE: ____________
Date: ____________

Examples of where good drawing details are paramount include the following:

- **Sanitary process piping** Isometric drawings (Figure 10.1) must clearly define valve orientations, fitting-to-fitting dimensions, supports, line slopes, tie-in locations, and elevations.
- **HVAC systems** Duct routings must be verified for interferences with structural steel or architectural components. The location of returns must provide proper air movement. Installation details for HEPA filters must show proper attachment and sealing details.
- **Architectural finishes** Details require clear representations of transitions, sealing requirements, dimensional "stack up," and materials of construction.

Inspection Philosophy

The criteria found in 21 CFR Part 211.42 basically address the end results that must be achieved through the design and construction process. This section states that "Any building or buildings used in the manufacture, processing, packaging, or holding of a drug product shall be of suitable size, construction, and location to facilitate cleaning, maintenance, and proper operations." [2] But the FDA does not specify how these end results are to be achieved.

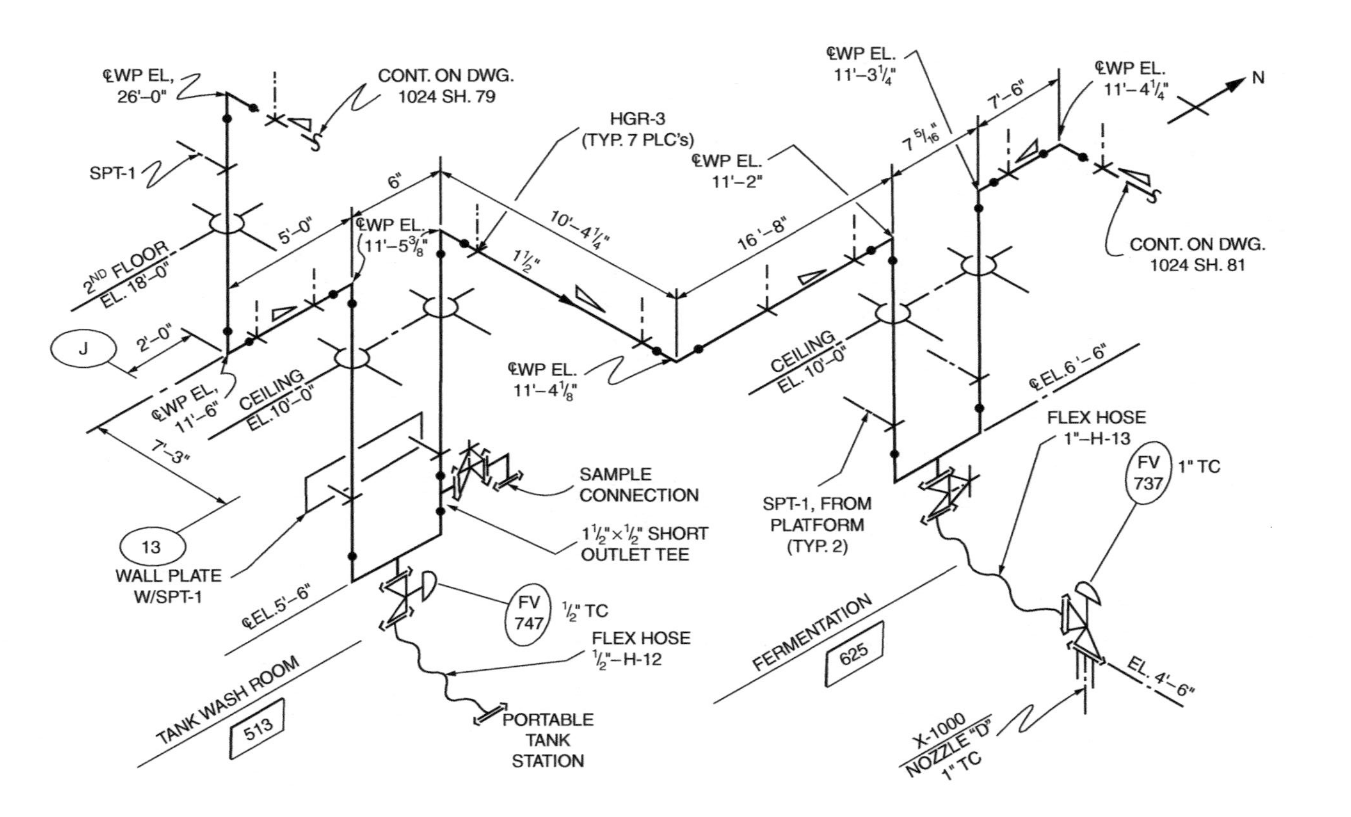

Figure 10.1 Isometric drawing for sanitary process piping.

Therefore, it is important for Owners and designers to address the overall philosophy that they will implement for inspections and to specify how they will apply it to each discipline, system, or area. These decisions will ultimately impact project cost, construction execution, and validation execution.

Standard industry practice and previous project experiences play an important role. The amount of inspection required varies from organization to organization. Table 7.3 (see page 200) shows various inspection approaches and their potential consequences. These approaches can apply to all areas and disciplines, for activities involving sanitary process piping, welding, passivation, HVAC system testing, instrumentation checkout and calibration, Programmable logic controller (PLC) loop checks, concrete inspection, and soils testing.

Determining the right approach requires consideration of process requirements, product requirements, and regulatory issues. The selected approach must be carried out by trained professionals experienced in the particular type of inspection program to be implemented. Proof of this experience should be a part of the IQ documentation.

Construction quality

Quality in construction anticipates the need for high-quality workmanship and installation, sound inspection and testing practices, and documented results. It begins early in the life of the project, from the development of contracts and the procurement of materials, and continues through final turnover and commissioning of the facility. It requires a sound program of management and a commitment to excellence from all project participants.

Program implementation

What makes a successful construction quality program? Focus and commitment: a focus on the attributes that are critical to the success of the project and a commitment from every contractor to implement the necessary controls to assure that every job is done right the first time.

The goal of a construction quality program is to prevent errors—errors that cause project costs to increase due to the time required to track them and the time required to correct them. The program must focus on conformance with design requirements and provide the documented proof that the facility has been constructed to meet these requirements. The language of the program must be included as a part of all trade contracts and bid documents so that uncertainty can be eliminated early when defining the scope of a contractor's work.

Project management must provide support for the quality effort. There must be action behind the words. Empowerment to take the necessary actions regarding the program's implementation should never be questioned. This holds true for all project participants.

Program attributes

The following discussion focuses on some important attributes of a construction quality program for a biotech facility. The attributes of the program would be applicable to any sterile product facility.

GMP walk-downs

While there is a tremendous focus on GMP compliance during the execution of project design, many times that fervor is not carried into the field. The assumption that "if it is built per the drawings, it will comply with GMP" may not always hold true. It is good practice to conduct GMP walk-downs of systems and rooms as work nears completion. Conditions that would cause operational or maintenance problems may be found. These walk-downs also give operations personnel the opportunity to see their work areas in advance.

Requirements meeting

Prior to the mobilization of any contractor to the job site, a meeting should be held to review the quality plan for that scope of work. Procedures for the notification of tests, the witness of inspections, the submittal of documentation, and the coordination of work should be discussed. Designated contacts for questions and/or issue resolution should be identified in writing so that there is no misunderstanding of key personnel.

Benchmarking

Certain types of construction activities will require the definition of acceptable standards of workmanship above what is written in a specification or on a drawing. Examples of such activities would be the application finish on epoxy flooring, the installation and sealing of insulation and jackets for sanitary piping drops, the weld finish on stainless steel work platforms in clean rooms, or the weld finish on sanitary process piping systems.

Inspection teams comprised of representatives from the Owner, construction, and design teams will review and comment on workmanship samples provided by the executing contractor. Acceptable results are documented and establish the basis for judging all future work of a similar type.

Mock-up

There are many examples in a sterile product facility where the use of mock-ups can eliminate confusion of drawing interpretation, eliminate congestion, and provide for better operational and maintenance access. Some examples would include the fabrication/installation of transfer panels, the layout of sampling stations, the layout/fabrication of utility stations, the orientation of filter banks, floor drain closure in clean rooms, or the sealing of wall penetrations into clean rooms.

Again, an inspection team will review the mock-up condition and determine acceptance for the project. All future work of the same type would then be judged against that first mock-up.

Closure inspection

Areas that will become permanently sealed or inaccessible after construction should be inspected and photographed for future reference. These areas include below-grade systems that involve leak detection for containment purposes, wall sections prior to final closure, or above ceiling areas not accessible by access doors.

Punchlist

Most biotech and pharmaceutical projects are executed based on an effort to complete the majority of the work focusing on a building or area work schedule, with a shift of emphasis near the end of the project to a system focus. In this scenario, system walk-downs are performed at approximately the 85–90 percent completion level, and incomplete items go onto a punchlist for completion. Sometimes, these punchlists can contain hundreds of items that must be addressed and completed before start-up and commissioning.

One way to avoid this scenario is to maintain a list of work items that have been noted as incomplete, missing, or not meeting design requirements from early in the project. Through regularly scheduled walk-downs of systems, teams can identify where work is incomplete at a time when correction can prove more efficient and cost-effective. The sooner the focus on a system can begin, the easier turnover will be.

Quality in work execution

The easiest way to discuss quality issues in the construction of sterile product facilities is to review some actual results of poor work execution, where quality programs were not implemented early and followed during field execution. Examples are catagorized by discipline.

Architectural

- Hollow metal door frames were purchased without proper sealing of head and jambs. Air leaked through openings, placing a classified area in noncompliance.
- Access doors in hard ceilings did not have gasketed doors to prevent air leakage.
- Interior viewing window frames did not have sloped sill closure pieces, creating an area to collect dust and debris.

- Exposed anchor bolt threads at the base plate of structural platform members in a clean room required cover plates to keep dirt from threads.
- Improper treatment of cove application for epoxy flooring led to cracks at cove base.
- Improper sealing of exposed GWB surfaces allowed fluids to contact the material, providing an area for mildew growth.
- Improper priming of platform surfaces led to peeling paint above reactor; paint chips found in nonsterile raw material.
- Interior wall width did not allow room for specialized process drain plumbing inside the wall.
- Floor slope requirements were not maintained during epoxy flooring installation, creating pockets where water would not flow to drain.
- Architectural details did not account for dimensional differences between dissimilar materials at transitions; rework required to remove voids and ledges caused by nonuniformity of materials.

Sanitary piping

- No markings on installed material; material found to be of incorrect type, requiring removal.
- Bent tube from field bending used in lieu of fittings; minimum wall thickness not met for that particular tube diameter. Bends had to be cut out and replaced.
- Contaminated welding gas used in welding operation; welds had to be cut out and rewelded.
- Contractor did not follow procedures in maintaining daily weld inspection records. Search of records found incomplete weld data; welds had to be removed and reworked.
- Dimensions on isometric drawings not field verified prior to fabrication of spools. Fitting dimensions used as basis for design were not manufactured by the same vendor as what was purchased. Equipment connections were incorrect.
- Welds found with no weld numbers; welds cut out.
- Fabrication made to incorrect drawing revision.
- Metal chips from deburring found inside process line.
- Lines installed with slope in wrong direction.
- Improperly supported lines sagged during functional checks.
- Improper gasket installation in tri-clamp caused leakage.
- Valve orientation incorrect for self-draining application.
- Final orientation of vessel to process line would not allow for removal of vent filter; housing was installed in a binded condition between tri-clamps.

- Boroscopic inspection of weld not performed at proper time. When realized, a blind weld condition was created due to installed position of line.

HVAC

- Holes field drilled into HEPA filter housing for installation caused air leakage.
- Area between HEPA filter housing and hard ceiling sealed with non-compatible caulking material.
- Soft drink can found inside duct.
- Temporary filters not installed at diffusers during system blowdown. Dust and dirt from duct blown into process suite.
- Galvanized duct material installed in stainless steel system.
- Leakage test documentation not maintained for installed sections of system; entire system required retesting.

Electrical

- Improperly installed light fixture lenses allowed air leakage through fixture.
- Exposed conduit in process area renovation did not have watertight bulkheading.
- Electrical boxes purchased with improper surface finish; did not use epoxy paint.
- Outlets in process area were not watertight.
- Boxes installed with knockout plugs instead of solid boxes; air leakage through knockouts.

Purchase orders

- Vendor equipment drawings issued as standards, not certified vendor data; actual field location of nozzles not reflected. Piping design had to be revised.
- Purchase order for sanitary process tubing allowed unpolished fittings for certain material classifications. Reorder of rejected material caused schedule delays.
- Overseas equipment ordered and received without UL rating of electrical panels. Panels had to be reworked for compliance before city inspector would allow issue of occupancy permit.
- Purchase orders for equipment manufactured overseas did not have utility service checklist; required services not available at site.
- Operation/maintenance manuals issued with metric vs. English units.

Lessons learned

An unknown project manager once said, "If you make a mistake once, it is learning. If you make that same mistake twice, it is stupidity." In the haste to complete a project, many project teams fail to take advantage of a valuable opportunity to improve quality on future projects. While members of the project team are still familiar with all that has occurred, it is good practice to plan a "Lessons Learned" session to review the good and not-so-good aspects of the project, with the goal of learning from your mistakes.

The focus of a "Lessons Learned" session is not to lay blame or to point fingers; it is to assess the team's performance in meeting the goals of the project and see where improvements could be made. In these sessions, discussions of what worked well should go hand-in-hand with issues that needed improvement.

By documenting all of the items that are addressed and the solutions offered as improvements for future reference, companies can create a valuable database of information that will benefit future projects. The following are excerpts from actual sessions that will give some examples of merit.

- **Staffing of the project** Due to accelerated schedules and a desire from the client to keep costs down early in the project, the staffing plan never met what would be considered an acceptable level of size or qualification. It is important that individuals assigned to a sterile facility manufacturing project have the necessary experience to deal with the review of design drawings, the resolution of field questions, and the support of validation activities. There must also be recognition of the fact that adequate review of preliminary design documents can save tremendous amounts of time and money by catching issues early. But to do so you must have qualified personnel available early.
- **The amount of time and money spent on the project due to construction cleanliness, or lack of it** More time should be spent with the trade contractors in reviewing cleanliness requirements for a drug manufacturing facility project. A formal education program for all contractors was recommended for future projects.
- **The status of design documents that were sent out for bids for mechanical systems** A better job of coordination between the A/E firm and the construction manager must occur if risk is to be minimized. The A/E firm must understand the building philosophy and contracting plan of the construction manager early so that design packages can reflect the bidding plan and sequence of need based on the construction schedule.
- **Problems with the submittal of information required for IQ** The difficulties experienced in contractor submittals were traced to a late start in defining what documents would be required and how they would be submitted. There was no standardization of forms; very specific time frames established for the submittal of construction-gen-

erated test documentation did not exist. The validation/commissioning group should begin early development of documentation requirements for contract bid packages that define the numbers of documents, type of forms, submittal time frame, type of review, and required approval.

References

1. *American Heritage Dictionary,* 2nd college edition. 1985. Boston: Houghton Mifflin Company.
2. *Code of Federal Regulations,* Title 21, Part 211.42. 1990. Washington, DC: U.S. Government Printing Office.

chapter 11

Establishment licensing

For facilities that will ultimately produce biological products, licensing the facility is a complicated process that has changed in recent years. In 1996, CBER moved away from the dual licensing model that utilized the Establishment Licensing Application (ELA) and the Product Licensing Application (PLA). The use of the ELA/PLA model was ended as of October 20, 2000. Now, a single licensing model is used, as defined in the Biological Licensing Application (BLA). This shift was implemented to allow CBER and CDER to harmonize their respective application formats under the single form 356h (Figure 11.1).

This change required some revision to established regulations and also led to the development of a series of Chemistry, Manufacturing, and Control (CMC) guidelines. These guidelines address format and content issues related to the new BLA. There are a number of CMC guidance documents for separate product classes. They include:

- Content and Format of Chemistry, Manufacturing, and Controls Information and Establishment Description for a Vaccine or Related Product
- Guidance for Industry: For the Submission of Chemistry, Manufacturing, and Controls and Establishment Description Information for Human Plasma-Derived Biological Products, Animal Plasma, or Serum-Derived Products
- Guidance for Industry on the Content and Format of Chemistry, Manufacturing, and Controls Information and Establishment Description Information for an Allergic Extract or Allergen Patch Test
- Guidance for the Submission of Chemistry, Manufacturing, and Controls Information and Establishment Description for Autologous Somatic Cell Therapy Products*

* Food and Drug Administration; CBER and CDER

DEPARTMENT OF HEALTH AND HUMAN SERVICES
PUBLIC HEALTH SERVICE
FOOD AND DRUG ADMINISTRATION

APPLICATION FOR ESTABLISHMENT LICENSE FOR MANUFACTURE OF BIOLOGICAL PRODUCTS

Form Approved: OMB No. 0910-0124.
Expiration Date: April 30, 1997.
See OMB Statement on last page.

DATE SUBMITTED:

NOTE: This report is mandated by Section 351 of the Public Health Service Act, the Federal Food, Drug, and Cosmetic Act, Section 502 and Title 21 CFR Part 600. No license may be granted unless this completed application form has been received.

GENERAL INSTRUCTIONS

Type or print legibly in ink. Complete all items. Items which are not applicable, enter "NA". If more space is needed for any item continue on an 8½x11 inch sheet, reference the entry by item number, and attach. Allow 1 inch top margin for filing purposes. Submit the original and yellow copies of the completed application. Assemble and staple each set, including all attachments. The application forms must be dated and signed by the Responsible Head. Return the application to DHHS/PHS, FDA/Director, Center for Biologics Evaluation and Research (HFM-1), 1401 Rockville Pike, Rockville, MD 20852-1448.

I. GENERAL INFORMATION

A. NAME AND ADDRESS OF MANUFACTURER FOR WHICH U.S. LICENSE IS BEING MADE INCLUDE A COPY OF THE CERTIFICATE OF INCORPORATION *(Without financial details)*.

CHECK ONE:

☐ New Application

☐ Revised Application

TELEPHONE NUMBER

( )

B. NAME, ADDRESS AND REGISTRATION NUMBER OF EACH LOCATION OF THE ESTABLISHMENT WHERE ACTUAL MANUFACTURING, INCLUDING TESTING, LABELING AND STORAGE, TAKES PLACE. IDENTIFY THE MANUFACTURING PERFORMED AT EACH LOCATION. IF THIS IS A SHARED OR CONTRACT MANUFACTURING ARRANGEMENT DESCRIBE THE FUNCTIONS PERFORMED AT EACH LOCATION.

C. NAME, ADDRESS AND PHONE NUMBER OF RESPONSIBLE INDIVIDUAL *(Responsible Head 21 CFR 600.10)* TO WHOM ALL OFFICIAL CORRESPONDENCE SHOULD BE DIRECTED AND WHO IS AUTHORIZED TO DISCUSS THIS APPLICATION.

D. NAME AND ADDRESS OF OTHER RESPONSIBLE OFFICIAL OR AGENT TO WHOM COPIES OF CORRESPONDENCE SHOULD BE DIRECTED.

E. NAME OF INDIVIDUAL(S) DELEGATED RESPONSIBILITY FOR RELEASE PROTOCOL SIGN-OFF BY THE RESPONSIBLE HEAD *(If it is not the Responsible Head)*.

F. A BRIEF DESCRIPTION OF THE PRODUCT OR PRODUCTS COVERED BY THIS APPLICATION. PLEASE INCLUDE A MANUFACTURING FLOW DIAGRAM INDICATING LOCATIONS WITHIN THE FACILITY WHERE EACH MANUFACTURING STEP IS PERFORMED.

FORM FDA 3210 (5/94) PAGE 1 of 18 PAGES

Figure 11.1a Application for establishment license for the manufacture of biological products (18-page form).

I.	GENERAL INFORMATION *(Cont'd)*

G. GENERAL OVERVIEW OF ALL MANUFACTURING LOCATIONS INCLUDING:

1. SITE PLAN
2. INDICATE IF THIS IS A MULTI PRODUCT FACILITY AND DESCRIBE ALL PRODUCTS *(Drugs, biologics, and medical devices or any combination)* MADE AT EACH SITE
3. OTHER INDUSTRY IN THE AREA
4. DISTANCE FROM FARM ANIMALS
5. SQUARE FOOTAGE
6. PEST CONTROL

H. PROVIDE FLOW DIAGRAMS FOR THE MOVEMENT OF RAW MATERIALS, PRODUCT *(Including in-process product)*, PERSONNEL, EQUIPMENT AND WASTE WITHIN THE FACILITY AND BETWEEN LOCATIONS *(If possible)*.

II.	WATER SYSTEM*(S)*

A. WHAT TYPES OF WATER SYSTEMS ARE PRESENT IN THE FACILITY? INCLUDE INFORMATION ON THE QUALITY OF INCOMING SOURCE WATER, CONNECTION TO THE SEWER AND BACKFLOW PREVENTION FEATURES.

B. WHAT TYPES OF WATER ARE USED IN PRODUCT MANUFACTURE AND SUPPORT AREAS OF THE FACILITY?

C. DESCRIBE THE PRETREATMENT SYSTEM*(S)*.

D. DESCRIBE THE SYSTEM*(S)* USED TO PRODUCE PURIFIED WATER.

FORM FDA 3210 (5/94) PAGE 2 of 18 PAGES

Figure 11.1b Application for establishment license for the manufacture of biological products (18-page form).

What is the BLA?

The BLA form identifies the required sections of the application. It also provides a general description of the data that will be required for submission. Included in Section 15 is the Establishment Description, per 21 CFR, Part 600. In this section, the term *establishment* refers to the facility where the drug product will be produced. Sections 1–20 on the BLA form constitute a checklist of the required information for the submission.

The information contained in Section 15 includes a description of the physical establishment, equipment, and animal care facilities. This information includes

- A definition of location, including all buildings, appurtenances, equipment, and animals used, and personnel engaged, by a manufacturer within a particular area designated by an address adequate for identification
- A definition of process referring to a manufacturing step that is performed on the product itself, which may affect its safety, purity, or potency
- A description of work areas
- A description and listing of equipment*

Establishment information

The information that should be contained in the BLA will provide investigators with a road map concerning the design and construction philosophy taken by the project team during execution of the project. This information can easily fill volumes, containing thousands of pages vital to the licensing process.

The BLA submission must include documents in a specific folder or section, including the following as applicable:

- Water systems
- HVAC systems
- Contamination and cross-contamination
- Formulation and filling
- Computer systems

In addition, floor plans and flow diagrams must be provided.

* 21 CFR, Part 600.11

General information

Information regarding the name and location of the manufacturer must be provided. If manufacturing operations for a given product will occur at more than one location, similar information for every location must also be provided.

The type of manufacturing that will occur at the defined locations must be provided. This would include operations for testing, production, filling, labeling, or storage if these occur at one or more locations.

This section also provides the name and address of the individual who is deemed the responsible head or the official to whom correspondence should be directed regarding all licensing and compliance activities. In most cases, this individual is the plant manager.

Buildings and facilities

One of the most important elements of the establishment information is the detailed description of the facility's areas that are directly involved in the manufacturing process. In this section, the design and construction features of each manufacturing area are described in detail for the reviewers.

Architectural treatments for surface finishes should be described for all floors, walls, and ceilings, as illustrated in Example 11.1. Important information regarding the design of utility systems, such as HVAC and water systems, should include a description of the system (Example 11.2) and a general diagram outlining the system design (Figure 11.2).

The flow of personnel, materials and product, and equipment through the facility is another important piece of information that must be included. Descriptions of flows accompanied by diagrams (Figures 11.3, 11.4, and 11.5) showing the interrelationships of areas within the manufacturing space should be clear and easily legible, whether in single or multiple diagrams.

There should also be a description of the various area classifications defining cleanroom areas. Diagrams that show classification boundaries, as in Figure 11.6, provide an easy means of visual interpretation.

Example 11.1 Architectural Finishes

The facility will house controlled class 100,000 and class 10,000 process and support areas. Fit and finish in these rooms will include cleanable/durable surfaces, coved floors, seamless walls and ceilings, and sealed penetrations.

Finishes will include epoxy painted gypsum wallboard walls and ceilings. High traffic areas will be protected with sheet vinyl wallcoverings as required. Floors will be chemical resistant epoxy with coved base of same material as floor. Door and window frames will be of hollow-metal construction with epoxy painted finish. Doors will also be of hollow-metal construction with an epoxy painted finish.

Example 11.2 Water for Injection System (WFI)

System Description: Hot WFI is produced in a multi-effect WFI still and drains by gravity into a 5,000 liter WFI storage vessel. The WFI is maintained at 80 ± 2°C by plant steam in the storage vessel jacket. WFI is recirculated to the use points through a stainless steel distribution loop that is fully drainable, and returned into the tank through a back pressure control valve and static (fixed) spray balls. During periods of extended recirculation, the recirculation pump heat rise is removed by a WFI loop cooler (double tube heat exchanger) using cooling tower water. Backflow prevention of cooling tower water is accomplished through lowering the cooling tower water pressure below that of the WFI loop pressure. The WFI tank and loop are designed for SIP with clean steam when the system is drained or as required.

Cartridge-style WFI heat exchangers are provided to cool WFI to ambient temperatures when required. These are located in the following areas:

- Buffer Preparation
- Tank Wash
- Purification
- Media Preparation

This section must also include a description of the relationship between the facility's manufacturing areas and nonmanufacturing areas, and with any surrounding buildings on the site. It is best to include a general facility layout and a general site plan that defines the locations of buildings and areas according to their functionality.

Animal areas

Information regarding the use of animals in a manufacturing process is important. This section must provide the specific location of any animal husbandry and quarantine facilities that are part of the facility, including information regarding design and construction features.

Descriptions of cage-washing systems, containment systems, environmental systems, food storage and handling systems, surgical areas, bedding facilities, and waste storage and removal systems should be provided.

If any animal testing is performed at another manufacturing location, a description of the testing arrangement must also be included in this section.

Equipment

The establishment information must identify each piece of equipment that is used in the manufacturing process and the location of that piece of equipment within the facility. This includes not only process equipment, such as vessels, chromatography skids, or centrifuges, but also any equipment used for material storage (freezers, incubators), filling, or labeling. It would also include certain utility equipment, such as WFI stills and clean steam generators, equipment used for sterilizing components, such as autoclaves or depyrogenation ovens, and equipment used for cleaning, such as CIP

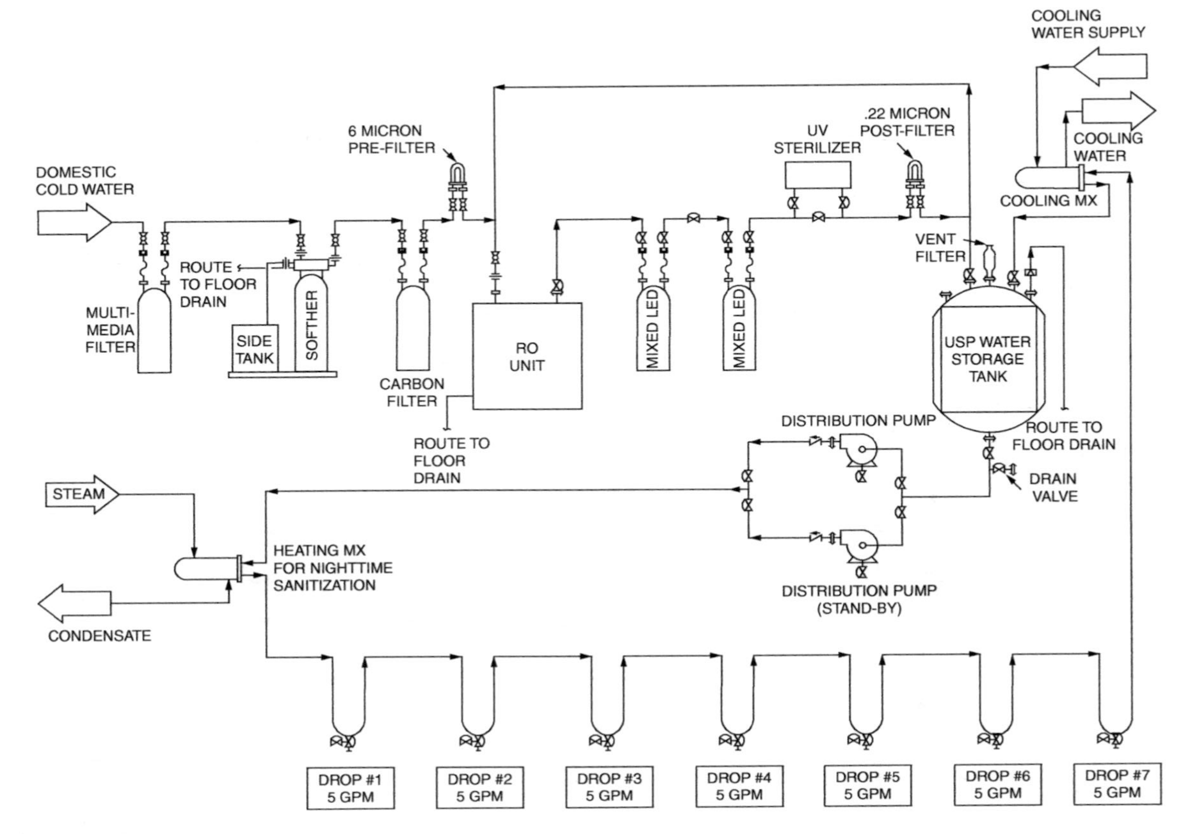

Figure 11.2 P&ID of RO/DI with nighttime sanitization.

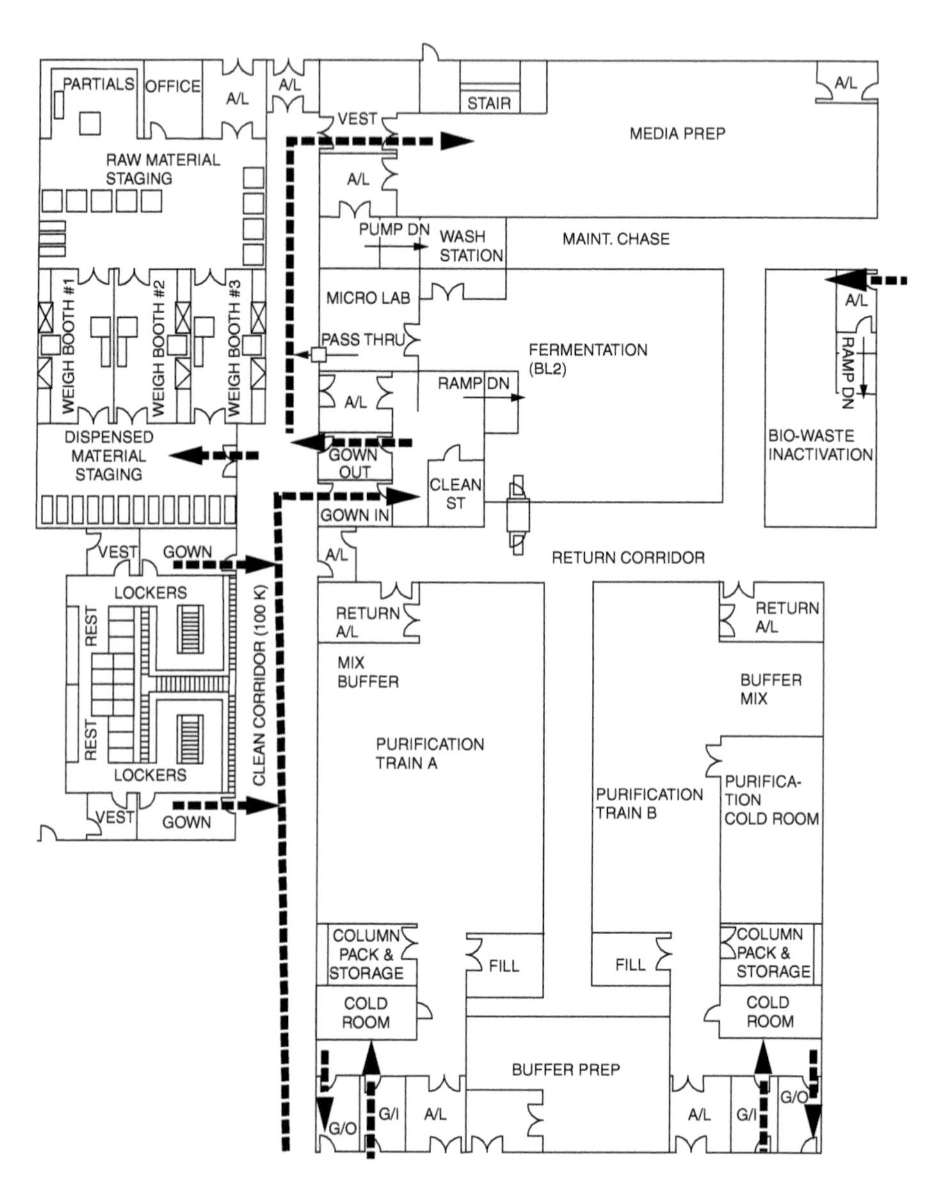

Figure 11.3 Personnel flow.

systems. Typically, general arrangement drawings, along with an equipment matrix or table, are used to provide the necessary information.

The methods used to clean and sterilize equipment must be defined, along with data to support the conditions and parameters chosen. An outline of all validation protocols and SOPs is normally provided, along with a summary of validation data to support the methodologies chosen. This would include the frequency of cleanings and the methods used to test for cleanliness and sterility.

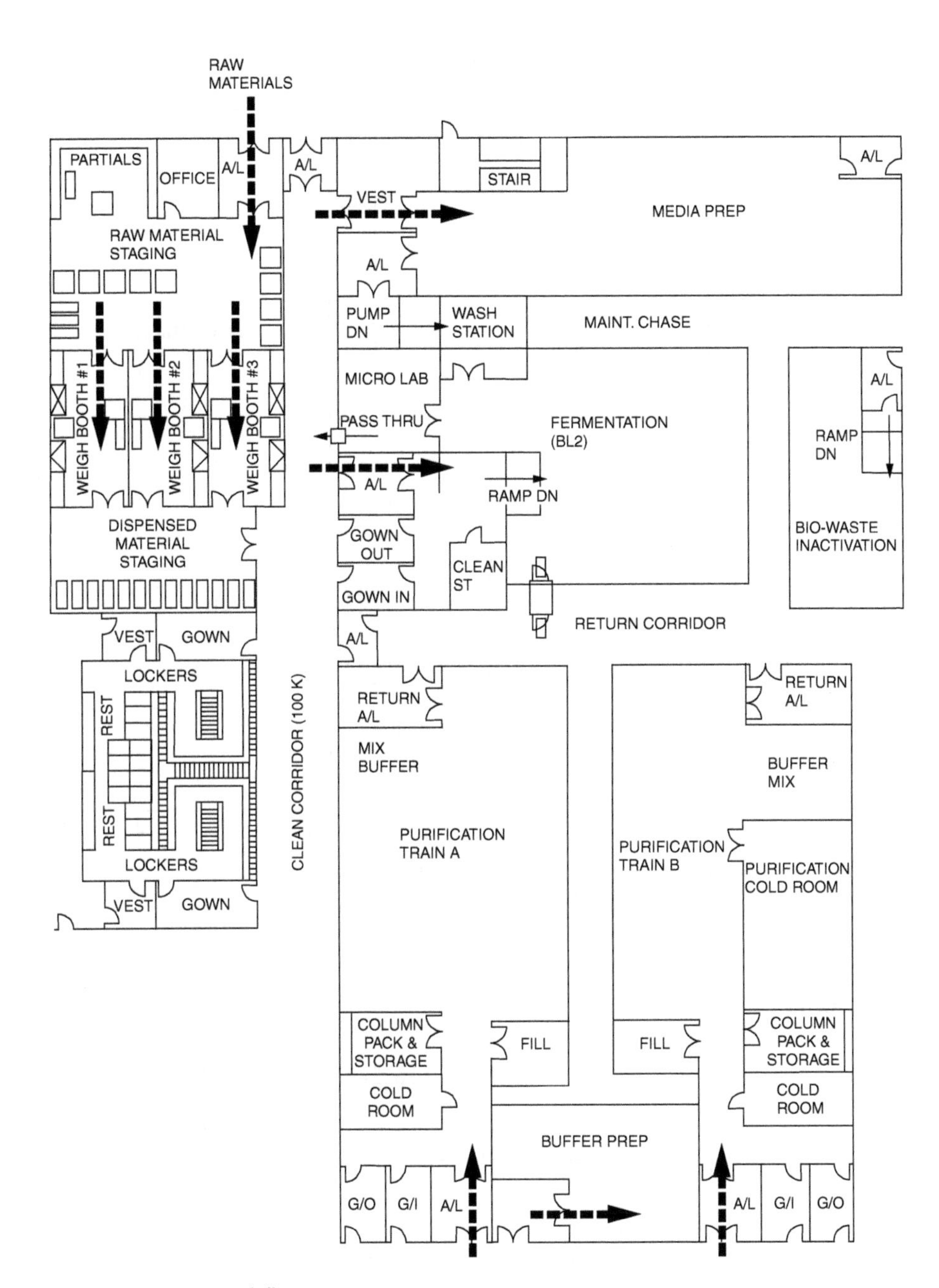

Figure 11.4 Material flow.

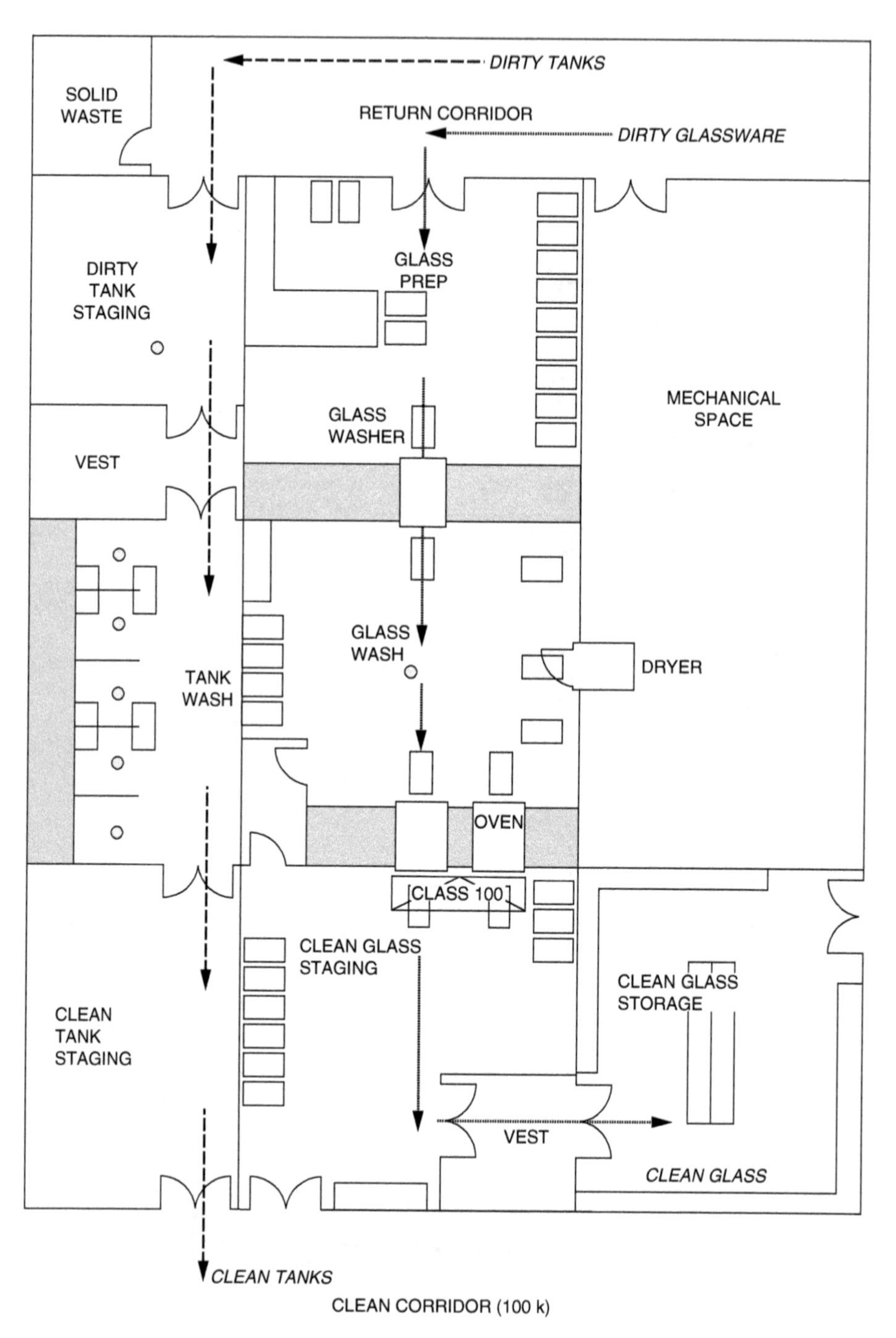

Figure 11.5 Clean flow.

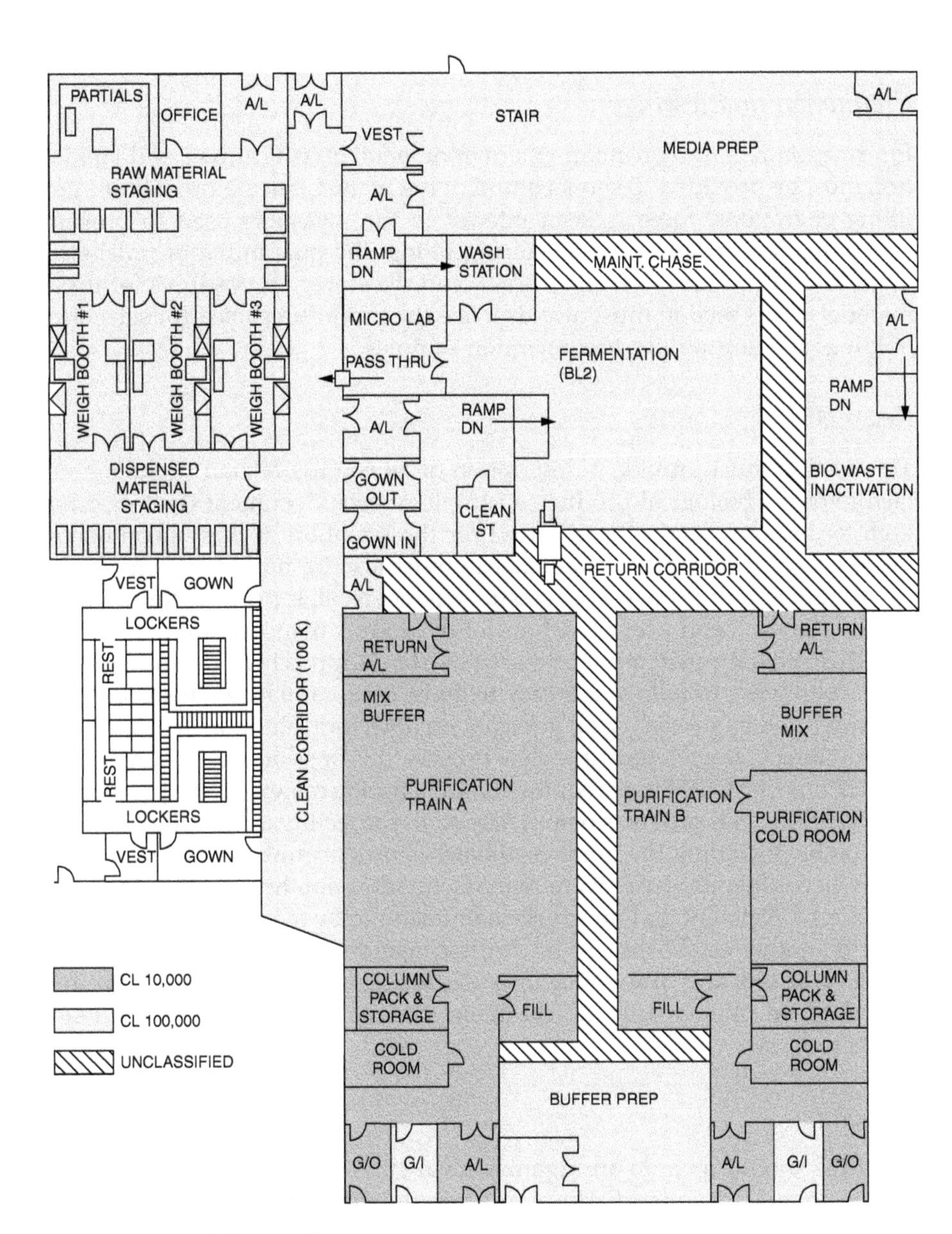

Figure 11.6 HVAC zone drawing.

Process automation and control systems should also be described, along with the calibration methods used to record data and to ensure compliance with predetermined parameters.

Production and testing

Information regarding sanitation, environmental classifications, and validation must be provided to show control of the manufacturing operations. For filling operations, there is keen interest in the measures used to prevent contamination of final product during filling, the quarantine of unlabeled and finished product awaiting release, and the storage procedures for these materials. This section must also describe the procedures used for selecting, holding, and storing product retention samples.

Records

This section must provide a description of all records related to the manufacture of the biological product, a sample of each document described for each lot or batch, and a time period for the retention of these documents. Procedures for document maintenance must also be provided, confirming to CBER that, if necessary, a successful product recall is possible through the tracking of each and every batch or lot of product manufactured.

This section must provide a step-by-step approach to documentation that is both easy to follow and easy to audit for compliance. This is usually accomplished by providing an information flowchart, along with procedural descriptions of the document review process. It is possible during prelicense inspections that CBER may require a demonstration review of this procedure.

A listing of all products manufactured in the facility during the previous 12 months, including the number of batches or lots, must also be provided. If any licensable products were manufactured at another location but sold by the establishment, or if products manufactured by the establishment were sold to another establishment for further manufacturing by the firm, these must be listed and the conditions of the material or product transfer described. In other words, if you made it there or sold it there, the FDA wants to know.

Personnel

The BLA should include an organizational chart, with titles and names of key personnel, for individuals involved in the manufacturing process. This should be a comprehensive look at the organizational structure and those individuals responsible for the day-to-day operations of the facility.

Since the qualification and training of personnel is an important issue to the FDA, there must also be documentation supporting the qualification and training of individuals and a description of the training philosophy used.

The pre-BLA meeting

Holding a pre-BLA meeting is a pivotal milestone in the FDA's managed review process. Its main purpose is to determine whether a submission of the application is appropriate. The pre-BLA meeting is held to describe for CBER reviewers the general information that will be submitted.

It is good practice to open dialogue with CBER/CDER early in the design process to ensure that design philosophy will meet regulatory intent. This reduces risk associated with design decisions that might cause issues later, during actual BLA review. As design issues arise that are open to interpretation, having review input from the regulators may help to resolve the issue at a time when the cost of making changes is minimal: minor drawing changes and engineering person-hours.

Inspection

The inspection of a facility for BLA approval need not be made until the facility is in operation. This is another reason that the pre-BLA meetings are valuable. A designated inspector from the Division of Manufacturing Product Quality conducts this inspection. Specifically, the duties of the inspector will include examination of the details of location, construction, equipment, and maintenance of areas used in any part of the manufacture of the product.

Once a facility is licensed, follow-on inspections will be made every two years; these can be made with or without notice during normal business hours, unless otherwise direct*. The goal of these inspections is to ensure that the facility remains in compliance with the parameters defined in the BLA. If, during these inspections, the inspector determines that aspects of GMP are not being met, a form 483 (Figure 11.7) will be issued.

* 21 CFR, Part 600.2

<table>
<tr><td colspan="2">DEPARTMENT OF HEALTH AND HUMAN SERVICES
PUBLIC HEALTH SERVICE
FOOD AND DRUG ADMINISTRATION</td><td colspan="2">DISTRICT ADDRESS AND PHONE NUMBER

Sample</td></tr>
<tr><td colspan="2">NAME OF INDIVIDUAL TO WHOM REPORT ISSUED TO:</td><td>PERIOD OF INSPECTION</td><td>C.F. NUMBER</td></tr>
<tr><td colspan="2">TITLE OF INDIVIDUAL</td><td colspan="2">TYPE ESTABLISHMENT INSPECTED</td></tr>
<tr><td colspan="2">FIRM NAME</td><td colspan="2">NAME OF FIRM, BRANCH OR UNIT INSPECTED</td></tr>
<tr><td colspan="2">STREET ADDRESS</td><td colspan="2">STREET ADDRESS OF PREMISES INSPECTED</td></tr>
<tr><td colspan="2">CITY AND STATE (Zip Code)</td><td colspan="2">CITY AND STATE (Zip Code)</td></tr>
<tr><td colspan="4">DURING AN INSPECTION OF YOUR FIRM (I) (WE) OBSERVED:</td></tr>
<tr><td>SEE REVERSE OF THIS PAGE</td><td>EMPLOYEE (S) SIGNATURE
VOID</td><td>EMPLOYEE(S) NAME AND TITLE (Print or type)</td><td>DATE ISSUED</td></tr>
</table>

FORM FDA 483 (5/85) PREVIOUS EDITION MAY BE USED. **INSPECTIONAL OBSERVATIONS** PAGE OF PAGES

The observations of objectionable conditions and practices listed on the front of this form are reported:

1. Pursuant to Section 704(b) of the Federal Food, Drug and Cosmetic Act, or
2. To assist firms inspected in complying with the Acts and regulations enforced by the Food and Drug Administration.

Section 704(b) of the Federal Food, Drug, and Cosmetic Act (21 USC374(b)) provides:

"Upon completion of any such inspection of a factory, warehouse, consulting laboratory, or other establishment, and prior to leaving the premises, the officer or employee making the inspection shall give to the owner, operator, or agent on charge a report in writing setting forth any conditions or practices observed by him which, in his judgement, indicate that any food, drug, device, or cosmetic in such establishment (1) consists in whole or in part of any filthy, putrid, or decomposed substance, or (2) has been prepared, packed, or held under insanitary conditions whereby it may have become contaminated with filth, or whereby it may have been rendered injurious to health. A copy of such report shall be sent promptly to the Secretary."

Figure 11.7 Form FDA 483.

Example 11.3 Points to Observe in Preparing a Meeting Agenda and Presentation Materials for a Pre-BLA Meeting

1. A paragraph that highlights the organization, size, location, product, and purpose.
2. A paragraph describing the purpose of the meeting.
3. A description of the product(s) and a brief description of the manufacturing process.
4. A product process flowchart.
5. A floor plan with manufacturing process flow described.
6. A floor plan with personnel flow described.
7. A brief description of the water system, with schematic.
8. A brief description of the HVAC system, including
   - A floor plan describing air handling units.
   - A floor plan showing pressure differentials for containment/protection.
   - A diagram showing air quality classifications of rooms and areas.
9. Multiuse facilities:
   - Product changeover procedures
   - Segregation of product, personnel
   - Cleaning validation and product residual testing plans
10. A brief description of validation procedures (IQ, OQ, PQ), including Validation Master Plan if available.
11. A description of the waste handling system, including biohazardous materials, with a waste flow diagram.
12. A timeline of scheduled submissions to FDA.
13. Outline of any unique licensing issues (e.g., shared/cooperative manufacturing).
14. Other issues, such as discussion of how the product moves through the facility, gowning requirements, inactivation procedures, and so on.

chapter 12

Containment basics

Because biopharmaceutical products are derived from living organisms, it is necessary to contain the biologically active agents that form the product in order to protect employees, the product, and the environment. Modern biological containment practice accomplishes this protection through programs involving special facility design, monitoring, and validation.

The containment of biological processes must begin with a risk assessment of the potential harm to employees and the environment posed by each component of the system. For effective containment design, there are three concerns:

1. The biological agent
2. The product
3. The process hazards

The overall containment strategy must be built on an integrated control of each of these concerns.

Good containment practice requires the following:

- Providing the appropriate level of containment while maximizing the accessibility for routine facility operation and maintenance
- Maintaining containment areas at lower room pressures than the external environment so that normal air movement is inward during operation and in failure conditions
- Providing easily cleanable surfaces that can withstand the required cleaning regiment
- Providing efficient means for the transfer of materials and equipment into and from the area that minimizes the possibility of cross-contamination
- Minimizing the generation of airborne contamination

Containment of biological agents used in biopharmaceutical processes is a two-phase approach. First, there is the concept of biological containment.

In most current biological processes, a microorganism is chosen that has a very low level of risk, a history of safe industrial use, or is unlikely to survive outside the production setting.

The second phase involves the utilization of the appropriate design criteria for creating facilities and systems that allow for the safe handling of the biological agent. This phase is known as physical containment. We will define physical containment as those procedures, systems, and structures designed to prevent the exposure of personnel to biological hazards, and to prevent the release of hazardous materials into the environment. The level of physical containment required for a biotech process is dependent on the potential hazard of the organism or agent being used. Such hazards include the potential to cause disease or allergic reactions in workers or toxic effects on the environment. Physical containment is the focus of this chapter.

Regulatory requirements

Physical containment levels have been established by the National Institutes of Health (NIH) in the Guidelines for Research Involving Recombinant DNA Molecules. [1] These guidelines center around good laboratory practice for experiments that are funded under the NIH. Because the FDA references them as a basis for the design of containment for biotech production operations, they are recognized as the design guidelines that all engineers must follow.

There are four levels of containment defined in the NIH guidelines for large-scale research or production. In this context, large scale is defined as volumes greater than 10 liters. These levels set the containment conditions, based on a review of the degree of hazard posed to the health of workers or the environment, due to the organism in use. The review of containment level assignment and contamination control philosophy is established by an Institutional Biosafety Committee (IBC), with representation from the Research, Manufacturing, Engineering, Safety, Regulatory, and QA/Validation Services groups.

The three main levels of containment that are found in biopharmaceutical manufacturing are designated Biosafety Level 1–Large Scale (BL1–LS), Biosafety Level 2–Large Scale (BL2–LS), and Biosafety Level 3–Large Scale (BL3–LS). [2] Table 12.1 provides a summary of the basic requirements posed by each level. Each level builds upon the previous requirements, with BL3–LS having the most stringent requirements. Each sequential level also has a compounding effect on process and building design.

It is important to remember that these are guidelines. While establishing requirements, they do not define how the containment should be accomplished or how many different ways it can be accomplished. Every facility should be designed according to the product, hazards, risk, and the best interest of the company.

Table 12.1 Physical Containment Requirements for Large-Scale rDNA Fermentations

	Applicable Regulatory Level		
Description of the Regulatory Requirement	BL1–LS	BL2–LS	BL3–LS
Closed vessel	■	■	■
Inactivation of cultures by validated procedure before removal from the closed system	■	■	■
Sample collection and addition of material in a closed system	■	■	■
Exhaust gases sterilized by filters before leaving the closed system	■	■	■
Sterilization by validated procedures before opening for maintenance or other purposes	■	■	■
Emergency plans and procedures for handling large losses	■	■	■
No leakage of viable organisms from rotating seals and other mechanical devices		■	■
Integrity evaluation procedures—monitors and sensors		■	■
Containment evaluation with the host organism before introduction of viable organism		■	■
Permanent identification of closed system (fermenter) and identification to be used in all records		■	■
Posting of universal biohazard sign on each closed system and containment equipment when working with viable organism		■	■
Posting of universal biohazard sign on entry doors when work is in progress			■
Operations in a controlled area:			
Separate specified entry			■
Double doors—air locks			■
Walls, ceiling, and floors permit ready cleaning and decontamination			■
Utilities and services (process piping and wiring) protected against contamination			■
Hand-washing facilities			■
Shower facilities in close proximity			■
Area designed to preclude release of culture fluids outside the controlled area in the event of an accident			■
Ventilation—movement of air, filtration of air, and so on			■

This table compares the NIH physical containment guidelines to the levels of containmen- required.

Containment design

In designing facilities where potential biohazards are to be processed, it is important for the designer to have a good appreciation of the design considerations relevant to the level of containment applicable to the facility. In designing contained facilities, there are two basic considerations: primary and secondary containment.

Primary containment

Primary containment involves the isolation of selected systems and/or pieces of process equipment within a facility in order to prevent exposure to biohazards. The principle methods of maintaining primary containment are the proper use of "closed" system equipment design principles, the use of validated methods for the inactivation of organisms before removal from the system, and the inactivation of effluents.

Equipment design

The basis of "closed" system design is the utilization of process equipment that does not expose the product to the open environment. A "closed" system (Figure 12.1) will have closed vessels of welded construction and hard, cleanable surfaces; hard-piped connections, with no threaded components; and a validated means of adding and/or removing materials from the system. "Closed" system design allows for the addition of WFI, clean steam, or compressed air during process operations; however, these materials are carefully controlled with respect to their quality, quantity, and method of addition.

Transfers are defined as any addition or removal of material from a "closed" system. This includes any sampling that is performed for quality testing purposes. The containment guidelines require that the design of equipment and/or components allow for sterilization to inactivate organisms before removal from the primary containment. This is usually accomplished through the use of a steam-lock addition port. In this configuration, the port can be decontaminated or resterilized prior to breaking the connection. This inactivates any organism in the immediate area of the connection.

Good containment practice requires that organisms be inactivated by sterilization, utilizing validated procedures, before opening a closed system for maintenance or any other purpose. The design of the containment system should also minimize the number of times that connections are made and/or broken. Current practice is to utilize steam-in-place (SIP) systems to accomplish sterilization of closed systems.

In a typical fermentor system, batch sterilization is accomplished via jacket heating and/or direct clean steam injection. The SIP system is normally automated to control the time required for fermentor batch sterilization heat up and cool down. System operating parameters for sterilization would typically be in the 15–20 psig range, with temperatures held at approximately at 120°C for a duration of 30–40 minutes. This must be a validated kill.

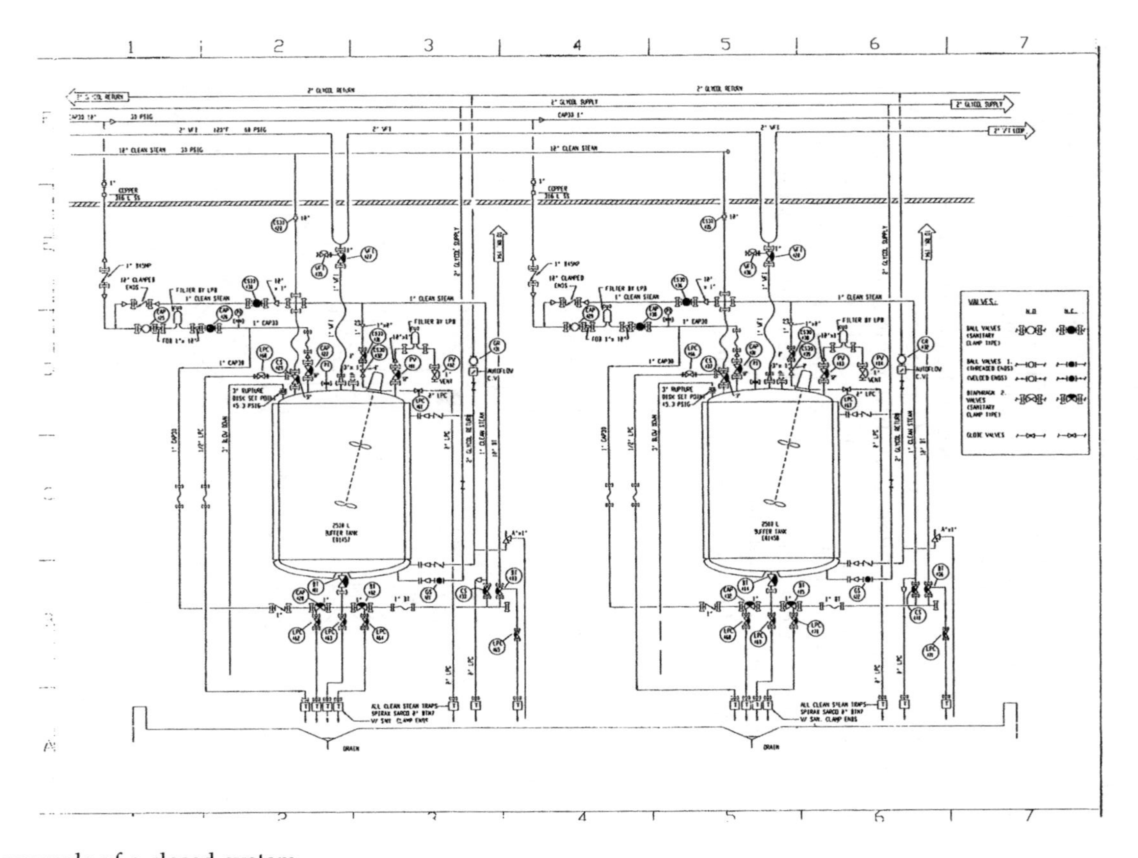

Figure 12.1 An example of a closed system.

Many of the components used to produce a product from a living organism have the potential for producing contaiminated aerosols. Agitators, pumps, and centrifuges all have rotating or sliding parts that are a concern. Current design practice employs the use of rotating seals to prevent the release of aerosols into the work environment. In Figure 12.2, a double mechanical seal design for an agitator shaft allows for flushing the chamber between the two seals with clean steam to inactivate any organisms. The contaminated condensate would then be piped to the biowaste system for further inactivation prior to release. This is important if a seal failure occurs, resulting in a leak of the vessel into the seal chamber. The integrity of the "closed" system is maintained.

Exhaust gases from "closed" systems must be treated to prevent the release of any viable organisms. Sterile 0.2 mm hydrophobic vent filters are commonly used for this purpose. These filters and their piping components should be designed to allow for sterilizing up to and through the vent filter. When this type of filter is specified, the design should allow for the in-place integrity testing of the filter in accordance with SOP. The major difficulty with vent filters is their tendency to foul and plug when wetted by moisture. This problem can be alleviated somewhat by using steam tracing of the filters.

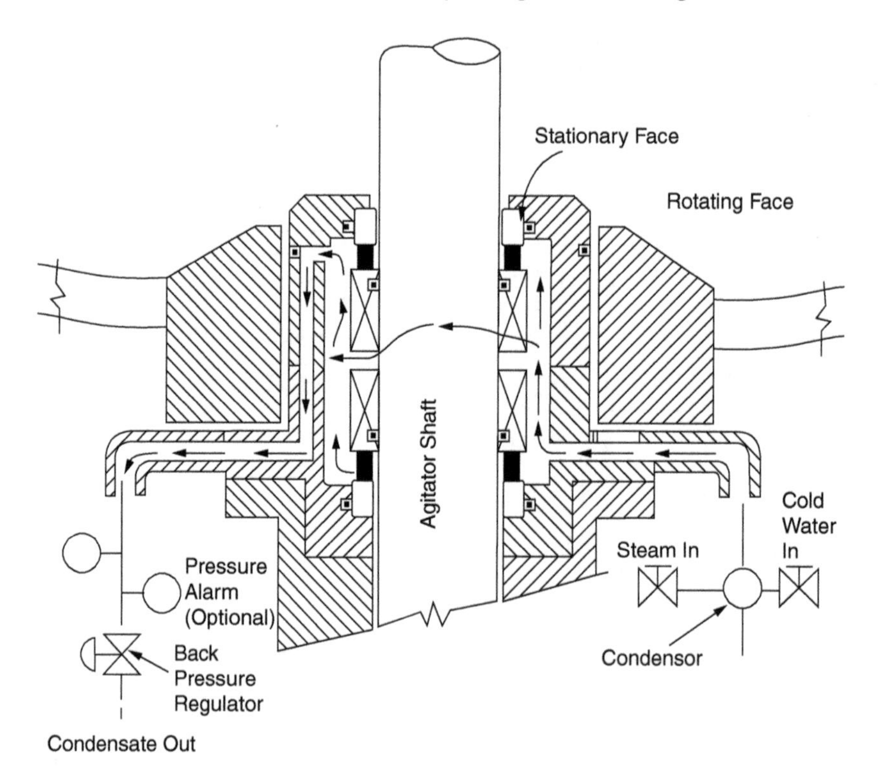

Figure 12.2 Contained double mechanical seal. Source: Miller, S., and Bergmann, D. 1993. Biocontainment design considerations for biopharmaceutical facilities. *Journal of Industrial Microbiology* (November).

Double O-ring seals will also be required for all flanged joints. This includes equipment and piping. Sanitary piping connectors provide proper sealing of mechanical pipe joints and can also be equipped with steam injection ports.

"Closed" systems and primary containment components must have monitoring devices incorporated into their design to evaluate the integrity of the system(s) during operation. Pressure-sensing devices on fermentors and process vessels, leak detection capability such as seal leak detection, and integrity test ports on filters are examples of containment integrity monitoring devices.

Drainage and effluent treatment

Discharges containing viable organisms from a containment process must be inactivated by a validated inactivation procedure prior to release from the "closed" system. These systems, referred to as inactivation, deactivation, or "kill" systems, collect contaminated or potentially contaminated liquid waste from a variety of discharge points, including fermentor drains, centrifuges, autoclaves, and floor drains, and hold contaminated waste for inactivation. The two most common techniques used to effect biological inactivation are thermal inactivation and chemical inactivation. These techniques are applied in either batch or continuous system designs.

Thermal Systems. Thermal inactivation is the most common method used for large-scale biological processes. In a thermal system, contaminated waste is heated to a sufficient temperature to ensure inactivation of recombinant organisms.

Chemical Systems. Chemical systems utilize chemicals, such as hypochlorite (bleach), to inactivate waste. While effective, this type of system has some inherent drawbacks, including the safe handling of chemicals, the corrosive nature of the process, and disposal of the now chemically contaminated effluent.

Batch Systems. Since the majority of biologic processes involve batch fermentation, batch systems match well with waste flows generated from the process. Figure 12.3 shows a batch, thermal inactivation system. In this system, waste drains by gravity from the process operations areas into a decontamination waste receiving tank. Once the tank volume reaches a predetermined level, the contents are shifted to the other collection tank. With the waste now isolated, mixing of the waste begins, and steam is applied to the jacket of the tank. The contents of the tank are heated to and maintained at 120°C for 30 minutes. Steam is also applied to the tank inlet line to sterilize it and to purge the air in the head space above the collected waste. After the inactivation cycle is complete, the waste is cooled to approximately 60°C by the addition of cooling water through the jacket. The waste is then discharged to the facility waste treatment system and the tank is put in standby mode for the next batch.

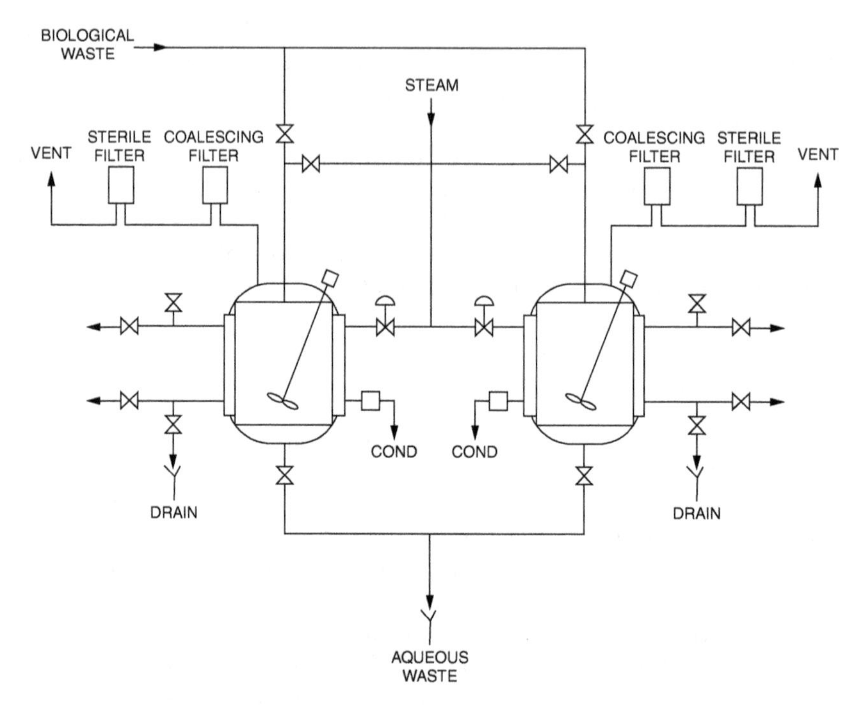

Figure 12.3 Batch biological inactivation system.

In designing a batch system, certain factors should be considered, including the following:

- Level of automation desired for control and documentation
- Number of tanks required based on the daily discharge from containment areas; do not overlook any of the waste streams
- Desired number of decontamination cycles per day
- Appropriate materials of construction
- Odor control
- Utility supply capacity of the facility
- Capital cost

While batch systems allow greater flexibility of operation, they are more capital intensive and require a greater amount of facility space, since the area housing the system components must meet the same biosafety level containment requirements as the area generating the effluent.

Continuous Systems. When the character of waste streams is more consistent and predictable, continuous inactivation systems are typically used. In a continuous system (Figure 12.4), biowaste can be heated with an appropriately sized heat exchanger and discharged directly to the facility waste treatment system. The concept is similar to the process commonly used for in-line media sterilization. Continuous systems offer less flexibility

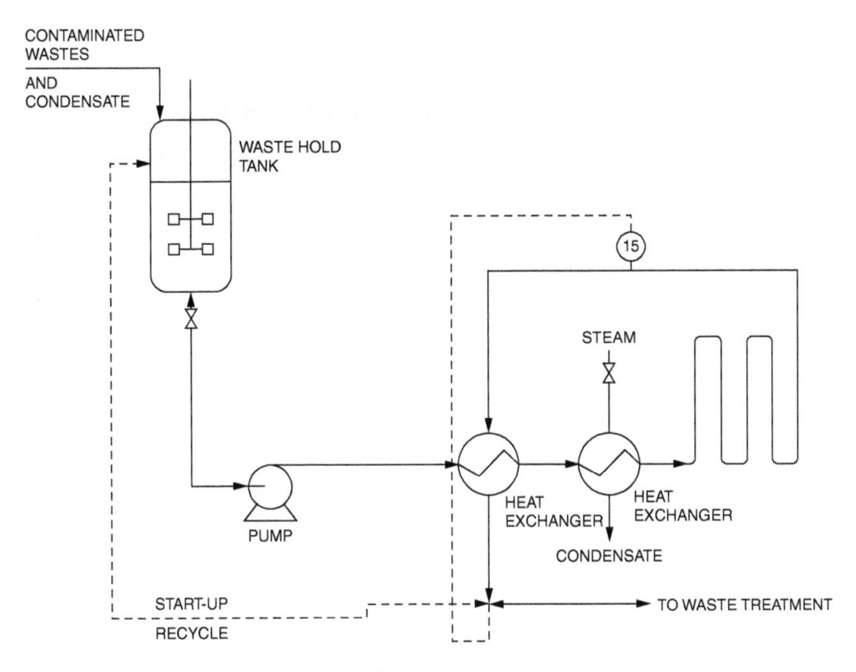

Figure 12.4 Continuous biowaste kill system.

than batch systems, but are also less capital intensive in equipment costs and require less facility space.

In addition to the liquid waste generated in biological processes, solid waste must also be decontaminated before leaving the containment area. A dedicated decontamination autoclave is used for this purpose. By properly locating the autoclave to minimize the distance from processing and to avoid crossing waste streams with incoming processing materials, GMP will be met and containment integrity maintained.

Secondary containment

Secondary containment is defined as protection of the external environment of the entire "containment area" to prevent the release of infectious materials should primary containment fail. It involves the use of rooms and structures (Figure 12.5) to support primary containment. The key elements of secondary containment design include

- Layout and building design
- HVAC
- Operational procedures

Layout and building design

The layout and design for containment areas should provide building features and arrangements that not only enhance the ability to contain biohaz-

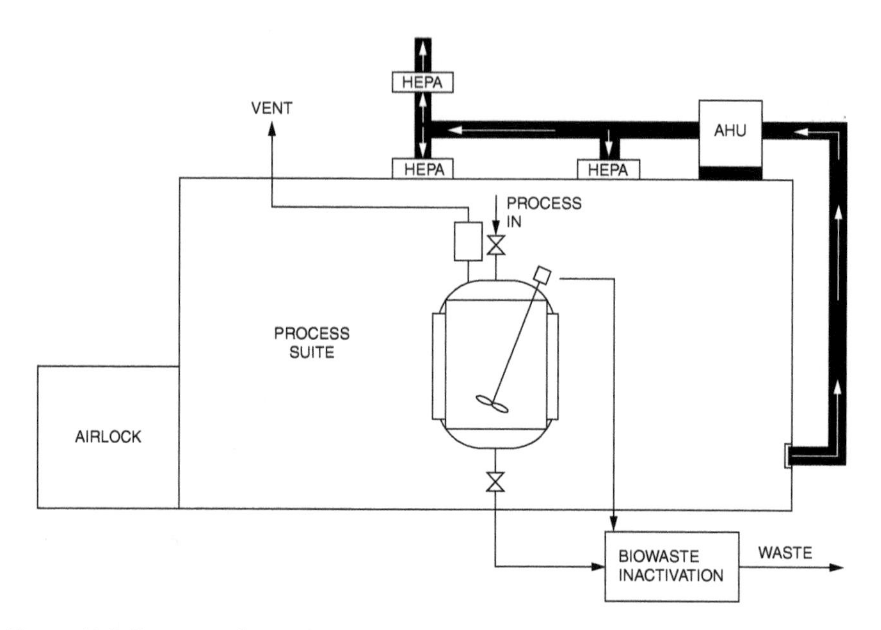

Figure 12.5 Rooms and structures to support primary containment.

ards, but also follow GMP guidelines relating to segregation, cross-contamination, and cleanability. Once the appropriate biosafety level requirement has been established and proper facility programming techniques have been applied to establish the overall facility requirements, it becomes the designer's responsibility to focus on the facility issues that center around containment.

Architectural Finishes. All containment area surfaces must be easily cleanable. In addition, many cleaning and decontamination regimes may require that these surfaces be resistant to a variety of solvents, disinfectants, acids, or alkalis. Cleaning practice as defined in the SOPs will identify which features are of most concern, the methods used for cleaning, and the frequency of the operations.

Surfaces must be sealed and have a texture that will optimize cleaning operations. Design considerations must be given to the process operations performed and the equipment utilized. If portable equipment is used, the floor surfaces must be hard to resist damage, and wall surfaces must be protected to prevent damage by the movement of equipment. Equipment surfaces must be corrosion resistant, and the equipment configuration must provide a waterproof seal.

Segregation. The containment area should be located in a manner that removes it from the general circulation flow of personnel. Personnel air locks and material air locks are generally used to isolate and control movement. These passages are usually controlled by an electronic security access system. Material, product, and personnel flows should follow the principles of GMP relating to cross-contamination, utilizing a unidirectional flow as much as possible.

It is desirable to minimize the maintenance activity within any operating area, especially a containment area. Utility services and major mechanical system components may be located outside the area adjacent to, but isolated from, the operating area. Figure 8.2 on page 220 shows a "gray space" conceptual layout that incorporates this philosophy. While this concept minimizes personnel exposure and helps keep environmental contamination low, it greatly increases facility cost.

Personnel Protection. Containment level designation and the processes employed will define the level of gowning and personnel protection that is required. SOPs will define the protective clothing required to enter the containment area and the method in which the clothing is to be applied and removed. Appendix 12.1 contains a sample gowning/degowning SOP. It may be appropriate for personnel to wear safety glasses or face shields in addition to the protective clothing defined in the SOP.

Doors leading into containment areas should contain glass panels to allow for viewing and have the universal biohazard sign clearly posted. Washing facilities may be required for personnel to disinfect their hands, depending on the risk of exposure to the organism.

Spill Management and Response. FDA inspection guidelines for biotech manufacturing facilities require that equipment and controls be in place to handle spills that could occur within biocontainment areas. [3] Physical features of the facility, as well as procedures for evacuation, spill containment, decontamination, and cleanup are procedures that the FDA looks for and scrutinizes.

Containment areas are normally designed to allow accidental spills and cleaning fluids to be contained within floors recessed below adjacent, noncontainment areas. Floors will be sloped to drains. Floor drains will not be connected to the standard drainage system, but will be piped to the appropriate biological waste treatment system for inactivation. Floor drains used in containment services must prevent backflow of waste or contaminated vapors from the waste system. Plug drains should be connected to traps that have an extra depth, a full two inches of water gauge greater than the maximum static pressure that can be generated by the HVAC system supply or exhaust. This will prevent waste fluids from rising from the drains in the event of an HVAC system malfunction. Facility SOPs must address methods to prevent microbial growth in the trap.

The installation of a check valve in the drain line will also provide assurance that back pressure from the inactivation system will not compromise the containment area. Whatever backflow prevention is used must still allow for unimpeded gravity flow.

It is also good practice to provide dikes around large pieces of stationary process equipment. The dike should be designed to contain the largest possible volume of process fluid that could be released in the event of a spill in addition to any chemical additives for inactivation.

SOPs should address the procedure for responding to and decontaminating a spill inside containment. An emergency checklist, similar to that shown

Table 12.2 Emergency Procedures Checklist for Spills

Small Spill (less than 10 L):	
1.	Immediately surround the area with absorbent spill pillows to contain the spill. Flood the contained area with the appropriate disinfectant.
2.	Discard used spill pillows into a bag for autoclaving.
3.	Wipe area with distilled water to remove the inactivation agent.
4.	Perform an appropriate microbiological recovery exercise with swabs.
5.	Wait until the recovery exercise demonstrates negative finding before resuming activity in that area.
6.	If microbiological growth occurs on test plates, decontaminate the room with approved decontaminating agent.
7.	Retest after second treatment.
Large Spill (greater than 10 L, where the maximum spill volume is no larger than the working volume of the largest fermentation vessel):	
1.	Move liquid into sump area with a squeegee.
2.	Spread solution of approved inactivation agent over entire floor area and let stand according to validated time requirements.
3.	Wash down entire area with distilled water, DI water, or WFI.
4.	Conduct microbiological tests in minimum of four critical locations in the room.
5.	If microbial growth occurs on test plates, decontaminate entire room with approved decontaminating agent.

in Table 12.2, should be reviewed by all operating personnel and made a part of their required training for working inside the containment area.

HVAC systems

Ventilation systems play an important part in containment design and represents a significant portion of the cost associated with secondary containment. Air pressure differentials are used to supplement the segregation and integrity characteristics of the overall containment system.

Standard GMP practice for HVAC system design is to establish air pressure differentials so that critical processing areas are at a higher pressure, with less critical areas at progressively lower pressures. These differentials are fractional in nature (specified in gradations of 0.025 inch water gauge) and can be established in two ways: passively by airflow balancing and by active control.

Containment areas are normally designed to operate at negative pressures relative to surrounding areas. This reduces the spread of organisms in case of a failure of the primary containment system. Personnel and material air locks are used to maintain the separation of the containment area from the rest of the facility. In Figure 12.6, airflow balances indicate the differential pressures from room to room.

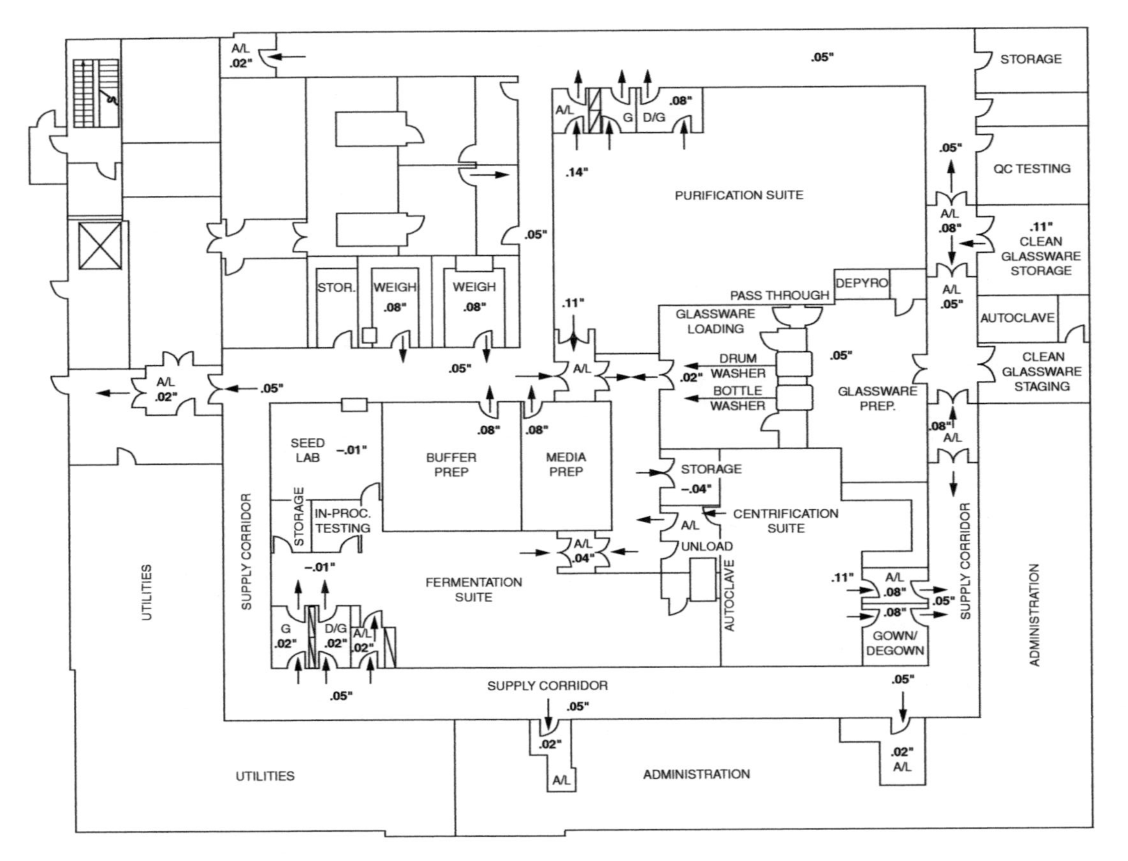

Figure 12.6 Layout showing air locks to separate containment area from rest of facility.

GMP operations are normally carried out under positive pressure; if a containment area is located within this GMP space, a zone of negative air pressure will be required. If these two air zones come in contact, the GMP zone would become depressurized, contaminating the area. HVAC zoning must minimize the disturbances seen from material and personnel flow between these zones.

The air supply for containment areas should come from air handling units that service only these areas. Exhaust air must not be recirculated to other rooms and should be exhausted away from the intakes of other process areas. Frequently, when aerosol generation is a concern, exhaust air is HEPA filtered just as supply air to remove any airborne contaminants.

In many cases, biological safety cabinets are present inside containment. These cabinets are often used in the preparation of fermenter inoculum. The exhaust gases from these cabinets should be HEPA filtered before exhaust from the containment area.

Operational procedures

Good containment design is only as good as the operational procedures that support good containment practice. Validated procedures, and an ongoing program of employee training, are as much a part of containment as any design feature.

Procedures for gowning, material transfer, cleaning, sterilizing, maintenance, and testing all are integral to sound containment practice. Operations personnel must know, and follow, these procedures if the containment foundation is to remain solid. Like the legs of a stool, failure of any one causes failure of the entire system.

Project impact

As engineers and scientists evaluate the risk associated with a particular process, there must be an understanding of the impact of selecting the proper biosafety level that is appropriate for the process operations and the organisms involved. There is a tendency to take a conservative approach to that risk. This can result in electing to design and construct a facility that has a higher containment level than is required. While issues relating to flexibility, confidence of the process or facility, or revisions to the regulations must be reviewed, overdesigning has a direct impact on the capital cost of the project.

References

1. *Guidelines for research involving recombinant DNA molecules.* 1994. Washington, DC: Centers for Disease Control and National Institutes of Health.
2. Ibid.
3. *Guideline for the inspection of biotechnology manufacturing facilities.* 1991. Rockville, MD: Food and Drug Administration, Division of Field Investigations (HFC-130).

Appendix 12.1

Sample gowning/degowning procedure

The purpose of this procedure is to define the gowning and degowning requirements for the ABC Biotech, Inc., manufacturing facility in Smalltown, USA. This facility is classified as restricted access.

This procedure is applicable to all persons entering or exiting the facility, including employees, contractors, and visitors. This procedure is to be strictly adhered to for the protection of both personnel and product.

1.0 Materials and equipment

The following materials and equipment will be provided as required for the specific area's gowning requirements:

- Tyvek® coveralls with or without hood. Designated alternates are polyester coveralls or equivalent paper coveralls
- Haircover and/or bonnet
- Beard cover
- Face mask
- Sterile surgical-type gloves
- Rubber boots
- Shoe covers
- Safety glasses
- Sterile sleeves
- Alcohol foam or liquid degermer

2.0 Definitions and notes

Any of the requirements defined in this procedure may be waived only by the Director of Manufacturing (or designate) by written notice.

For this procedure, Street Clothes are defined as pants, shirts, jackets, and other garments of clothing worn by persons and exposed to the outside environment. Undergarments and socks are not included. Shoes must be appropriate for use with shoe covers (heels no more than one inch high, closed-toe shoes preferred). Large and loose fitting jewelry is not permitted. The use of heavy makeup is prohibited.

In the event of an emergency or evacuation drill, gowning and degowning requirements may not apply. Discretion will be with the individual in such cases. Adherence to postevacuation protocols is mandatory.

3.0 Procedure

3.1 Facility Gowning and Degowning Requirements

3.1.1 Gowning:

Access to the manufacturing facility requires personnel to remove street clothes and put on a plant uniform, captive shoes, and safety glasses. Newly laundered uniforms will be provided. If a plant uniform was used during the current day and is still available, it may be reused.

Visitors to the facility that will not directly handle product (vendors, repair personnel, inspectors, dignitaries) are only required to wear clean Tyvek® coveralls or equivalent, shoe covers, and hair covers when entering the facility.

Captive shoes will be sprayed with a disinfectant once a week as a minimum.

3.1.2 Degowning:

Upon completion of work in the manufacturing facility, proceed to the appropriate locker room and change back into street clothes. Store captive shoes in locker and place uniform in laundry hamper provided.

If additional work is to be performed in the manufacturing facility, proceed to the appropriate locker room and change back into street clothes. Store shoes and plant uniform in locker.

3.2 Gowning and Degowning Requirements for Biocontainment

3.2.1 Gowning:

Upon entry into Gowning Room #111, put on disposable Tyvek® coveralls. Waterproof boots shall be placed over captive shoes. Hair cover, beard cover, and safety glasses must remain on person.

Hands shall be washed with alcohol foam or other liquid degermer before putting on sterile gloves. Check gloves for holes or tears prior to entering wash area.

CIP operations will require the use of a chemical safety apron over Tyvek® coveralls. Full face shields are also required for personnel involved in the mixing of CIP chemicals (acid and base).

3.2.2 Degowning:

Personnel completing work in Biocontainment area must proceed to Degowning Room #156. Step into disinfectant boot bath to decontaminate boots. Remove Tyvek® coveralls to approximately mid-thigh and sit on bench facing the direction of room entry.

Remove boot on one foot and the Tyvek® coveralls from the same side. DO NOT PLACE FOOT BACK DOWN ON CONTAMINATED SIDE OF FLOOR! Lift foot over bench to opposite side. Repeat for the other foot and the remainder of the Tyvek® coveralls.

Remove hair cover, beard cover, and gloves and place in waste receptacle marked for that purpose. Clean hands with alcohol foam or other disinfectant.

Put on new hair cover and, if applicable, beard cover.

chapter 13

Multiproduct facilities for biologics

Many biotech companies have elected to follow the approach of large, traditional pharmaceutical manufacturers and build dedicated facilities for the manufacture of a single product. While this approach has obvious advantages related to inventory control, staffing, and regulatory compliance, it can put a tremendous financial burden on young companies with numerous products in the clinical development pipeline and a limited cash flow to support capital expenditures. It also limits both the company's ability to respond to the continually changing needs of the healthcare industry and advances in its own product research and development.

In order to contain costs and make products available to the market in a timely manner, many biotech companies elect to develop multiuse manufacturing facilities, where two or more products can be processed within the same, or shared, manufacturing space. CBER has specifically addressed multiuse facilities, and the practice is recognized and accepted in the industry. While accepting the multiproduct concept, CBER requires proof that companies can effectively design and operate multiuse manufacturing facilities.

Multiuse facilities may be considered for several biological manufacturing steps, including culture preparation, fermentation, initial and final purification, formulation, filling, and lyophilization. Activities that are generally found in multiuse facilities may include the manufacture of a single product in multiple lots, at various stages of production within a common area; the manufacture of multiple products at similar stages of production in a common area; or the use of a facility for different products on a campaign basis.

The important issues in designing and operating multiuse manufacturing facilities are engineering design, procedural or temporal separation of production activities, the use of validated closed systems, and the implementation of a validated cleaning/changeover program [1].

Many factors have allowed industry to rethink its position toward multiuse facilities. Some of the most important are the development and utilization of sensitive analytical methods for validating cleaning/changeover procedures; more efficient purification methods that significantly reduce the amounts of process-related impurities in the final product; and improved engineering and construction methods that allow for more effective process segregation and facility cleaning. Properly implemented, these allow a manufacturer to minimize the potential for cross-contamination of products while demonstrating to regulatory agencies that there is acceptable control over the manufacturing operations required for GMP compliance.

Regulatory background

Biological products are regulated in the United States under Section 351 of the Public Health Service Act and the Food, Drug, and Cosmetic Act. Here lies the authority to license and inspect both the product and the manufacturing facility. General facility requirements for the manufacture of biological products are found in the biologic establishment standards, Part 600 [2], and the Current Good Manufacturing Practices for Finished Pharmaceuticals, Part 211 [3]. The FDA has also published several "Points to Consider" documents that address the manufacture and testing of biologicals, reflecting current thinking within the agency based on its regulatory experience [4, 5, 6]. The NIH's *Guidelines for Research Involving Recombinant DNA Molecules* [7] is another document that has significant design ramifications for any biotech manufacturing facility, especially multiuse facilities.

As with any government-issued regulation, interpretation is a key issue. Because biologicals are derived from living organisms, biopharmaceuticals are, in many respects, quite different from pure, chemically derived drugs. These differences have a tremendous impact on the design of a biotech facility. New products and new delivery systems only add to the issues facing manufacturers, especially in a multiuse facility, where microbial contamination is a concern from the earliest steps in manufacturing.

Definitions

When discussing the issues surrounding the manufacturing of biologicals in a multiuse facility, it is important to understand the terminology that is used. These definitions have been adapted from the Pharmaceutical Manufacturers' Association's (PMA) white paper on multiuse manufacturing facilities for biologics [8].

- **Dedicated equipment/facility** Equipment/facility permanently used for processing only one product at a particular stage of production.
- **Campaign manufacturing** Processing of more than one product in the same facility, equipment, or both in a sequential manner. Only one product is present in the facility at a time. Documented changeover procedures that have been validated to minimize potential cross-contamination are completed between products. Generally,

campaigning involves using the facility, equipment, or both for the manufacture of multiple batches of one product over an extended period of time (weeks or months) before changeover to the next product.

- **Concurrent manufacturing** Concurrent manufacturing in a multiuse facility is characterized by simultaneous production of a number of different products in segregated areas within the same facility. Segregation may be achieved by spatial separation or validated closed systems. The equipment or facility used may be dedicated on a product-specific basis or may be used interchangeably for different products following a validated changeover procedure. Concurrent manufacturing of multiple batches of the same product at a particular stage of processing is considered an acceptable industry practice. This assumes that all GMP regulations are followed to avoid mix-up between batches.
- **Changeover procedure** Documented cleaning or decontamination procedures that have been validated to remove product constituents, process chemicals, and host cell materials from facilities and product-contact equipment, as appropriate. This includes any required equipment assembly and disassembly procedures necessitated by the nature of the product or products in question.
- **Closed system** Process equipment in which the product is not exposed to the immediate environment. In many closed systems, materials (e.g., filtered air, WFI, clean steam, etc.) may enter or leave the system, but the quality of these materials is carefully controlled. In addition, the manner in which these materials are added and removed from the system (e.g., filtration, aseptic connection) is carefully controlled. A closed system should be validated to prevent the escape of product and the entry of contaminants from the external environment into the product.
- **Infectious agent** Microorganism or virus considered to be pathogenic to humans or animals.
- **Open system** Process equipment that is not designed to prevent exposure of the product to its immediate environment. Any system must be considered open unless it has been validated as closed.
- **Segregation** Separation of one process or process step from another through appropriate use of facilities, equipment, or procedures to prevent product cross-contamination.
- **Suite** A functional manufacturing area consisting of one or more rooms with shared air handling and personnel access that is segregated from the rest of the facility. A suite contains a separate air supply and exhaust system, separate personnel access and egress, and separate process equipment. It does not necessarily include a separate supply of water, compressed air or gases, or steam, provided that suitable engineering controls are in place to prevent product contamination from these systems.

Engineering design

CBER's primary concern with multiuse facilities is the potential for cross-contamination of the product at any stage of the manufacturing process. Facility design, engineering controls, and operating procedures of the facility must minimize this potential. These three areas can be seen as the legs of a stool: all are required to minimize risk, and each is as important as the others.

CBER sees the key elements for the proper design and operation of a multiuse facility as no different from those that support successful dedicated facilities. These elements are as follows [9]:

- Good facility design includes proper separation of manufacturing and testing activities; proper material, product, and personnel flows; and adequate space and layout for manufacturing and support operations.
- Good plant systems are required for HVAC, water, sterilization, and waste treatment.
- A good commissioning program covers all direct-impact systems, procedures, and testing activities.
- Good personnel training is needed in the areas of GMP compliance and work skills.
- Good environmental monitoring includes both personnel and air and water quality, as required.

Facility design

The first key to successful multiuse facility design is to know what products will be produced in the facility. It is important to evaluate the potential impact of each product on all operations of the facility, whether the product is in development, early clinical stages, or full-scale production. This evaluation should include the host system or systems, the potential for cross-contamination, biocontainment requirements, biological effects, toxic or infectious agents, and cleaning validations.

Process description

There is no document that is more important in the development of a multiuse facility design than the process description, which defines the products being manufactured and the processes being utilized. The process description should include the name of the product or products, the design basis for the annual output of the product, a detailed description of each process step (fermentation, media prep, purification), a description of the services required for the process (glass wash, tank wash, CIP), and the utilities necessary to support the process (WFI, clean steam, waste neutralization).

The process description should also have, either as part of the document or as a separate attachment, an operational analysis to review all multiproduct operations. This will ensure that adequate time is provided for each of the manufacturing unit operations that make up the process, that all necessary materials and equipment are available where and when they are needed, and that all the necessary documentation records are identified to document the manufacturing process. This is discussed later in this chapter.

The process description becomes the basis for the development of the PFDs and the P&IDs. These drawings, normally done during the early stages of engineering design, become the basis for detailed design drawings of individual systems and rooms within the facility.

Layout

Good facility design should provide for the movement of equipment, personnel, raw and waste materials, and product through the facility while minimizing the interaction between staff and process streams from different stages of the process. Because biopharmaceuticals are derived from living organisms, it is important to determine the flow and control of personnel and to identify areas where physical containment of the manufacturing processes is required and areas where containment is not required. Of the three primary reasons for containment — personnel protection, product protection, and environmental protection — protection of the product in a multiuse facility may be the most important.

Figure 13.1 gives an example of a manufacturing layout for fermentation operations, showing the flow of materials and personnel. In general, the flow should be unidirectional where feasible, so that materials and equipment enter contained areas by means of an airlock and leave through either a double-door autoclave or an air lock. When necessary, dedicated areas or suites are established for some process unit operations in order to provide an additional level of segregation and thus reduce the risk of potential contamination from environmental sources.

Clean equipment and supplies should not cross the path of dirty equipment and supplies. *Dirty* in this context refers not only to equipment or supplies that could be contaminated, but to components that have been used in the process and have not gone through a validated cleaning procedure.

Since people are the major source of contamination within any biotech facility, the flow of personnel, from the time workers enter the facility until they reach their work destinations, must be taken into consideration. Only those individuals involved in specific operations, such as fermentation or purification, should be allowed into these areas when these work operations are in progress. One proven approach to ensuring proper segregation of operations and personnel is shown in Figure 13.2, where personnel enter through a central gowning area and then proceed to additional gowning areas for their specific work operations. Access to individual suites is controlled via an electronic card access security system. When processing operations are not in progress and cleaning is underway, access is again controlled via the electronic card access system.

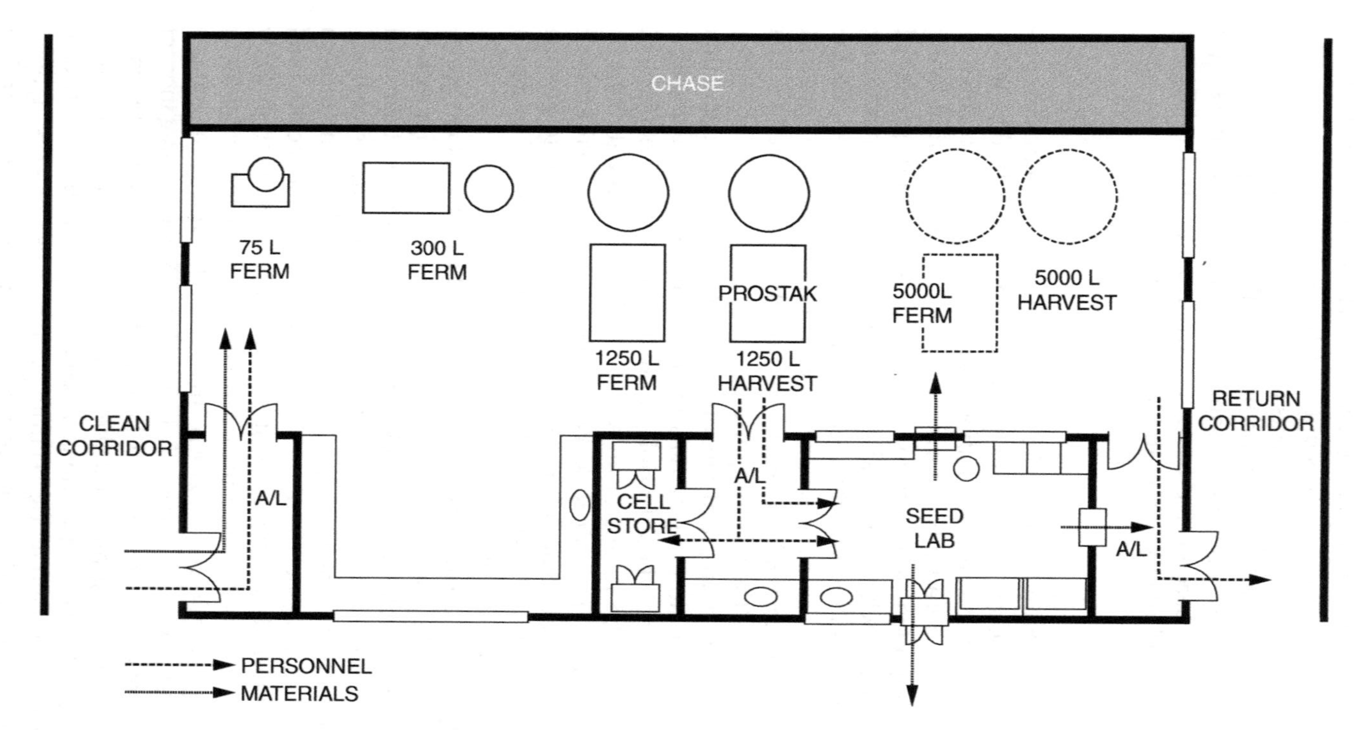

Figure 13.1 Fermentation suite for a bulk biopharmaceutical facility.

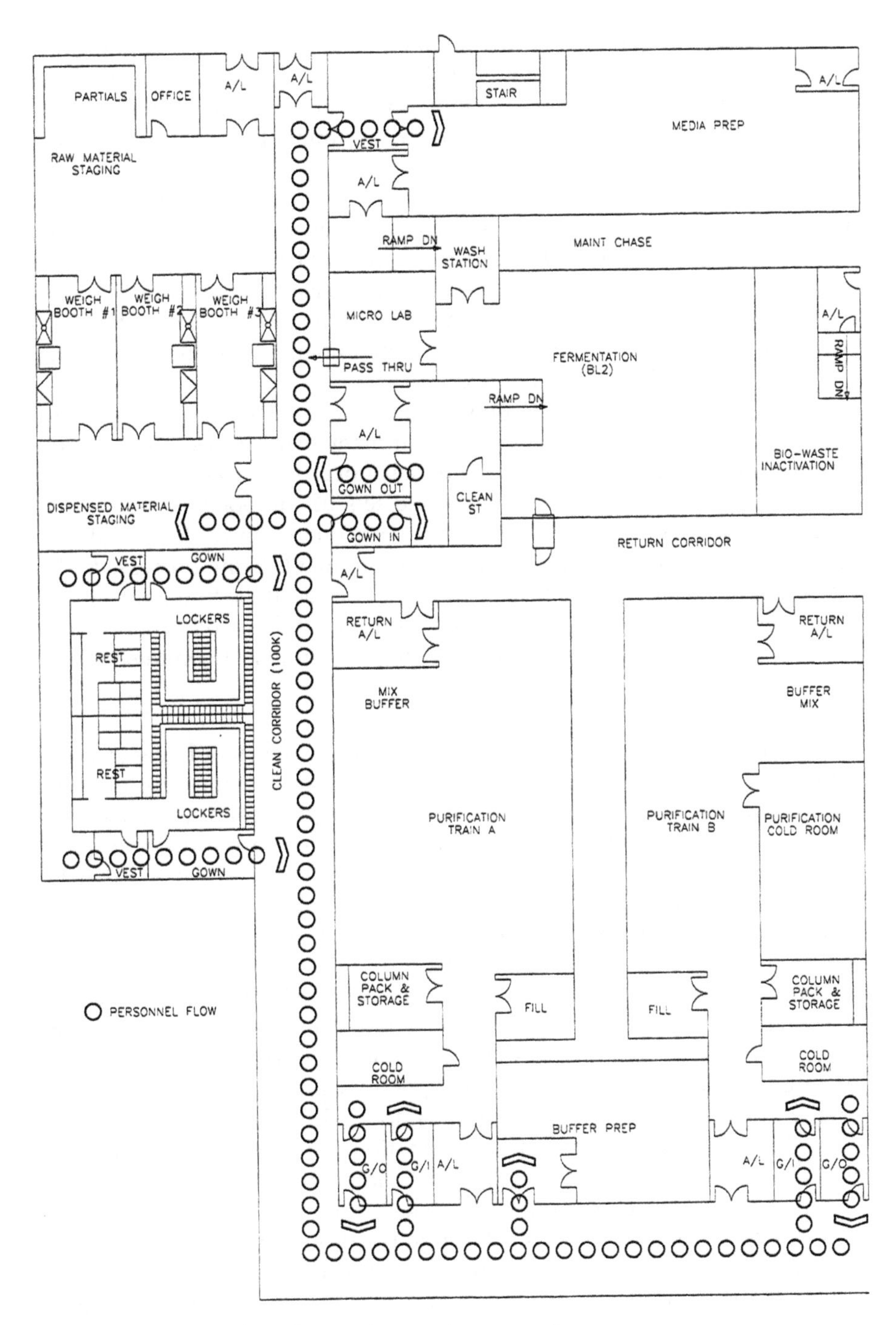

Figure 13.2 Segregation of operations and personnel.

As with equipment, the facility layout must also avoid having personnel cross the path of dirty components during manufacturing operations. Many times, personnel flow diagrams will reveal areas of question, usually involved with the staging and movement of equipment to and from cleaning operations.

In multiuse facilities, it is important that the facility layout allows for the segregated flow of components, such as portable tanks or glassware, that are leaving process areas to be cleaned. If a central tank and glass wash area is utilized, the flow of dirty components in and clean components out must be separate, as shown in Figure 13.3. Dirty components should not be transported through clean corridors where contamination potential is greater, unless they are isolated or contained.

It is also important that adequate space be provided for each operation and for the storage of supplies and equipment that are necessary for the execution of the process, but that may or may not be utilized for each separate product. If it is necessary to use multiple sets of components that will be dedicated only to one product each, these components, once cleaned, should not be stored in the room where a different product is being processed. This viewpoint has been documented by the FDA in questions to manufacturers regarding how segregation of product equipment from other equipment (such as laboratory equipment in a fermentation suite) is accomplished [10].

HVAC systems

In any sterile product manufacturing facility, it is important to control both viable and nonviable airborne particulates. This fact is critical for multiuse facilities, where particulates that cause concern as physical contaminants can also become vehicles for introducing microorganisms from different products or product lots. The quality of the environment must ensure that the individual product stream is protected from microbial and particulate contamination during any open processing steps, such as fermentor inoculation, centrifugation, or column chromatography.

When multiple products are manufactured under a concurrent manufacturing philosophy, either in a single facility or in single areas of environmental control, the main risk is in cross-contamination via recirculation of air through the HVAC system. If the operation philosophy is one of campaigned manufacturing, then the risk is not real-time cross-contamination through the HVAC system, but contamination from product residue in the HVAC system from the previous product campaign.

A proven way to minimize the risk of cross-contamination between different processing operations is through isolation. Isolation is best accomplished by using separate air handling systems to establish isolated manufacturing areas within the facility. Operations such as fermentation, purification, and product filling should have separate air handling systems that supply HEPA–filtered air only to the particular room or suite where the operation is performed. Figure 13.4 illustrates separate HVAC zones for a facility layout for different process operations.

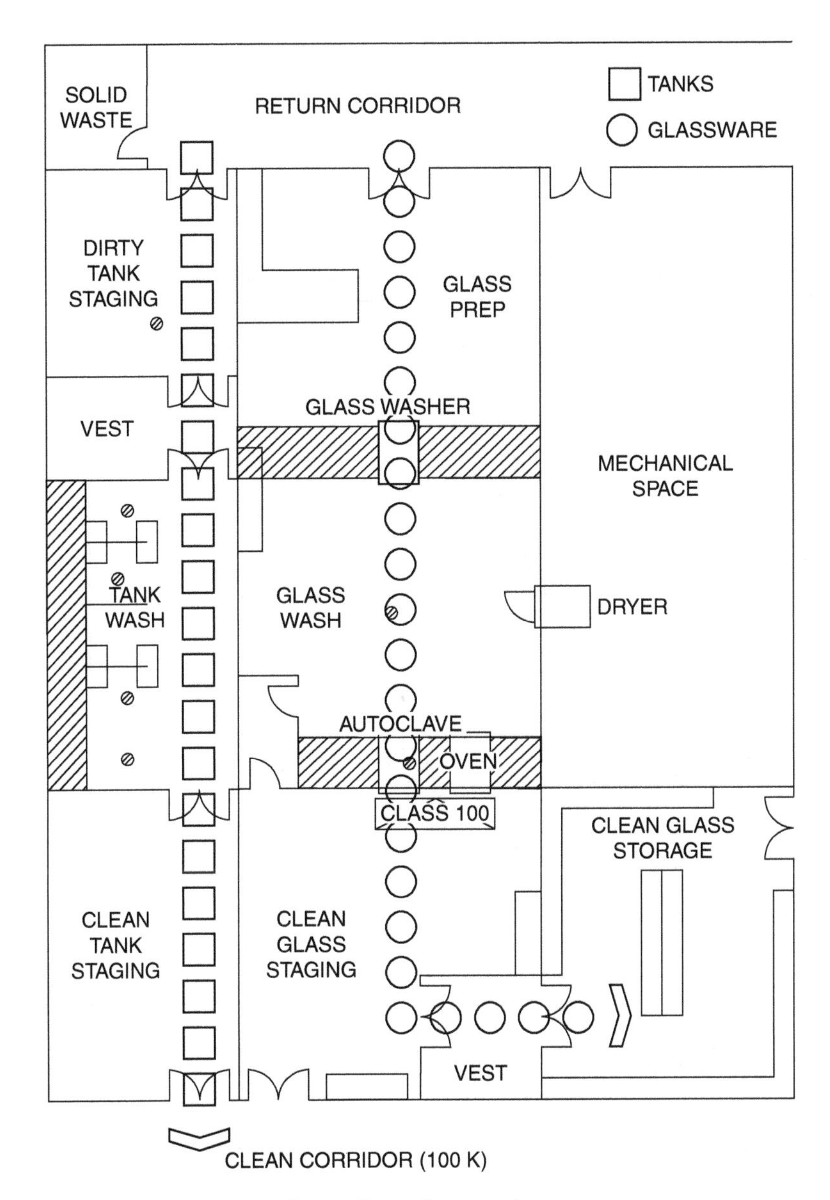

Figure 13.3 "Dirty" versus "clean" flow of components.

Effective contamination control also requires air pressure differentials between different manufacturing areas. The areas with the highest allowable particulate load (class 100,000) will be maintained at a lower pressure than areas with less allowable particulate loads (class 10,000). Generally, as the purity of the product increases through the process, the flow of the product moves through a series of rooms or suites that have increasing pressure differentials and increasing air quality.

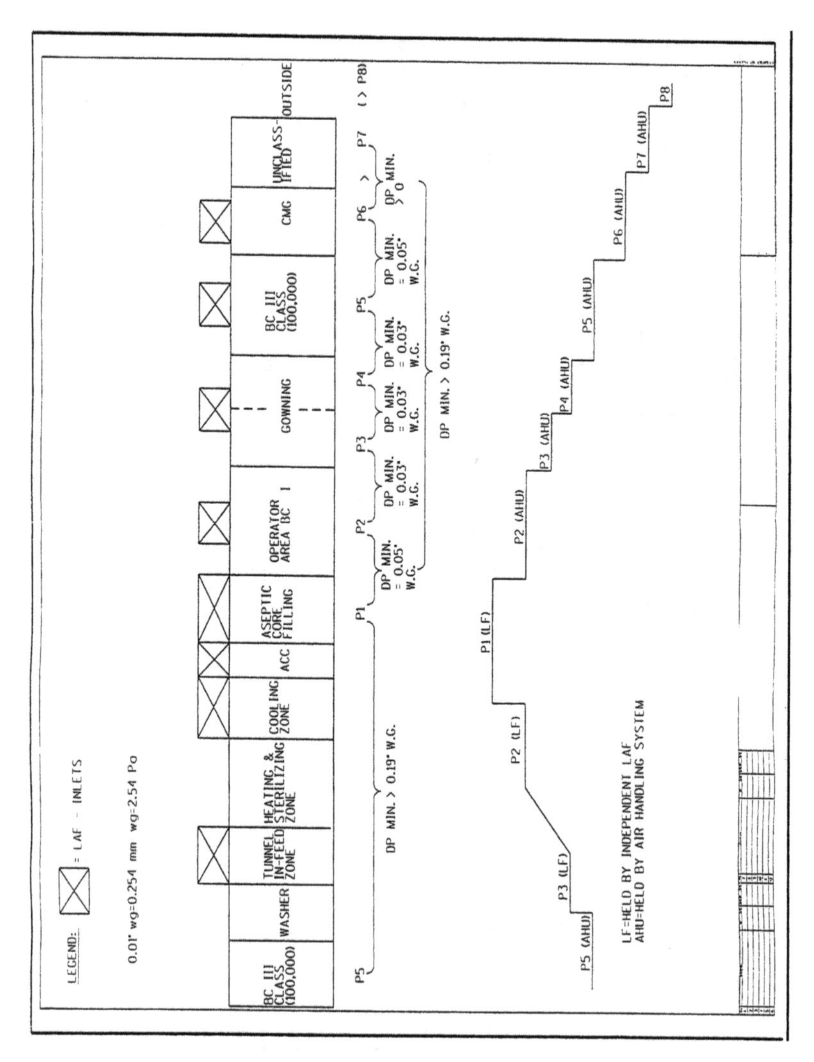

Figure 13.4 HVAC zone drawing for classifications.

Contamination control also depends on the control of particulates found on personnel and equipment that must continually pass through the various areas of any multiuse facility. The use of airlocks for equipment and personnel, and pass-throughs for materials and components, can prevent airborne cross-contamination of products and raw materials between operating areas.

Sterilization systems

A multiuse facility must include proper provisions for the cleaning and sterilizing of equipment and supplies that enter the containment and processing areas of the facility. All materials and supplies leaving containment should pass through a decontamination autoclave. Glassware, components, and waste generated during processing operations should all be sterilized prior to leaving the containment area boundary. Sterilized materials should be allowed to cool under a controlled, HEPA–filtered air source to help reduce the chance of microbial contamination.

The primary issue related to facility layout should be that clean and dirty component pathways must never cross. Separate autoclaves for sterilization and decontamination should be provided to ensure that these flows remain distinct.

Open vs. closed systems

There is not a true regulatory definition for closed systems or processing. A closed process is a process step where the product is not exposed to the immediate room environment. The actual definition of the closed system is the responsibility of the manufacturer. A closed system must prevent the escape of product into the environment and prevent entry of contaminants from the external environment into the product.

Multiuse facilities can utilize both open and closed systems for manufacturing operations. However, it is important to understand how the manufacturing philosophy will impact system design.

A specific processing area or suite can be used for a number of products on a campaign basis, but only a single product should be in the suite at a given time during the process. In open systems, the processing of different products or different steps of the process for the same product, such as formulation or multiple purification steps, should be campaigned and include validated changeover procedures. It is also recognized practice for any production or purification process conducted in an open system scenario to be performed under class 100 air [11].

Processing of different products or different steps of the process for the same product in validated closed systems can occur simultaneously, allowing for concurrent manufacturing of products [12]. Closed systems provide greater assurance of product protection and allow for more thorough cleaning through the use of CIP systems that can be easily validated. A key issue here is the overall integrity of the closed system and how operations, such as aseptic transfers, are performed.

Scheduling

The development of a logical manufacturing schedule as a part of the process description is a key element of multiuse facility design. Manufacturers must prove to the FDA that adequate time is provided for every manufacturing operation; that all of the necessary materials, equipment, components, and personnel are available in the required areas at the proper time; and that all the required records are issued to validate or document the manufacturing process.

Many factors influence the development of a manufacturing schedule. Production requirements, batch size, process yields, and the desired plant operating schedule must all be addressed. In multiuse facilities, particularly those that operate on a campaign basis, the changeover between products is a critical schedule attribute. The schedule must provide adequate time for the tracking of equipment, the generation of required documentation, and the cleaning operations following completion of a product run; it also must allow adequate time for the introduction of the next product entering the process.

In order to have a working schedule, you must clear an operation through a piece of equipment or system before the next one comes along. This must include adequate time for any QA/QC inspections or holds and for the necessary cleaning operations that are defined in the SOPs. Each unit operation and its duration must be identified in a timeline analysis similar to the one shown in Table 13.1.

Gantt charts can then be developed to represent graphically the entire production turnaround (Figure 13.5).

Changeover

Changeover is defined as a series of operations that, when performed, provide assurance that the equipment, facility, and personnel used to produce a biopharmaceutical product have been prepared for the introduction of the next product into the equipment or system.

- **Equipment changeover** Equipment changeover requires written procedures to ensure that all the necessary steps for proper cleaning and validation documentation are followed. Typically, the changeover of equipment involves automated and/or manual cleaning, sterilization, sampling for verification, and, in many cases, the replacement of items classified as *soft goods* (e.g., gaskets, tubing, filters). Appendix 13.1 contains an example of a generic changeover procedure.
- **Facility or area changeover** The room or suite that holds the process equipment must also go through a complete changeover between

Table 13.1 Time Analysis of Operations and Their Duration

Task Name	Duration (Hours)
Production Batch #1	144.0
Media Preparation: Total Time	34.2
Main Medium: Total Time	23.3
Media Makeup	3.7
SIP Media Prep Tank	1.5
Receive/Cool WFI	0.8
Add Peptone (Tri-Blender)	1.5
Charcoal Filtration	0.5
Additional Components (Tri-Blender)	1.0
Ultrafiltration	7.0
SIP Permeate Tank	1.5
Ultrafilter/Collect Permeate	2.0
Dispose of Retentate	0.5
CIP Media Prep Tank	3.0
QA/QC Media Hold	2.0
Media to Seed Fermentor	1.0
Media to Main Fermentor	1.0
CIP UF Permeate Tank	1.0
Dextrose Makeup	16.3
Receive/Cool WFI	0.3
Add Dextrose	1.0
SIP Dextrose/Port Tank	3.0
QA/QC Hold	1.0
Move to Main Fermentor	1.0
Feed Dextrose	8.0
CIP Portable Tank	1.0
Carbonate Makeup	15.8
SIP Portable Tank	1.5
Receive/Cool WFI	0.3
Add Carbonate to Tank	1.0
QA/QC Hold	1.0
Move to Main Fermentor	1.0
Use Carbonate	8.0
CIP Portable Tank	1.0
Deoxycholate Makup	7.2
Obtain Xfer Can	0.5
Receive/Cool WFI	0.3
Add Deoxycholate	1.0
QA/QC Hold	1.0
Move to Lysis Vessel	1.0
Add Deoxycholate Solution	1.0
Remove Xfer Can from Area	0.5
Acetic Acid Makeup	7.7
Obtain Xfer Can	0.5
Receive/Cool WFI	0.3
Add Glacial Acetic Acid	1.0
QA/QC Hold	1.0

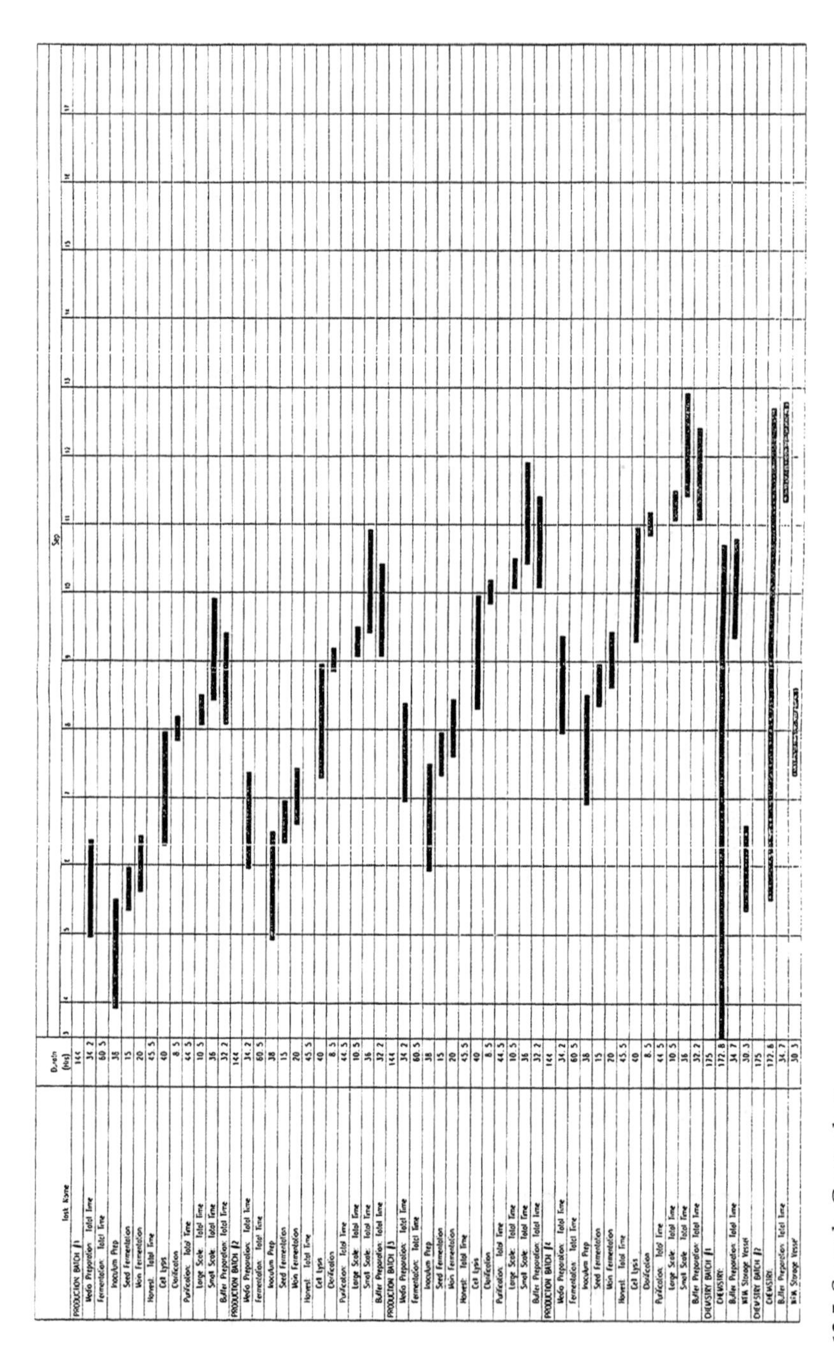

Figure 13.5 Sample Gantt chart.

products. As with equipment, written procedures are required to document the cleaning of the area. The cleaning of the area may involve wipe down, wash down, and sanitization.

- **Personnel changeover** GMP requires that all personnel engaged in the manufacture of a drug product shall wear clean clothing appropriate for the duties they perform [13]. For proper changeover in a multiuse facility, this may involve a gowning differentiation. Gowning and degowning procedures must specify requirements for cleanliness.

Personnel

In a multiuse facility, it is the manufacturer's responsibility to prove to the FDA that proper measures to prevent cross-contamination due to personnel movement between processing areas (fermentation, purification, formulation) have been implemented. This is generally accomplished through the training of personnel, controlling access and movement of personnel within the facility, and enforcing validated gowning procedures to prevent personnel from carrying contaminants from one area to another.

Personnel training, which should be documented as part of the overall facility validation program, should emphasize that access to certain processing areas is restricted to only those individuals working in the area. A manufacturer may choose either to dedicate personnel to a single area or to develop multitask crews that follow a product through the process. In either case, personnel may require gowning changes before leaving one area and entering another. The appendix in Chapter 12 gives an example of a generic gowning and degowning procedure.

Equipment

Multiuse facilities can utilize various types of equipment that fall into one of the following classifications: fixed, portable, dedicated, nondedicated, disposable.

- **Dedicated equipment** is equipment that is dedicated to one specific process step and to only one product. It may be equipment that is permanently fixed in place or that is portable, able to be moved from room to room. The decision to dedicate a particular piece of equipment will usually center on the ability to validate the cleaning of the equipment. Each manufacturer must evaluate the economics, effort, and feasibility of validating the cleaning operation against the cost of dedicating equipment to a single product. Manufacturers will generally dedicate chromatography columns and/or resins, ultrafiltration units, and filling equipment parts.
- **Nondedicated equipment** may be reused for different products or different product applications. This classification of equipment can

only be utilized between different processes after a validated cleaning/changeover procedure has occurred. Items such as fermentors, process vessels, and centrifuges are normally classified as nondedicated equipment. Some equipment that does not come into direct contact with the product, such as media and buffer vessels, can be used for more than one product, provided that validated cleaning occurs between batches.

- **Disposable equipment** is used only once, then discarded. Some types of glassware, tubing, and filters fall into this category.

All of the processing equipment used in a multiuse facility must be designed to facilitate cleaning and allow for visual inspection and verification of this cleaning. Following standard GMP, surfaces that contact product or in-process materials should be smooth and nonreactive. And the equipment must be of adequate size and properly located within the facility to allow for its proper utilization, cleaning, and maintenance.

Environmental monitoring

Because of the concerns about cross-contamination in a multiuse facility, an environmental monitoring program is a critical aspect of the proper engineering controls that govern operations. The FDA reviews environmental coverage as a part of regular GMP inspections. The manufacturer must provide data that documents its ability to isolate the various operating areas of the facility and prove that these areas and their operations are under control.

The monitoring program will vary, depending on operations. In many cases, air quality and room pressurization, viable and nonviable particulate counts, and bioburden on surfaces and personnel are recorded under both static and dynamic conditions to give inspectors a clear picture of the effectiveness of cleaning and sanitization procedures during processing. Such data is collected regularly, with trends developed to show the profiles for each area at any point in time.

The inspection will determine that the program is designed to verify that organisms are subject to appropriate biocontainment practice. Equipment and controls described in the BLA as part of the biocontainment and waste processing systems are validated to operate to the standards — the equipment is in place, it is operating, and it is properly maintained. SOPs for sampling, isolation, counting, and reporting of results will be reviewed as a part of this inspection.

Cleaning/changeover validation

Cleaning validation is one of the major challenges in the successful operation of a multiuse facility. *Cleaning validation* is defined as a system designed to provide a high degree of assurance that a specific cleaning or cleaning/

changeover procedure will consistently clean a particular piece of equipment to a predetermined level of cleanliness [14]. The procedures and methodology that a manufacturer uses to demonstrate equipment fitness for use are scrutinized very closely by the FDA. In particular, inspectors examine the detail and specificity of the procedure for the cleaning of the process being validated and ensure that the proper documentation is available to support the claims of the manufacturer.

Defining clean

What is *clean*? Defining this condition has become very difficult due to the extreme sensitivity of new analytical techniques for sampling. It is common today for these techniques to show that some trace materials can be found even under the most stringent conditions. The FDA recognizes this fact in current inspection guidelines for biotech manufacturing facilities [15].

It now becomes necessary to define *clean* in a manner that is practical for the products being manufactured and processes being utilized. The PMA has approached this definition by saying that the equipment or facility must be adequately cleaned in order to serve its functional purpose [16]. Guidelines established by the Environmental Protection Agency (EPA) and toxicologists set this level at an amount that has been shown to not have any harmful biological effect in the most sensitive animal system known (e.g., no effect) [17]. This permissible level of residue must be justified by the manufacturer. Validation data must then verify that the cleaning process removes contaminants to this level.

Different levels of cleanliness may be required for a multiuse facility. This level is dependent on process operations and manufacturing approach. For concurrent manufacturing, product-to-product cross-contamination is a concern. During a product campaign, the concern is the control of residual materials from previous lots adding back potential impurities that contaminate product.

As a rule, the more purified the product becomes during the manufacturing process, the more stringent cleaning becomes. The cleaning process must be sufficient to keep bioburden to a minimum, remove any residual cleaning agents, and provide a baseline of cleanliness that allows successive product lots to be manufactured without any adverse effects.

Cleaning of systems in a multiuse facility is typically done with CIP systems. CIP systems are hard-piped to maintain containment and have a distinct advantage over manual cleaning methods in their ability to be automated, thus eliminating human intervention and the potential for inconsistent results, contamination, and errors. A typical CIP cycle is a rinse, followed by an acid/caustic wash, followed by three consecutive rinses, with the last rinse using WFI–quality water. Example 13.1 gives a general CIP cycle description.

Cleaning validation

The validation of a cleaning process in a multiuse facility is based on defining *clean* and developing and validating analytical methods to ensure proper sensitivity, accuracy, and reproducibility. In multiuse facilities, there are two basic strategies for cleaning validation:

1. Prove that the cleaning procedure is effective for each product.
2. Prove that the cleaning procedure is effective for the least soluble or worst-case product.

Written procedures must define a method to establish the worst-case condition for equipment and facilities in order to challenge the cleaning process. The FDA recognizes two acceptable methodologies to access cleaning: surface sampling and rinse solution sampling. The use of a pseudoproduct or placebo is generally not accepted by the FDA [18].

Surface sampling typically involves the swabbing of product-contact surfaces with a dry or solvent-saturated swab and visual examination of those surfaces [19]. The advantages of this method are that areas of equipment that are the hardest to clean and yet easily accessible can be evaluated and that any residue that has dried out or is insoluble can be sampled or physically removed.

Rinse sampling has the advantage of covering a large surface area and the ability to sample inaccessible systems. Caution must be exercised where residue may not be soluble or may be physically obstructed by components of the equipment. Rinse water analysis should include residue-specific assays, such as cleaning-agent and endotoxin assays, and residue-nonspecific assays for water quality analysis and TOC.

Validation focuses on the details of the cleaning procedure: what should be done and the materials to be utilized, the capability of the operators performing the cleaning, all pertinent parameters that must be checked, and the written documentation that validates each.

In addition to the cleaning of equipment and facilities, changeover requires line clearance. This is an inventory that is conducted to ensure that all materials, equipment, and documents have been properly reconciled and removed from the area or stored in designated locations. This can be done by utilizing a checklist format, as shown in Example 13.2.

Manufacturing considerations

Fermentation

Fermentation areas in multiuse facilities are not required to be dedicated to only a single product. The manufacturer must provide validated proof that the area can be cleaned and sterilized, and that system integrity is maintained. If the fermentor is a closed system and all effluent properly contained

Example 13.1. A General CIP Cycle Description

The facility will be provided with three CIP skid-mounted recirculation units to clean process equipment and piping. The first unit will be dedicated to the fermentation area. The second unit will serve purification and media preparation operations. The third unit will be dedicated to the central tank wash area. Each unit will be equipped with two tanks; one for wash and recirculation solutions, the other dedicated for WFI storage. The WFI tank will be designed to allow periodic SIP with clean steam.

The CIP sequence will be automatically from the CPI controller. Electronic checking of the circuit to be cleaned will be performed by the controller prior to beginning the sequence.

The following is the typical CIP cycle design:

Pre-rinse	3–30 second bursts
Air blow	1 minute
Caustic recirculation	20–30 minutes
Air blow	1 minute
Acid sanitizer recirculation	5–10 minutes
Air blow	1 minute
Final rinse	3–30 second bursts

The cleaning solutions will be prepared automatically in the circulation tank on each CIP skid. Skid-mounted feed pumps will withdraw caustic detergent and acid sanitizer a central chemical feed station and feed into the recirculating piping loop to/from each wash tank.

in accordance with the applicable biosafety level, then it is possible to process a number of products simultaneously. However, validated changeover procedures must be in place in order to use a fermentor for more than one product.

The process of inoculating the fermentor requires special precautions in order to minimize the time that the inoculum is open to the environment. A typical fermentor system would have inoculum transferred from an aspirator bottle to a transfer can under a class 100 hood. The inoculum transfer from the can to the seed fermentor would be a steam-locked air transfer. The transfer from the seed fermentor to the main fermentor would be via a hard-piped, closed connection.

Purification

Large- and small-scale purification operations should be performed in self-contained areas of the facility, where more stringent gowning, environmental monitoring, and cleaning may be appropriate. Large-scale purification operations (Figure 13.6) using closed systems and hard-piped connections may permit the simultaneous processing of more than one product in an area.

Small-scale purification (Figure 13.7) is typically conducted in a classified area where more dedicated equipment may be required for process operations if product campaigning is used.

Example 13.2 Standard Operating Procedure Checklist

STANDARD
OPERATING PROCEDURE

CHANGEOVER PROCEDURES IN THE MICROBIAL MANUFACTURING FACILITY	Number
	Supersedes

Attachment A
FORMULATION CHANGEOVER CHECKLIST

Changeover from ____________________ Lot# ______

Formulation Rm. 100 NA ☐

__________ Nothing from previous product is in room. __________
__________ Room Cleaned per SOP-101 __________

Formulation Tank T-3109

__________ 1. Equipment log checked __________
__________ 2. Status card checked __________
__________ 3. Product designation label checked __________
__________ 4. P0409-02 checked and attached __________
Manufacturing check/initial/date QA check/initial/date

Growing Rm. 200

__________ Nothing from previous product is in room. __________
__________ Room Cleaned __________
Manufacturing check/initial/date QA check/initial/date

Filtration Rm. 250

__________ Nothing from previous product is in room. __________
__________ Room Cleaned __________
Manufacturing check/initial/date QA check/initial/date

Formulation Rm. 300 NA ☐

__________ Nothing from previous product is in room. __________
__________ Room Cleaned per SOP-101 __________
Manufacturing check/initial/date QA check/initial/date

COMMENTS:

____
____
____

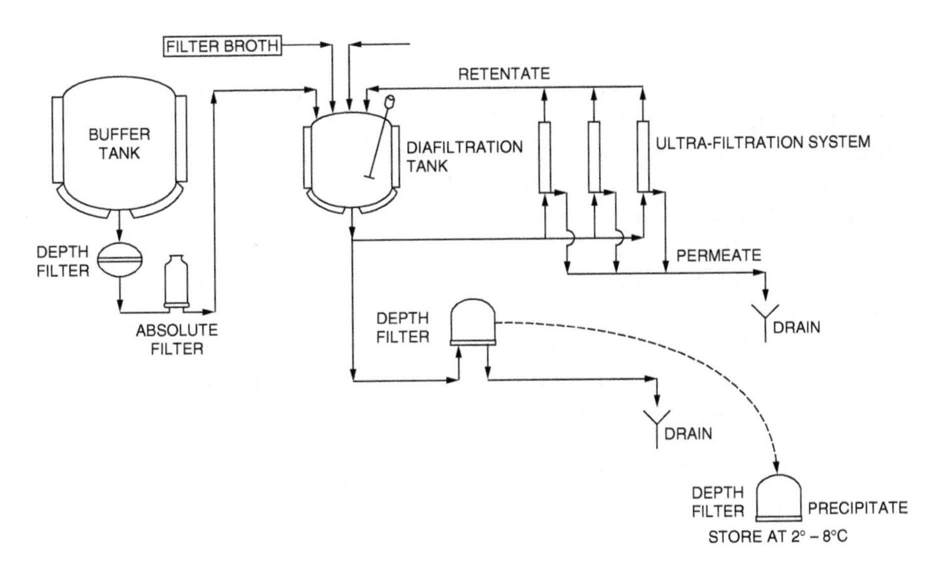

Figure 13.6 Large-scale purification.

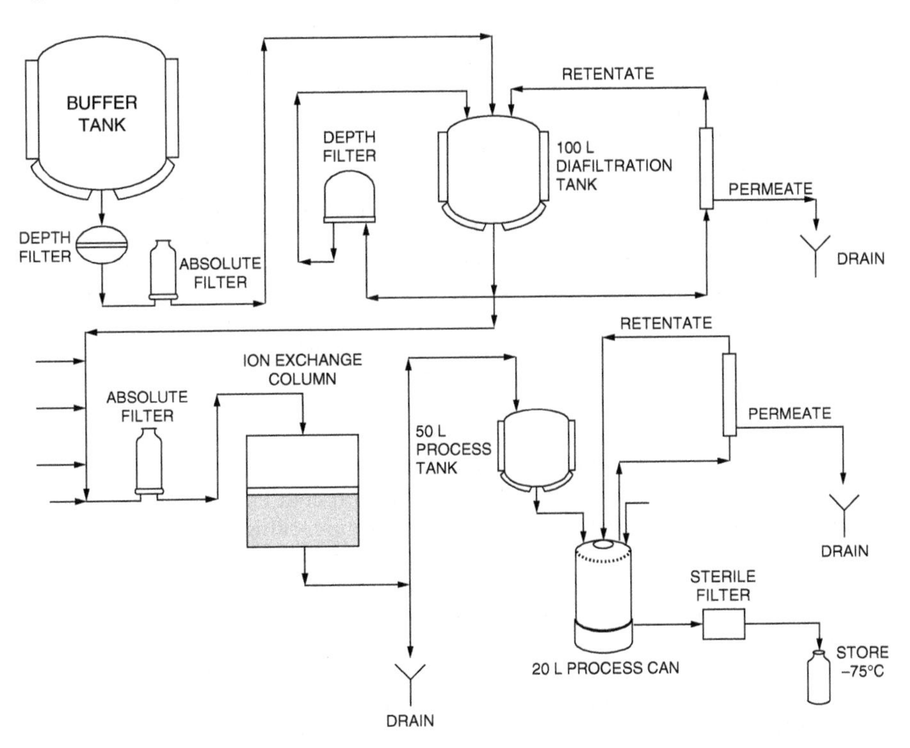

Figure 13.7 Small-scale purification.

Filling

Filling operations in a multiuse facility should be considered a dedicated operation. The facility layout should create a separate area for filling, isolated from other production operations. Areas for containment and aseptic filling operations cannot have a common hallway or shared routes of access by personnel and equipment.

Since biopharmaceutical products require a small batch size, there is less automation and more operator involvement in the fill line. Because of this involvement, there is a greater concern for contamination by the operator. Environmental monitoring programs may require an evaluation of microbial levels on the people working on the fill line, through surface monitoring of gloves, gowns, and personnel.

Dedicated equipment, components, and glassware should be required for each function. Again, a separate air-handling system with positive pressure directional air is required.

References

1. PMA white paper on multiuse manufacturing facilities for biologicals. 1992. Washington, DC: Pharmaceutical Manufacturers' Association.
2. *Code of Federal Regulations*, Title 21, Parts 600–680. 1992. Washington, DC: U.S. Government Printing Office.
3. *Code of Federal Regulations*, Title 21, Part 211.42. 1990. Washington, DC: U.S. Government Printing Office.
4. Food and Drug Administration. 1985. *Points to consider in the production and testing of new drugs and biologicals produced by DNA technology.* Rockville, MD: FDA Congressional and Consumer and International Affairs Staff.
5. Food and Drug Administration. 1987. *Points to consider in the manufacture and testing of monoclonal antibody products for human use.* Rockville, MD: FDA Congressional Consumer and International Affairs Staff.
6. Food and Drug Administration. 1987. *Points to consider in the characterization of cell lines used to produce biologicals.* Rockville, MD: FDA Congressional Consumer and International Affairs Staff.
7. National Institutes of Health. 1986. *Guidelines for research involving recombinant DNA molecules.* Washington, DC: U.S. Government Printing Office.
8. PMA white paper on multiuse manufacturing facilities for biologicals. 1992. Washington, DC: Pharmaceutical Manufacturers' Association.
9. PMA white paper on multiuse manufacturing facilities for biologicals. 1992. Washington, DC: Pharmaceutical Manufacturers' Association.
10. PMA white paper on multiuse manufacturing facilities for biologicals. 1992. Washington, DC: Pharmaceutical Manufacturers' Association.
11. *Federal Standard 209E, Cleanroom and Work Station, Controlled Environments.* 1992. Washington, DC: General Services Administration.
12. PMA white paper on multiuse manufacturing facilities for biologicals. 1992. Washington, DC: Pharmaceutical Manufacturers' Association.
13. *Code of Federal Regulations*, Title 21, Part 211.28. 1990. Washington, DC: U.S. Government Printing Office.

14. PMA white paper on multiuse manufacturing facilities for biologicals. 1992. Washington, DC: Pharmaceutical Manufacturers' Association.
15. *Guideline for the inspection of biotechnology manufacturing facilities.* 1991. Rockville, MD: Food and Drug Administration. Division of Field Investigations (HFC-130).
16. PMA white paper on multiuse manufacturing facilities for biologicals. 1992. Washington, DC: Pharmaceutical Manufacturers' Association.
17. Perez, M. 1977. Human safety data collection and evaluation for the approval of new animal drugs. *Journal of Toxicology and Environmental Health* 3.
18. Cleaning validation. 1992. *Mid-Atlantic region inspection guide*. Rockville, MD: Food and Drug Administration.
19. Hardec, S. W. 1984. The validation of cleaning procedures. *Pharmaceutical Technology* 8 (5).

Appendix 13.1

Sample changeover cleaning procedure

The purpose of this procedure is to describe the changeover sanitization and cleaning requirements for the manufacturing and purification areas dedicated to multiple products.

1.0 MATERIALS AND EQUIPMENT
1.1 Changeover Cleaning Record, CC-431
1.2 HEPA–filtered canister-type vacuum
1.3 Nylon, nonshedding rag mop and replacement mop heads
1.4 Sponge mops and replacement mop heads
1.5 Plastic or stainless steel buckets
1.6 Plastic or stainless steel wringers
1.7 Floor signs: CAUTION WET FLOORS
1.8 Nonshedding cleanroom-type towels
1.9 Floor squeegee and replacement blades
1.10 Rubber gloves
1.11 Plastic trash bags
1.12 Approved disinfectants (XY-12 or LpH)
1.13 Isopropyl alcohol (IPA) (70%)

2.0 SAFETY

Personnel must follow all gowning requirements when entering designated areas for cleaning. See Material Safety Data Sheets (MSDSs) for hazard classification and safety recommendations for all cleaning agents. Use caution when working on or near wet floors and be sure to post areas for the safety of others.

3.0 PROCEDURE

3.1 Perform a changeover cleaning in areas that are not dedicated to the production of a single product and/or product component. Changeover cleaning shall be performed upon completion of a product or product component campaign prior to the start of the next product or product component campaign. Changeover cleaning of the area shall not be performed until completion of the changeover cleaning for process equipment in accordance with SOP #EC-321 (Changeover Cleaning of Process Equipment).

3.2 Cleaning Procedures: The use of dedicated mops and buckets is required for a changeover cleaning of designated manufacturing areas. The use of mops, buckets/ wringers, and disinfectants shall be in accordance with SOPs CC-323, 311, and 352, respectively.

3.3 Changeover Cleaning

a. Pick up all trash and debris from the floors and vacuum.
b. Obtain and prepare the appropriate disinfectant per SOP CC-388.
c. Using a clean sponge mop, carefully wipe the ceiling with the disinfectant. Rinse the mop as necessary. Wipe all lights and HVAC grills with disinfectant using nonshedding cleanroom-type towels.
d. Wipe down all walls from ceiling to floor, doors, and windows with disinfectant using downward strokes with a sponge mop. Rinse mop as necessary. Rinse windows and other surfaces with IPA.
e. Wipe down all work surfaces with disinfectant using nonshedding cleanroom-type towels and allow to air dry. Work surfaces include the following:

    Doors, windows, sinks, benches, and other furniture
    Interior and exterior cabinets
    All hoods
    Storage carts
    Shelving bins
    Major equipment and valves
    Instrumentation gauges
    Piping
    HVAC vents
    Control boxes and hardware

f. Rinse surfaces with IPA as necessary.
g. Use a squeegee to direct all excess water to the floor drains prior to mopping the floors. Wet mop the floors using a rag mop with disinfectant and allow to air dry.
h. Dispose of all nonshedding towels and sponge mop heads in the trash receptacle.
i. Clean all mops, buckets, and wringers in accordance with SOP CC-601 (Use and Cleaning of Mops and Buckets). Hang all mop handles on designated racks and deposit used mop heads in designated containers located in the designated Cleaning Supply Storage Area (Room #188). Store all other cleaning supplies appropriately.
j. Empty all trash receptacles in the area and move the trash to an appropriate location for maintenance personnel to dispose. Clean the trash receptacle with disinfectant.
k. Carefully pour all used disinfectant solutions down the appropriate drains as described in SOP CC-622, Attachment II.
l. Record changeover cleaning on the Monthly Facility Cleaning Record (F-345). File the completed records in the logbook for each room.

chapter 14

Contract formulation and philosophy

In our present and increasingly litigious society, the importance of the contract in executing projects is becoming more apparent. Regardless of the size of the company or the size of the project, developing the right set of contractual documents is a key element to the overall success of a project. There are many different options available in structuring a contract. There is no right or wrong option. Success will come from the overall philosophy taken in developing the execution plan for the project and in the diligence of monitoring the contract for compliance.

What is a contract?

For design and construction contracts, a simple definition of a contract is a promise or a set of promises to furnish services and/or materials to design a structure, build a structure, or to improve real property for another party who promises to pay for the work performed. It must also be clearly understood that for a breach of these promises, the law gives a remedy and in some ways recognizes the performance of the promises as a duty.

The one common denominator in the execution of any project is the contract. But how the contract is formed, what it asks for, and how it defines the execution process can be as varied as the needs of the parties involved. There are numerous ways in which a contract can be structured in order to meet the project's goal. Some of the circumstances that can arise from certain contract formats must be recognized as potentially hazardous to the project's goals, be they cost or schedule related.

It must be understood that contract documents must be written clearly, fairly, and with the intent to protect the rights of the client and other executing parties. The fewer unknowns, the better. Fewer unknown or questionable items translates into fewer potential areas for claims due to interpretation of statements.

Format

As is true of contracts in general, the form of design and construction contracts varies greatly from contract to contract. While some are short and simply stated, such as for a simple feasibility study, others are long and detailed, such as the design and construction management of a sterile product manufacturing facility.

However, the repeated use of specific forms, particular terms, or clauses has resulted in some standardization of contract forms within the industry. One of the most often used formats is that established by the American Institute of Architects (AIA). [1]

Many companies also have standard contract forms that are implemented for many varying types of projects. In some cases, a company lacking a formal contractual model for the execution of design/construction work will revamp existing service contract formats to meet the needs of the new project. While certainly a viable option, be aware of the time required in developing these type of documents. If time will become a constraint in executing the formal agreement, be sure to start this process early.

Basic elements

Contracts normally consist of six clearly distinguishable, basic parts, as follows:

1. Bid form
2. Agreement
3. General terms and conditions
4. Special provisions
5. Documents
6. Addenda

The bid form, or proposal form, provides the bid form, the actual invitation to bid, and the instruction to bidders. Depending on contractual philosophy, it may include very detailed information for bidding purposes or it may be somewhat generic in providing basic data necessary to estimate costs and fee projections to be utilized at final negotiations. (See Example 2.4 on page 26 for a related example—the Request for Proposal.) It is interesting to note that the majority of design and construction contracts are not negotiated, but competitively bid, where the contractor has no say with respect to the terms or subject matter of the contract except for the price or prices. [2]

The agreement identifies the parties to the contract, the execution date, the contract price and commercial terms, the commitment of the contractor to perform the services required to execute the project in accordance with the referenced documents, and the signatures of both parties.

General terms, general conditions, or general provisions relate the specific clauses applicable to all projects. Typical general terms and conditions contain language pertaining to quality, timing, insurance, conditions, and payment terms, to name only a few. Again, the AIA has standardized these general conditions.

Special provisions or special conditions are specific clauses that relate specifically to the particular project. For example, the special provisions may identify certain activities that may be performed and/or provided by the Owner, such as third-party testing, passivation, or cleanroom certification.

The document section is somewhat self-explanatory; it lists applicable references, such as drawings, specifications, profiles, cross sections, or details applicable to the defined scope of work. It must include the revision and date of each document and be very descriptive where filed changes or as-built conditions are identified.

The final part of the contract identifies any addenda that supplement, modify, or amend any information contained in the contract negotiated during the bid cycle or prior to the execution of the contract. These should not be confused with change orders or supplemental agreements negotiated after the contract is awarded.

Execution formats

There are numerous ways in which an organization can choose to have a project executed. The chosen method of delivery will determine the best format of contract to be developed. It will also determine the bidding format for obtaining pricing.

Lump sum

Lump sum contracts are the most important and common form of fixed-price contracts. In a lump sum contract, the contractor agrees to perform the complete scope of work outlined in the contract for a price agreed to and fixed at the time the contract is awarded.

For most lump sum contracts, pricing is obtained through the competitive bid process. This allows Owners to know what their costs will be for any particular segment of work that is bid. It also allows an Owner the ability to easily track increases in costs through the issue of change orders to the original contract for scope of work additions.

Lump sum contracts for construction require that design drawings used for bidding be complete. Drawings that are not in an "Issued for Construction" state will lead to numerous changes to cover evolution of the design process. These can normally include changes due to the completion of certified vendor drawings requiring rerouting of sanitary process piping from equipment, rerouting of cleanroom HVAC duct to eliminate interferences, or revisions to architectural features due to final sizing of equipment skids in cleanroom areas. Drawing changes issued after the contract is awarded can lead to added costs due to rework, additional new work, changed conditions for performing work (e.g., added congestion in areas or increases in work height), or extended delays due to added material delivery time.

A lump sum contract price includes all costs that a contractor will incur for both direct and indirect labor, materials, rentals, overhead, and profit. This will ensure that the contractor will control the indirect job costs as best as possible to maximize profit. It also will provide an incentive to maintain the schedule since extended time in completing work will lead to more incurred costs.

One very important aspect to remember when selecting a lump sum contract format is the qualifications of the bidders. Biotech and pharmaceutical projects require a certain "acquired" experience in dealing with "industry special" attributes normally encountered (e.g., GMP compliance, cleanroom environments, and validation). Any design or construction organization not familiar with these attributes may (or will) not reflect the costs associated in their bids correctly, leading to a potential money-losing situation for the contractor and a potential cost and schedule overrun for the client.

In evaluating lump sum contract proposals, Owners must be careful to establish which proposals are within an acceptable control estimate range and which are unusually low. Lack of knowledge in the biotech/pharmaceutical industry, improper estimating practices, or desperation to obtain work can lead to abnormally low proposed costs from contractors. A careful evaluation of lump sum proposals is necessary to ensure that the scope of work being evaluated is a direct "apples-to-apples" comparison. Failure can lead to extreme cost overruns.

Other fixed-price contract forms

A variation of lump sum contracts for design and construction projects is the unit-priced contract, where the price per unit is fixed at the time the contract is awarded. While this price per unit is intended to remain unchanged during the execution of the project, the number of units may increase or decrease, which means the total compensation to the contractor may also change. The actual final contract price is not based on estimated quantities, but on actual expended or installed quantities.

The obvious problem for a unit-priced structure is that if there is a substantial increase or decrease in quantities, the contractor could experience significant cost overruns or underruns. If the contractor cannot make a good determination of the work to be performed, either by lack of experience or insufficient design information, then the estimated unit price is probably not valid. When this happens, the contractor or the Owner may feel they have been misled and may seek price adjustments or other claims.

It is extremely difficult to estimate finite quantities in construction unless design information is complete. In biotech and pharmaceutical construction the sequencing of architectural work, fabrication and installation of sanitary process piping, cleanroom duct installation, and validation costs are very difficult to estimate on a unit-price basis without extensive background experience. The large number of potential field interferences and rework situations further adds difficulty to the validity of unit-price values.

If a unit pricing format is chosen, the contract must contain a carefully drafted clause specifying the allowable quantity variance (increase or decrease) that will trigger a price adjustment. If this is not done, the Owner can expect to pay the agreed upon unit price for actual quantities since there is no other basis for evaluation.

Another variation of the fixed-price contract is the guaranteed maximum price or GMAX price. In this format, the Owner requires the design or construction contractor to guarantee a maximum price for the work to be performed, beyond which the owner is not obligated to pay. The only adjustments to the GMAX price will come if the Owner agrees to changes in the scope of work.

Cost-plus contracts

In a cost-plus contract the design or construction contractor agrees to perform the scope work specified at actual cost, plus an additional negotiated amount to cover the contractor's profit. The cost-plus contract can take on many variations, including cost-plus–fixed fee, where the amount of the fee is a fixed value; cost-plus–percent of cost, where the additional compensation is a negotiated percentage of the actual costs; or cost-plus–incentive fee, where the fee is a calculated shared savings value for underruns.

Contractors favor cost-plus contracts because they eliminate their exposure to problems that may be encountered. However, Owners should be very careful with cost-plus contract situations where new, inexperienced contractors to the biotech and/or pharmaceutical industry are involved. Errors committed or problems created by the contractor's lack of knowledge and understanding will be paid for by the Owner, through higher job performance costs resulting from low productivity, poor scheduling, or overestimating actual completed work.

Cost-plus contracts do not create incentive for a contractor to control indirect job costs. It will become a very time-consuming and costly exercise for an Owner or a construction manager to monitor and audit monthly indirect costs—where they are "hidden" and how they are distributed. Excessive indirect costs due to lack of knowledge in the biotech/pharmaceutical industry will be added as a form of "contingency" to protect the contractor. These costs are difficult to find and even more difficult to reduce.

Time and expenses, time and material

Time and Expenses (T/E) and Time and Material (T/M) contracts are frequently used for very small projects or to contract the performance of additional scope of work. T/E and T/M contracts are very basic: The client pays for the labor (time) expended to perform the work, the expenses incurred by the design firm, or the materials used by the construction contractor. In this contract format, all costs are covered.

Design-build vs. construction management

A design-build contract identifies the firm designing the plant to also perform the construction, either through self-performing the work or acting as the construction manager. Construction management contracts involve a firm, oftentimes not the design firm, acting as the Owner's representative in issuing subcontracts for the performance of project work activities, and managing the overall execution and control of these activities. Both scenarios are common, and usually reflect the experience of the Owner and the preferred method of structuring a project. There are many differences in the two approaches that should be discussed.

In the common design-build approach, a single firm has management and control responsibility for all design and construction activities. This firm may also perform some, if not all, of the construction activities with their own craft employees, although this is frequent in the industry. The emphasis placed on this approach is one of single point of responsibility—only one firm with which to deal. A single firm can better maximize schedule efficiency, provide lower costs through comprehensive management, and produce a quality end product under a single quality control plan.

Proponents of the separation of design and construction responsibilities view design-build contracts as the "fox in the hen house" approach. Problems with design will not be flushed out as well since the construction organization is less likely to criticize its corporate peers. Advantages of the construction management approach and its focus on owner advocacy include a second set of eyes for reviewing the design to provide better constructability and value engineering to the design. The construction manager does only one thing—build. They maximize contracting efficiency to bring total construction costs down, even though they charge fees equivalent to 2–5 percent of direct costs.

The arguments are strong for both approaches; the choice should be based on the needs of the project and the comfort level of the client in managing either approach. The experience of firms and the talents of the individuals proposed for the project are important in both approaches.

Scope of work

Any discussion of contracts is incomplete without emphasis placed on completion of a detailed scope of work. It is the scope of work that defines the work to be performed, the criteria it will be measured against, and the basis for which a price is formulated.

As with most projects, sterile product facilities become more complicated and more difficult to revise as their completion date draws near. Omissions or errors in scope definition become costly to correct, especially for those activities related to cGMP issues. As an example, assume that a definitive scope for providing sanitary process piping supports did not address the spacing requirements, or the need for compliance to facility cleaning procedures. The problem becomes the need to maintain proper slopes for self-drainage that requires more frequent placement of supports. Also, if some of the manufacturing areas see steam cleaning vs. chemical cleaning, the type of material in the support spacer will need to be of a different type, say Viton® (E. I. Dupont) vs. standard rubber. To correct both of these omissions will require time and money—ordering new material, reworking installed items, or possibly rerouting to provide space for added supports.

When defining the scope of work for a contract, remember that the goal is to have the necessary work completed in a timely manner, for the cost agreed upon, and to avoid any possible claims that could arise from the actual work varying from the scope definition.

- Specify all work required as provided for in the drawings and specifications.
- Ensure that the drawings and specifications represent the work to be performed.
- If work methods are specified, be sure they are followed and not allowed to vary unless approved.
- Identify any drawing or specification changes immediately and amend the contract accordingly.
- Provide sufficient details to perform the work. Do not require contractors to perform unanticipated engineering services.

Bid analysis and selection process

Bid documents, containing the detailed scope of work and all pertinent documents, should be issued to qualified bidders chosen from a process that should include the following:

- Review of experience
- In-house interviews
- Networking with previous clients

The list of qualified bidders should reflect firms that you are confident could successfully execute the project. It is recommended that the primary members of the project team visit the potential firms at their offices. Such a visit gives the project team a good sense of the corporate atmosphere, other projects in progress, and a general feel for the contractor's operation. For example, visits to a potential engineering designer can help the project team determine whether the staff has the professed specialized experience at the working level.

Since the process of requesting, preparing, and analyzing bids is a costly and time-consuming task, it is a good idea to create a short list of bidders who will receive the final bid documents. This short list should be limited to no more than three or four firms.

The decision regarding which firm to choose to perform the scope of work in question will be based on the following considerations:

- Experience of the firm in similar assignments
- Experience of the assigned personnel in biotech/pharmaceutical projects
- Cost
- Project control
- Schedule compliance
- Quality
- Past performance

For the selection of an engineering firm, cost should be lightly regarded when compared to the other factors. This reason, from experience, is that reputable engineering design firms usually have similar multipliers, bid philosophies, and methods of execution. Variations of less than 10 percent are not uncommon. All other factors being equal, cost then becomes a consideration.

Experience is the key factor. Knowing the industry-specific problems, familiarity with technologies, past performance with clients in the biotech/pharmaceutical industry, and experienced personnel are important. Repeat business and favorable recommendation from peers are also good indicators.

The use of a ranking system to evaluate proposals is a good method of selection. Any hidden favoritism or bias can be eliminated from the process by utilizing weighted factors for each element of the evaluation process. An example of such an evaluation can be found in the discussion on sanitary process piping (see Chapter 7).

References

1. American Institute of Architects, 1735 New York Avenue NW, Washington, DC 20006.
2. DeBenedictis, D., and R. McLeod. Construction contracts. In *Handbook of construction management*, p. 48. New York: Van Nostrand Reinhold.

chapter 15

Future trends

What does the future hold for the biotech and pharmaceutical industries? This is the question that was posed in the first edition of this book. Since then, it is an understatement to say that the biotech and pharmaceutical industries have changed. In many areas, the changes have been dramatic.

As I look back on the trends of the late 1990s, I can make some general observations about where the industry has progressed. Barrier technology has continued to see greater implementation but has not made the dramatic impacts that were predicted. In many instances, the development of improved process technologies has negated the need for barriers, due in part to a movement away from stringent cleanroom classifications. Outsourcing has expanded as predicted. Contract manufacturing, research, and testing companies have flourished, as many small and developing companies have decided that their limited capital resources are better spent on research than on infrastructure. The industry restructuring continues today. Many big-name mergers, such as IDEC/Biogen, Glaxo/SmithKline, and Pfizer/Pharmacia, have captured headlines worldwide. All of the famous industry prognosticators believe that the consolidation is not about to slow down soon. And the changes in the regulatory landscape continue, with new changes in the structure of the FDA, a review of GMP, and the expanded influence of more internationally focused regulatory guidance, such as the efforts of the International Committee on Harmonization (ICH) to produce standards.

So now, where do we go? Let's discuss the current trends in the industry that will, again, have a tremendous impact on facility design and operation as we move into the 21st century.

Manufacturing

Improved process technologies, a greater focus on economies of scale, and the emergence of new production alternatives will all play a major role in shaping facility decisions in the future. Particularly in the biotech industry, where predicted industry growth is expected to continue to increase in both the number of products in the pipeline and the financing of new ventures, the changes could be dramatic.

The current pipeline of drugs remains solid. Fueled by the discoveries coming from the mapping of the human genome and in the advancing fields of gene therapy, bioinformatics, and proteomics, the industry will continue to see many new targets developed and will see a decrease in the time required for drug discovery and development. This will not only decrease the costs to develop drugs but will also fuel the need for manufacturing capacity, based on the increase in drugs coming to market.

The increase in the use of monoclonal antibodies (MAbs) will continue, with projections of the global sales of MAbs surpassing $8 billion by the year 2005 [1]. This will have a significant impact on the industry, due to the need for production capacity involving large-scale facilities that carry very high capital costs, significant human resources to operate and maintain, and a significant amount of time to design, construct, and validate.

Manufacturing technologies

The biotechnology and pharmaceutical industries will continue to see new and improved applications of manufacturing technologies — better ways to "build the mousetrap." These will include not only the area of human therapeutics, but also industrial applications, agriculture, and medical devices.

The area of human therapeutics, and more specifically the production of proteins, is currently focused on cell-culture technology. The focus on transgenic technology implementation holds much promise to many within the industry. *Transgenics* is the science of producing human proteins within nonhuman systems, such as plants (corn, tobacco, wheat) and animals (chicken, sheep, cattle). A number of companies are working in this area, and many predict that the industry will see a transgenic product on the market within ten years.

The impact of this technology on facilities could be very dramatic. Under a transgenic production system utilizing plants, the upstream production of proteins would not involve 12,000L stainless steel vessels but would employ thousands of acres of tobacco or corn as the production means. Production scale-up could produce a dramatic reduction in cost of goods — and such a system could become a tremendous boost for depressed agricultural areas of the United States.

As of this writing, there is not an approved transgenic product on the market for human therapeutics. The hurdles include dealing with the issues of controlling glycosylation, the potential for contamination, a lack of regulatory guidelines, and public perception. But if the biotech industry remains focused on proving the technology, it is a good bet that products will be delivered. If this occurs, facilities will have a very different look in terms of unit operations on the front end of the process. Equipment for handling bulk agricultural products will need to be incorporated into a facility design. The site selection process will need to look more to agricultural areas than to the current biotech clusters.

The science of biotechnology is also moving into many industrial-focused applications. The same focus on cost effectiveness in the production of human therapeutics is also driving industry to look at applications for bio-based fuels, bio-remediation products, and materials such as enzymes (proteins composed of amino acids). The future holds the development of very large-scale production facilities for enzymes and materials that will ultimately be used in the production of pharmaceutical products and various medical devices.

Applications of emerging technologies will also play a significant role in advances in tissue engineering, wound care, and the development of materials for surgical products, such as bone graft substitutes and so-called "artificial skin" for the treatment of burns.

Industry restructuring

In the first edition of this book, I discussed the generally held belief that the industry would begin to look very different by the turn of the century. And the predictions were right. Strategic alliances, mergers and acquisitions, and organizational strategies each have played a significant role in shaping a new pharmaceutical industry.

We now have GlaxoSmithKline, IDEC Biogen, and a much larger Pfizer as industry leaders. We see such companies as Eyetech, Genta, and Amylin closing collaboration deals that are worth millions of dollars in revenue. At the end of 2002, we saw the value of pharma/biotech strategic partnering deals reach $6 billion.

We will also continue to see cash-rich firms pursue the smaller, developing companies as they struggle to survive tight capital markets while trying to maintain their drug discovery focus. This will be particularly true for early-stage drug discovery companies.

Will there be fewer facilities built? Or will there be more focus on increasing capacity as new organizations sense a greater stability in their ability to meet capital needs? Time will tell.

Regulatory changes

The changes that have impacted the regulatory landscape over the past five years may be only an indicator of the changes that are coming. The reauthorization of the Prescription Drug User Fee Act (PDUFA) will continue to allow the Agency to increase the investigator ranks to meet future demands for product and facility reviews. However, according to FDA statistics, at the end of 2002, the median approval time for a standard NDA filing increased to 15.3 months, the longest review time since 1996.

The consolidation of functions between CBER and CDER could improve communications between the Agency and potentially reduce review cycles.

However, there is some concern that this consolidation will lead to decreases in staff that could have the opposite effect for facilities related to vaccines, blood products, and cell therapies. Projects in those areas could feel an impact in terms of review times, as well.

The Agency has also announced that the first major review of GMP regulations will occur in 2004. The Agency is turning to a risk-based approach to good manufacturing practices that will help it predict where its inspections will most likely have the greatest impact. The extent of these changes could affect guidance documents that have an impact on facility design and delivery.

Internationally, the continued recognition and acceptance of ICH guidelines are bringing a new focus on facilitating a global view of good manufacturing practices. Differences between current FDA and European guidelines are being addressed through the Q7A document, and providing a more consistent approach to regulatory interpretation.

References

1. Burrill & Company. 2003. *Biotech 2003.*

Index

D

I

J

K

L

M

Q

R

S

T

U

V

W

Z

Milton Keynes UK
Ingram Content Group UK Ltd.
UKHW021822071024
449327UK00021B/1389